Physics, the environment and man

Physics, the environment and man

Isaac Turiel

San Francisco State University

Prentice-Hall, Inc., Englewood Cliffs, New Jersey

Library of Congress Cataloging in Publication Data

TURIEL, ISAAC.
Physics, the environment and man.

Includes bibliographical references and index.
1. Physics. 2. Man—Influence of environment.
3. Man—Influence on nature. I. Title.
QC28.T87 301.31 74-26781
ISBN 0-13-672402-7

10 9 8 7 6 5 4 3 2 1

Printed in the United States of America

Prentice-Hall International, Inc., London
Prentice-Hall of Australia Pty, Ltd., Sydney
Prentice-Hall of Canada, Ltd., Toronto
Prentice-Hall of India Private Limited, New Delhi
Prentice-Hall of Japan, Inc. Tokyo

Contents

Preface

This book has two aims: one is to inform the reader on some of the important environmental issues of our time; the second is to introduce some of the important principles and concepts of science in a nonabstract way. Whatever knowledge of physics, chemistry, and biology is necessary to gain a clearer understanding of these environmental issues is introduced as the book progresses, so a prior background in science is not necessary.

Many people are now aware of the complex web of interactions between man and his environment. Complex problems such as whether or not to build new nuclear power plants, or whether or not an electric automobile should be developed, require interdisciplinary approaches to their solution. For example, the use of electric automobiles will reduce photochemical smog in the cities, but electrical energy will be needed to recharge their batteries. If more fossil-fuel power plants are built to supply the needed energy, then we will be changing the distribution and composition of the air pollution. We must consider the health and other effects of the different air pollutants. If instead of fossil-fuel power plants, nuclear power plants are built, then we need to consider all of the problems associated with nuclear power plants. In this book we discuss the nature of man's interaction with his physical environment so that decisions, such as whether or not to build the SST or an electric automobile, can be seen in their full impact upon man and the environment.

I hope that participants in a course based on this book come away with a better understanding of their physical environment; in particular, how we influence the state of the environment by our activities and how the environment influences us. Another goal for this type of course is the improvement of the ability of those participating to understand newspaper and magazine articles on environmental issues and to vote knowledgeably on propositions of environmental impact.

Much of this book is based on a course that I taught at Indian Valley Colleges in California. There is more than enough material here for a one-semester course, and instructors and students should feel free to choose those parts of the book to study that jointly interests them most. The only exceptions are the first two chapters which lay a foundation for the rest of the book. I have attempted to make the other five parts of the text as self-inclusive as possible, with a minimum of repitition, so it would be possible to skip one or more parts if desired.

The most satisfying aspect of such a course as compared to the traditional physics or physical science course for nonscience majors is that there is an opportunity for students and teachers to engage in discussions and get to know one another. I hope this type of interaction will be possible for all who use this book in formal courses. Topics of discussion, such as energy conservation, mass transit versus the private automobile, and the effects of the energy crisis on our quality of life, are open ended and require little specialized knowledge for participation. The suggested reading at the end of each chapter is meant for those readers who wish to delve into certain subjects more deeply than is done in the text. I hope that both students and instructors will feel no responsibility to finish the whole book, and will add and subtract material freely, according to their combined interests.

I would like to take this opportunity to thank all those individuals who provided me with information and photographs for use in this text. For illustrations used on the chapter opening pages I wish to thank the U.S. Atomic Energy Commission, EPA–Documerica, and Steve Lux and Margaret McAbee of the Prentice-Hall staff. To all of those people at Prentice-Hall who helped in the publication of this text, particularly William Grimshaw, I wish to express my sincere appreciation. In addition, I would like to thank my family, Ellen, Erica, and Rachel, for their support, encouragement, and patience while this book was being written. I owe a special debt of gratitude to Ellen for the many helpful suggestions she gave upon reviewing the manuscript.

ISAAC TURIEL

Physics, the environment and man

Part One

Introduction to the properties and structure of matter and the biosphere

Chapter 1

Properties and structure of matter

Man as a reflection of his environment

Man is influenced in many ways by different environmental factors, both natural and man-made. In this book we discuss a number of interactions between man and his physical environment. The effects of such factors as temperature, humidity, radiation, noise, and air pollutants are considered. From time to time we also look at the effects man has had on the environment; in particular, from such activities as power production, mining, and transportation. In order to understand better the complex interactions between man and his environment, concepts and principles of physics, chemistry, and biology are introduced. The main emphasis, however, is on physics.

Table 1-1 ELEMENTARY COMPOSITION OF HUMAN BODY AND EARTH'S SURFACE

			Human body		*Earth's surface*†	
Element	*Atomic number*	*Atomic weight*	*Percent of body weight*	*Order of atomic abundance*	*Percent of surface weight*	*Order of atomic abundance*
*Hydrogen	1	1.0	10.0	1	5.4	1
*Oxygen	8	16.0	65.0	2	66.3	2
*Carbon	6	12.0	18.0	3	0.017	15
*Nitrogen	7	14.0	3.0	4	0.0023	20
*Calcium	20	40.1	1.8	5	1.84	6
*Phosphorus	15	31.0	1.1	6	0.059	13
*Sulfur	16	32.1	0.22	7	0.070	12
*Sodium	11	23.0	0.13	8	1.94	5
*Potassium	19	39.1	0.20	9	1.31	9
*Chlorine	17	35.5	0.15	10	0.96	10
*Magnesium	12	24.3	0.043	11	1.11	7
Fluorine	9	19.0	0.0090	12	0.035	14
*Iron	26	55.9	0.0057	13	2.50	8
Silicon	14	28.1	0.0020	14	13.9	3
*Zinc	30	65.4	0.0033	15	0.0033	26
Bromine	35	79.9	0.0020	16	0.0034	27
Rubidium	37	85.5	0.0017	17	0.0060	24
Aluminum	13	27.0	0.00014	18	4.1	4
Strontium	38	87.6	0.00020	19	0.023	17
*Copper	29	63.5	0.00014	20	0.0023	29
Boron	5	10.8	0.000014	21	0.0004	28
*Manganese	25	54.9	0.000030	22	0.050	16
Titanium	22	47.9	0.000021	23	0.22	11
Nickel	28	58.7	0.000014	24	0.0040	25
Lithium	3	6.9	0.000001	25	0.0015	18
Barium	56	137.4	0.000023	26	0.020	21

*Essential life elements.

†All of hydrosphere plus an equal mass of lithosphere.

(From *Environmental Physiology Syllabus*, Nello Pace, Design Enterprises, 1972, Berkeley, CA)

If we look at the abundance of the elements comprising the earth's surface and those making up the human body, we find a striking similarity in the two lists (Table 1-1). In terms of number of atoms present, hydrogen and oxygen are the two most abundant elements both in our bodies and on the earth. Moving down the list we can see that, with few exceptions (silicon and aluminum), the elements that are very abundant on the earth are also very abundant in the body. Both aluminum and silicon are strongly bound to oxygen in compounds found in the earth's crust and are unavailable to enter into chemical reactions in the body. Therefore, although they are very abundant in the earth's crust, aluminum and silicon are very scarce in the human body. The above comparison indicates that man's chemical makeup is a reflection of his environment.

The earth rotates once every 24 hours about an axis which passes through both poles and through the center of the earth. This 24-hour cycle of light and dark has affected man's bodily functions (probably over hundreds of thousands of years) in such a way that they, too, follow a 24-hour cycle

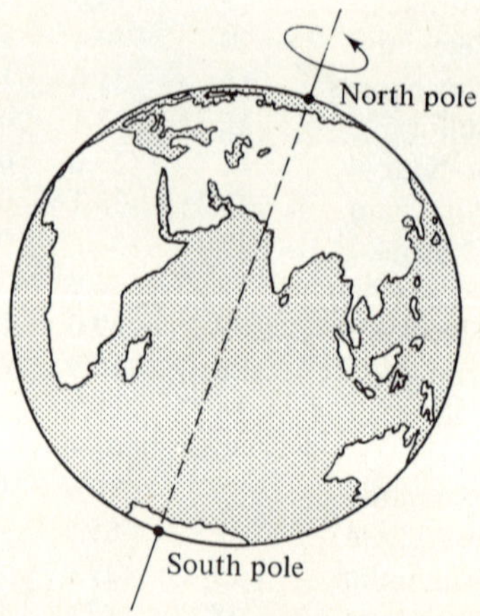

or circadian rhythm. The word "circadian" is derived from the Latin *circa* (about) and *dies* (day). Perhaps organisms (including man) that had biological rhythms close to that of the earth's rotation retained some advantage in survival (like greater efficiency in hunting during daylight hours); and they survived in greater numbers through a process of natural selection.

It appears that the period (24 hours) of this circadian rhythm is essentially unaffected by external stimuli, although the time of the maxima and minima of certain body functions can be shifted. Some examples of body functions which exhibit a circadian rhythm are body temperature, the volume of urine excreted, the reaction to a drug, and the number of dividing cells in any tissue. Figure 1-1 shows the effects on certain bodily functions when individuals were kept underground without any contact with the outside world. It can be seen that the 24-hour period remains essentially undisturbed, but that there is some shifting of the time of occurrence of the peaks.

If we take a jet plane from New York to Paris, our circadian clock keeps on going with New York time. When we arrive in Paris we are out of phase, by about 6 hours, with the local pattern of light and dark. Entrainment to

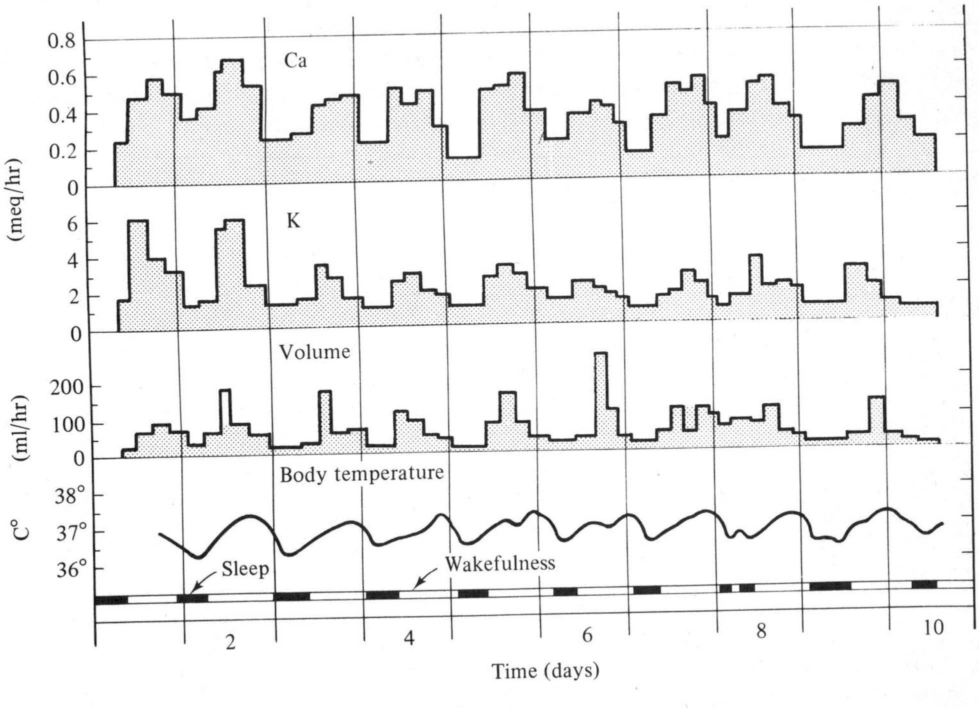

Figure 1-1 Circadian rhythm of urine excretion (calcium, potassium, and water), body temperature, and sleep-wake cycle in a human subject kept without timepiece in complete isolation from the outside world. (From Aschoff, J., "Circadian Rhythms in Man," *Science*, Vol. 148 (June 11, 1965), 1427–1432. Copyright 1965 by the American Association for the Advancement of Science.)

local time takes two to three days, during which we feel tired and less efficient. This observation (and the experiment described above) indicates that our circadian rhythm has an external and internal component. It appears that the 24-hour period is controlled internally and is not easily influenced by external stimuli, if at all. However, the phase, or time of occurrence of maxima and minima in bodily functions, is influenced by environmental factors such as ambient light.

Density

Matter has many different properties. For example, it has mass (or inertia), and it exerts forces upon other matter. A basic property of matter with which we are all familiar is density. A lead ball is much heavier than an aluminum ball of the same size because lead is more dense than aluminum. A cork will float on water since it is less dense than water.

Density is defined as the mass per unit volume of a substance. The density of any substance depends on the geometrical arrangement of the atoms comprising it (that is, how tightly they are packed together) and on the masses of the individual atoms.

$$\text{Density} = \text{mass/volume} = m/V \tag{1-1}$$

Water has a density of 1 gram/per cubic centimeter.* This means that 1 cubic centimeter (cm^3) of water has a mass of 1 gram (g), 10 cm^3 of water has a mass of 10 grams, and so on. The densities of several materials are listed in Table 1-2.

Table 1-2

Substance	*(Density grams/cm³)*
Aluminum	2.70
Iron	7.86
Lead	11.3
Gold	19.3
Wood	0.35–0.9
Bone	1.7–1.9
Water	1.00
Gasoline	0.66–0.69
Mercury	13.6
Air	1.29×10^{-3}

*One centimeter (cm) equals 0.394 inch. A mass of 1 gram weighs about $\frac{1}{30}$ ounce. The metric system of units is discussed more fully on page 43.

Example 1-1 About how many pounds does a quart of water weigh? A quart has a volume of about 1,000 cm^3. The mass in grams of a quart of water can be computed from Eq. (1-1);

$$d = m/V \qquad m = d \times V$$

$$m = 1 \text{ gram/cm}^3 \times 1{,}000 \text{ cm}^3 = 1{,}000 \text{ grams}$$

A mass of 454 grams weighs 1 pound. So

$$1{,}000 \text{ grams} = \frac{1{,}000 \text{ grams}}{454 \text{ grams/pound}} = 2.2 \text{ pounds}$$

Therefore, a quart of water weighs about 2.2 pounds and has a mass of 1,000 grams. (Later we discuss the difference between mass and weight.)

Atomic structure

The idea that matter is composed of indivisible particles called atoms dates back to the 5th century B.C. Democritus, a Greek, was the most well-known proponent of atomism. The concepts of a scientific method did not exist at the time of the natural philosophers, and the theory of atomism was justified by its simplicity and beauty rather than by experiments. Aristotle, another Greek philosopher, opposed the idea of atomism in favor of the concept of four basic elements: earth, air, fire, and water. Because of Aristotle's greater reputation, his ideas were accepted. Those of Democritus were dropped and not revived again for about 2,000 years.

In the 19th century, the concept of matter being composed of atoms was very useful in describing the details of a whole group of experiments involving chemical reactions. While these chemical reactions were being studied it was found that certain substances, called elements, were incapable of being decomposed by chemical means or created by chemical union. To date we know 92 elements that occur naturally on the earth (See inside back cover). There are also at present 13 man-made elements that have been created in nuclear reactors and in particle accelerators.

Of the 92 naturally occurring elements, only a small number comprise a great fraction of the weight of the earth's surface and of our bodies. Only four elements [hydrogen (H), oxygen (O), carbon (C), and nitrogen (N)] make up 96% of the body's weight and four elements [oxygen (O), hydrogen (H), aluminum (Al), and silicon (Si)] make up about 90% of the weight of the earth's surface (see Table 1-1).

The smallest particle of an element which can enter into a chemical combination is called an atom. An iron bar (Fig. 1-2) is pictured as being composed of individual iron atoms arranged in a particular geometric pattern. For the present we can consider an atom as a particle with a diameter of

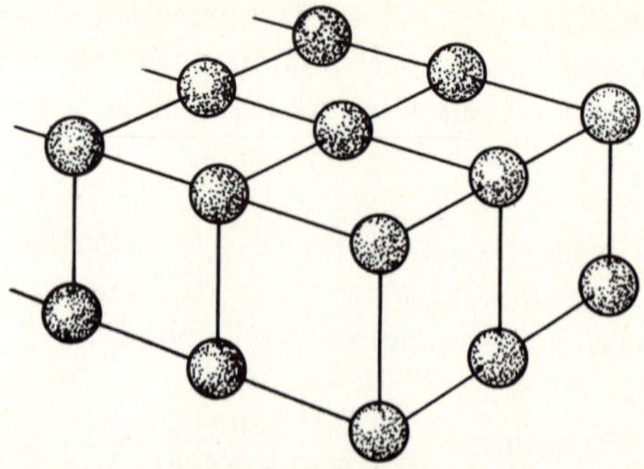

Figure 1-2 The geometrical arrangement of Fe atoms in a piece of iron

about 10^{-10} meter (or 0.0000000001 m, which is 1 preceded by 10 decimal places). (In Appendix A there is a description of how to manipulate powers of 10.)

All chemical compounds are formed of atoms. The difference between compounds is attributable to the nature, number, and arrangement of their constituent atoms. The smallest subdivision of any substance, still retaining the properties of that substance, is a molecule. Figure 1-3 shows how atoms

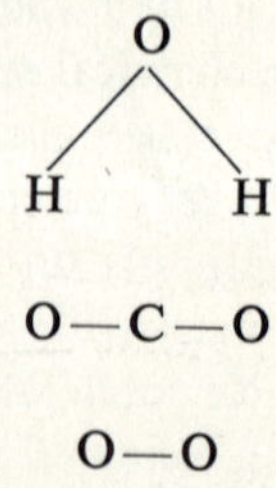

water—H_2O

carbon dioxide—CO_2

oxygen—O_2

Figure 1-3 A representative molecule of some simple compounds

combine in definite proportions to form molecules. Each water molecule is composed of two atoms of the element hydrogen and one atom of the element oxygen. All carbon dioxide molecules are made up of one atom of carbon and two atoms of oxygen, and all molecules of oxygen are composed of two atoms of oxygen.

Whether we are talking about solids, liquids, or gases the atoms and molecules which comprise them are continually moving about. In the case of a solid, such as the iron bar in Fig. 1-2, the iron atoms vibrate about fixed positions. In liquids, such as water, the molecules are also vibrating, but they can also wander throughout the liquid. The molecules which make up gases, such as carbon dioxide and oxygen, are completely free to move around and are in continuous motion. Solids, liquids, and gases are the three states of matter*.

By the early 1870s it was found that the elements could be arranged in the form of a table (See inside front cover) such that the elements in each column have many of the same physical and chemical properties. For example, the elements in the first column (excluding hydrogen) are silvery metals that combine violently with water to form basic solutions, and the elements in the last column are all chemically inert gases (helium, neon, argon, krypton, xenon, and radon).

Until the 20th century it was thought that atoms (the building blocks of the elements) were indivisible. Early in this century a simple model for the structure of the atom was developed. While this model is not exact, it does provide a very useful picture of the internal structure of the atom, and its use leads to many interesting and correct conclusions.

According to this model, atoms are composed of a small, heavy nucleus and of shells of electrons surrounding the nucleus with lots of empty space between them. The electrons circle the nucleus as the planets revolve about the sun. The nucleus of an atom consists of uncharged particles, called neutrons, and positively charged particles, called protons. Electrons, which are negatively charged, have an extremely small mass (about 1,800 times less massive than a proton or neutron). The elements are ordered in the periodic table by their atomic number (the number above the chemical symbol), which is equal to the number of protons in the nucleus. So if we know the number of protons in the nucleus of an atom, then we know the name of the element by looking at the periodic table.

In an electrically neutral atom, the number of protons is equal to the number of electrons (since protons and electrons have equal but opposite electric charge). The simplest atom is that of the element hydrogen (Fig. 1-4). An atom of ordinary hydrogen has one proton in the nucleus and one electron going around the nucleus. The second element in the periodic table is helium (Fig. 1-5).

An atom of helium has two electrons orbiting a nucleus that consists of two protons and two neutrons. As we go from lighter to heavier atoms in

*The plasma state is sometimes called the fourth state of matter. It is discussed in Chapter 11.

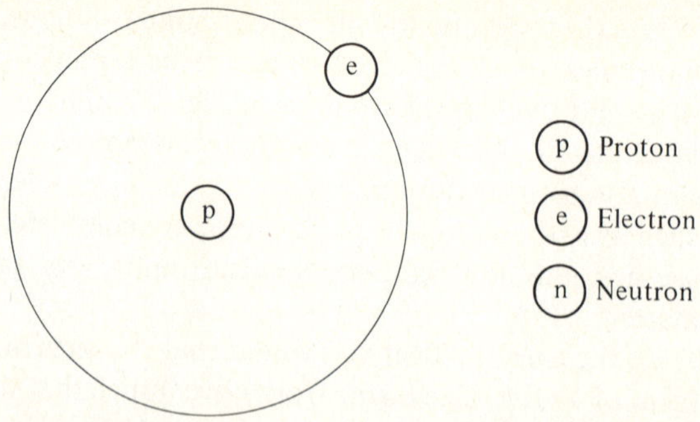

Figure 1-4 An atom of ordinary hydrogen

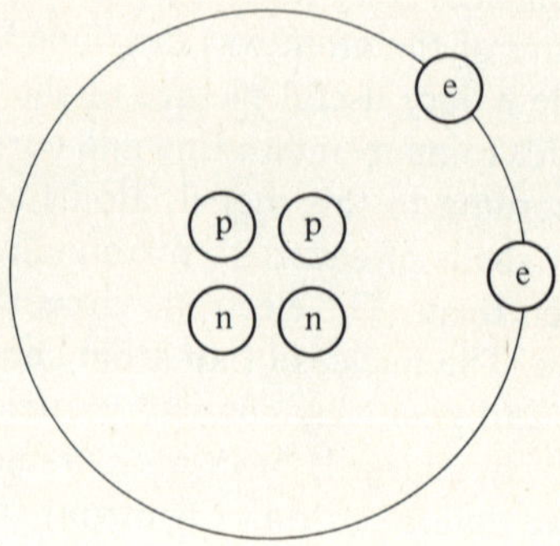

Figure 1-5 An atom of helium

the periodic table, we keep adding one proton to the nucleus to form an atom of the next element.

Example 1-2 How can we find out the physical and chemical properties of the two newest elements? These elements have the chemical symbols Rf and Ha. If we look at the periodic table, we see that Rf is in the same column as zirconium, titanium, and thorium. Therefore, if we could generate enough Rf to hold in our hands we could expect it to look like and have the properties of a metal. We could also look up the chemical properties of the elements in column IVb and assume that Rf has similar properties. The same procedure should be followed for Ha.

Energy levels and excitation

In the simplified model of the atom, which has just been discussed, the electrons that are pictured as orbiting the nucleus can be found only at certain distances from the nucleus. The electrons can move from one allowed orbit (or level) to another of higher energy only if their energy* is increased by the right amount (by absorbing light or by some other means). That is, in order to jump from level one to level two, the electron, which was originally in level one, must absorb enough energy to make up for the difference in energy between level two and level one. This process, whereby an electron moves from one energy level to a higher one, is called *excitation.* If enough energy is imparted to the electron to remove it entirely from the influence of the nucleus, then we say that the atom is *ionized.* Excitation is very important because an increase in the energy of an atom or molecule makes that atom or molecule more likely to enter into chemical reactions with other atoms or molecules.

Electrons can also move from a higher energy level to a lower one (as in

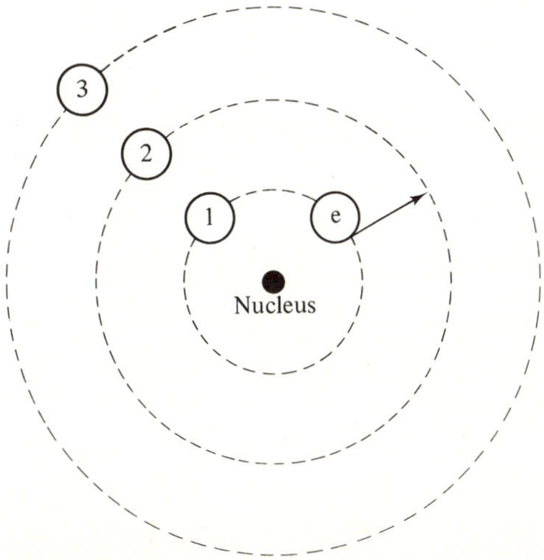

Figure 1-6 In this simplified model of the Bohr atom, named after the Danish physicist, Niels Bohr, who developed it in 1912, the electrons are found only in certain well-defined orbitals. The electron that is shown making a transition from energy level 1 to energy level 2 must absorb energy to do so.

*Energy comes in many different forms. Light and sound energy are two examples with which we are familiar. One definition of energy is the ability to do work. In Chapter 11 there is more about energy and work.

going from level two to level one). In this case, the energy of the electron decreases and the excess energy is emitted in the form of light or some other type of electromagentic radiation. We discuss electromagnetic radiation in Chapter 3.

Isotopes

Every element has two or more isotopes. An isotope exists where the atoms comprising an element all have the same number of protons but not the same number of neutrons in the nuclei of the atoms. Isotopes of an element have the same chemical properties since they have the same number of electrons in their outer shells (chemical properties depend on the number of electrons in the outer shell), but different physical properties. An example of a physical property which would be different for isotopes of the same element is mass, which in the metric system is measured in units of kilograms.

Hydrogen has three isotopes which are designated by the following symbols: $_1H^1$, $_1H^2$, and $_1H^3$. The superscript is the mass number, which is equal to the number of neutrons plus protons, and the subscript is the atomic number. The isotopes of hydrogen all have one proton in the nucleus but different numbers of neutrons (0, 1, or 2). (See Fig. 1-7.) One or more isotopes of each element are unstable, which means that they spontaneously change into other elements by emitting radiation until new stable nuclei finally are formed. This phenomenon is called radioactive decay and will be discussed in Chapter 5. Of the three isotopes in Fig. 1-7, only tritium is radioactive.

This simple picture of an atom's structure, with the electrons in well-defined shells around the nucleus, is not exact. According to the present theories of quantum mechanics, the locations of the electrons which are circling the nucleus are not exactly known, but can be discussed only in terms

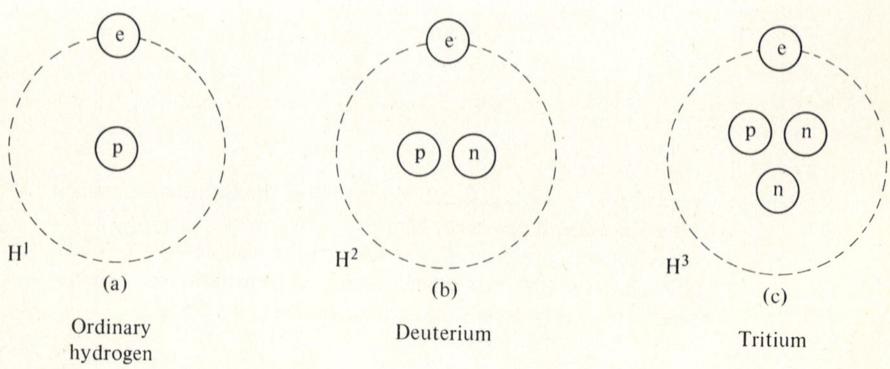

Figure 1-7 (a) Ordinary hydrogen; (b) Deuterium; (c) Tritium

of a probability of their being within a certain distance of the nucleus. Therefore, we really should consider the negative charge of the electrons as spread out in space and not as being concentrated at a point in space at any one time. Even though the simpler model of the atom (the Bohr atom) which has been described is not exact, it is easier to visualize than the quantum mechanical model, and we use it in this book.

Electric force

Charge is an essential property of matter responsible for all electric phenomena, it occurs in two forms arbitrarily designated as negative and positive. The fundamental unit of electric charge is that carried by an electron or proton. All electrons and protons carry the same amount of charge in terms of magnitude, but electrons are negative and protons positive in sign. Charged bodies can also be produced by removing electrons from neutral atoms or by adding electrons to neutral atoms. For example, if a hydrogen atom is ionized by irradiating it with ultraviolet light [Fig. 1-8(a)], then we no longer have a neutral hydrogen atom. Instead, we have two charged bodies which are called ions; they are the removed electron and the remaining proton.

There are two types of forces* between charged bodies. One is called an

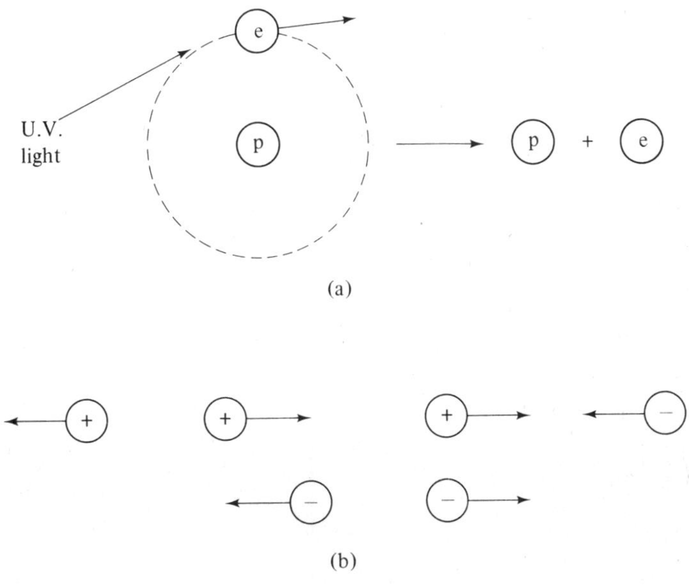

Figure 1-8 Ionization of a hydrogen atom

*Force is strictly defined as that which causes a body to accelerate. Force can usually be described as a push or a pull.

electric force (or Coulomb force, named for Charles Coulomb, the French engineer and scientist, who in 1788 experimentally determined the equation describing the force between two charged bodies). Coulomb forces exist whether or not the charged bodies are in motion. The force which is due to the motion of charges is called a magnetic force; it will be discussed in Chapter 2. Coulomb discovered that the electric force between two charged bodies is directly proportional to the product of their charges and inversely proportional to the square of the distance between them. There is no electric force between an uncharged body (such as a neutron) and a charged body. The electric force between unlike charges is attractive, and between like charges is repulsive. See Fig. 1-8(b).

Temperature and heat

The molecules of a gas are in constant random motion, as can be seen by looking at dust particles in a sunlit room. The dust particles can be seen to be moving about in a random way due to collisions with the various molecules and atoms which comprise the physical mixture we call air. If we apply heat to a gas-filled container (such as a tubeless tire), we know that the temperature of the gas will increase. On a molecular level, an increase in temperature corresponds to an increase in the average kinetic energy of the molecules; that is, the gas molecules move at greater speeds. Kinetic energy is defined as the energy of a body due to its motion. The temperature of a gas is proportional to the average kinetic energy of its molecules, which means that a doubling of the temperature results in a doubling of the average kinetic energy, a halving of the temperature results in halving of the average

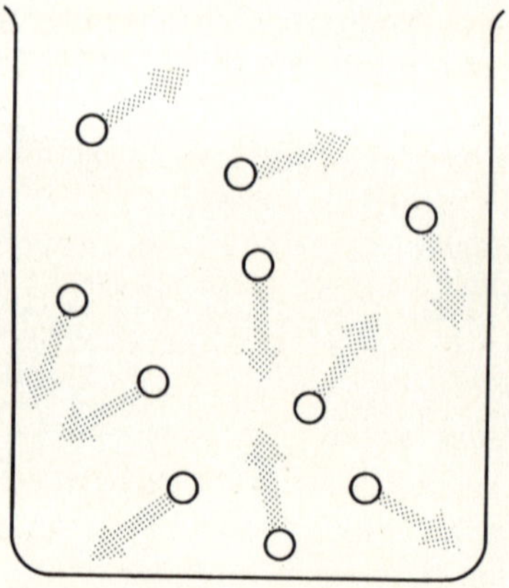

Gas molecules in random motion
inside a container

kinetic energy, and so on.* We must use the word *average* because the molecules of a gas move about at different speeds.

We can use the same definition of temperature for a liquid but not for a solid, since the molecules in a solid are not as free to move around as the molecules in a gas or liquid. In a solid, the molecules or atoms can vibrate only back and forth about some fixed position, and the vibrational energy is proportional to the absolute temperature.

We can summarize the previous discussion by noting that all atoms and molecules, whether incorporated into solids, liquids, or gases, are in constant motion. The energy associated with this motion is called thermal energy. The greater this motion in a substance, the greater the temperature of the substance. When we put our bodies into a bathtub full of hot (more than 98.6°F) water, thermal energy is transferred from the water to our skin, the temperature of our skin increases, and the temperature of the water decreases (although very slightly). Energy is transferred during collisions between water molecules and molecules making up our skin. When we go swimming in the ocean, thermal energy is transferred from our body to the cooler water (unless the ocean temperature is greater than 98.6°F). In these examples, there is a flow of thermal energy from a hotter body to a cooler one. This thermal energy, which is transferred from one body to another because of a temperature difference, is called heat. Heat will of itself always flow from a substance at a higher temperature to one at a lower temperature. This normal direction of heat flow can be reversed by an external input of energy, as is done in a refrigerator.

Temperature measurement Many instruments are used to measure temperature, but the most commonly used is a mercury thermometer. As the temperature goes up or down, the mercury in a thin glass column expands or contracts. The most widely used temperature scale is not Fahrenheit (which is used mainly in the U.S.), but the Celsius scale, where the temperature of boiling water is set at 100°C and the temperature of melting ice is set at 0°C. The interval between 0 and 100 degrees is divided into 100 equal divisions called degrees. As can be seen in Fig. 1-9, there are two other temperature scales in common use today, the Fahrenheit and Kelvin (or absolute) scales. The lowest temperature on the Kelvin scale is 0°K, which is known as absolute zero.

Even at absolute zero, molecules would still have a small kinetic energy, called zero point energy. The explanation of this energy involves quantum mechanics.

*We are talking about absolute temperature, which will be defined later.

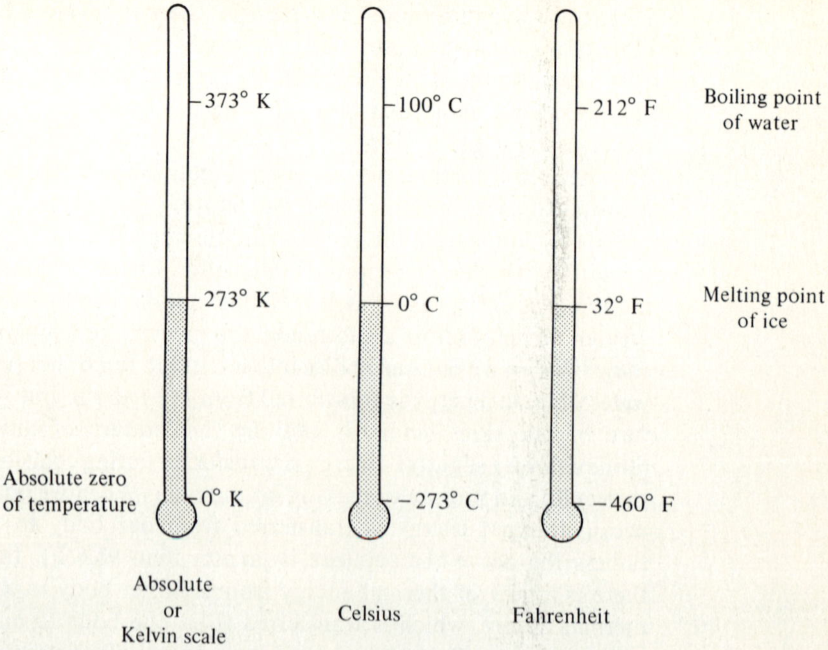

Figure 1-9 Comparison of three different temperature scales: Fahrenheit, Celsius, and Kelvin. Such scales are established by selecting as fixed points two temperatures which can be reproduced easily (boiling point of water and melting point of ice), assigning numbers to these temperatures, and then dividing the space between them into some convenient number of divisions.

Suggested reading

1 FREEMAN, IRA M., *Physics Made Simple*. Garden City, N.Y.: Doubleday, & Company, Inc., 1965.

2 ASCHOFF, JURGEN, "Circadian Rhythms in Man," *Science*, **148** (June 1965), 1427–1432.

3 PACE, NELLO, *Environmental Physiology Syllabus*. Berkeley, CA: Design Enterprises, 1972.

Some introductory physics texts

4 HEWITT, PAUL, *Conceptual Physics*. Boston: Little, Brown and Co., 1971.

5 MILLER, FRANKLIN, Jr., *College Physics*, 3d ed. New York: Harcourt, Brace & Jovanovich, 1972.

6 J. A. RICHARDS, F. W. SEARS, M. R. WEHR, and M. W. ZEMANSKY,

Modern University Physics. Reading, MA: Addison-Wesley Publishing Company, Inc., 1960.

A few environmental books

7 HARTE, JOHN, and ROBERT H. Socolow, eds. *Patient Earth*. New York: Holt, Rinehart and Winston, Inc., 1971.

8 WAGNER, RICHARD, *Environment and Man*. New York: W.W. Norton & Co., 1971.

9 HODGES, LAURENT, *Environmental Pollution*. New York: Holt, Rinehart and Winston, Inc., 1973.

10 McCABE, ROBERT H., and R. F. MINES, eds., *Man and Environment*. Englewood Cliffs, NJ: Prentice-Hall, Inc., 1972.

11 PRIEST, JOSEPH, *Problems of our Physical Environment*. Reading, MA: Addison-Wesley Publishing Co., 1973.

12 MORAN, JOSEPH M., MICHAEL D. MORGAN, and JAMES H. WIERSMA, *An Introduction to Environmental Sciences*. Boston: Little, Brown and Co., 1973.

Questions

1 List the 14 most abundant elements in the human body. Then list the 14 most abundant elements in the earth's surface (in terms of numbers of atoms). How many elements appear in both lists? What conclusion can be drawn?

2 Is there a correlation between body temperature and the sleep-wake cycle in man? (See Fig. 1-1.)

3 The full name of the element with the symbol Xe is __________. What physical and chemical properties does it have?

4 Which particle listed below is the smallest?
(*a*) a molecule
(*b*) an atom
(*c*) a nucleus
(*d*) a proton

5 How many protons and neutrons are in the nucleus of the most abundant isotope of carbon (C^{12})? How many electrons are found around the nucleus of a neutral carbon atom?

6 (*a*) How much mass does the air have in a room which is a cube 3 meters on a side? One meter equals 100 centimeters. (*b*) What is the weight of this air in pounds?

7 (*a*) The electric force between a proton and electron is __________.
(*b*) The electric force between a neutron and proton is __________.
(*c*) Is there any electric force between the nucleus of an atom and its orbital electrons? Explain.

8 Is there more thermal energy in a cup of hot coffee or in a bathtub full of cold water? Explain. What happens (in addition to becoming angry) if you drop the hot coffee into the bathtub?

9 When a thermometer is placed in a pot of boiling water, the mercury in the thermometer rises. Energy is transferred from the water to the mercury, causing the mercury to warm up and expand. What are the steps in this energy transfer?

Chapter 2
The biosphere

Introduction

The biosphere is that portion of the earth and its atmosphere that contains living organisms. In this book we are mainly concerned with man's activities in the biosphere. However, on occasion we also discuss some properties of other regions such as the interior of the earth or outer space. The properties of these regions can affect events taking place in the biosphere. The biosphere is usually classified into three sections:

1 Lithosphere—The solid portion of the earth which forms a 10-mile thick layer around the surface of the earth.
2 Hydrosphere—All the liquid bodies on earth.
3 Atmosphere—The gaseous envelope that surrounds the earth and is retained by its gravitational field.

The earth and its magnetic field

The earth The interior of the earth has three distinct zones: the crust, mantle, and core (Fig. 2-1). The outermost zone of the earth, called the crust, is composed of solid rock and is 20 to 30 miles thick. Below the crust is the mantle which extends approximately 1,800 miles into the interior of the earth. The mantle is also solid, but its composition is different. The third zone, the core, extends from the 1,800-mile limit of the mantle to the

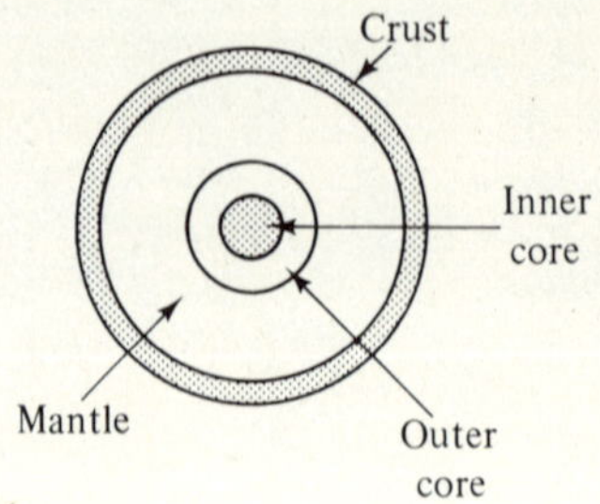

Figure 2-1 The main zones of the earth's interior

center of the earth, about 3,950 miles from the surface. The earth's core probably has two parts: an outer core of iron and some material of appreciably lower density, and an inner core of nickel and iron with some denser materials. There is good evidence that the core is liquid. Man's greatest penetration into the earth's interior is a gold mine in South Africa which is 12,000 feet (still within the crust) below the earth's surface. Temperature, pressure, and density all increase with increasing penetration of the earth's interior.

Magnetic field The first magnetic phenomena to be observed were probably those associated with natural permanent magnets, compounds of iron found near the ancient city of Magnesia. These natural magnets are able to attract unmagnetized iron to them and to cause such iron to acquire the properties of the natural magnet. If magnetized iron is in the shape of a rod, and if it is freely suspended about a vertical axis, then it will line up in a north-south direction. The use of magnets as aids to navigation can be traced back at least to the 11th century.

In the early part of the 19th century it was discovered that moving electric charges (as are found in wire carrying an electric current) can produce a magnetic field in a region of space. A magnetic field is characterized by the existence of a detectable magnetic force at every point in the region in which the magnetic field exists. That is, a charged particle, a piece of iron, or a magnet may experience a force if brought into the region of a magnetic field. It is now known that natural magnets derive their special properties from moving charges. In this case, the electrons in the iron atoms produce a magnetic field as they circle their respective nuclei. Therefore, all magnetic fields owe their origin to moving electric charges. The earth's magnetic field is thought to arise from the motion of charges in the liquid part of the earth's core (the outer core) (Fig. 2-2). Some scientists think that the earth's rotation sets these charges in motion. This hypothesis is supported by the following facts: Jupiter, which has a 10-hour day, has a much stronger magnetic field than the earth; and the moon, which rotates once in 27 days, has a very weak magnetic field. Another possible source of energy which may cause the motion of charges in the earth's outer core is heat (from radioactive decay) from the earth's inner core.

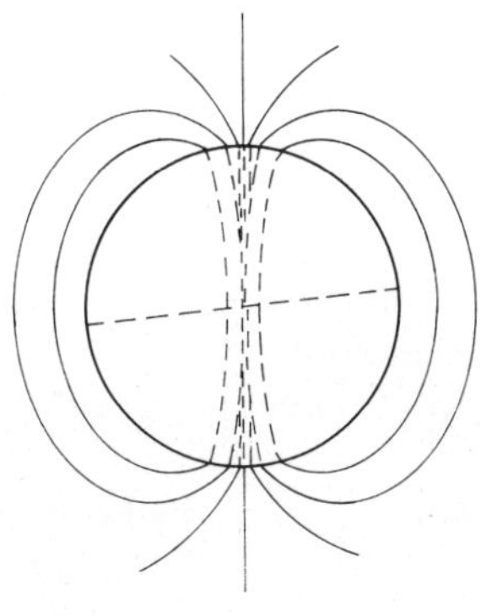

Figure 2-2 The magnetic field of the earth is similar to that of a bar magnet. The magnetic poles of the earth do not coincide with the geographical poles, and the axis of the magnetic field does not line up with the earth's rotational axis.

Whatever the source of the earth's magnetic field, it is not stable, but has wandered throughout the past. We know this from looking at natural compass needles that have been frozen into rocks. These needles are grains of magnetic iron oxide minerals, such as magnetite (FeO_4). At high temperatures iron atoms tend to align themselves with the magnetic field of the earth. Upon cooling and solidifying, after a lava flow, for example, their magnetization can no longer be appreciably influenced by changes in the earth's magnetic field. Therefore, the grains are magnetic fossils, recording the direction of the earth's field at the time the rock was formed. As samples of rock from different strata formed throughout geologic time are tested, the magnetic field of the earth can be charted as a function of time. When this is done it is found that the earth's magnetic field has fallen to zero at times and then reversed its direction. The last known reversal of the earth's magnetic field took place 700,000 years ago. Prior reversals are dated at 870,000 and 950,000 years ago.

Charged particles are deflected by a magnetic field, with a force that depends on their speeds and on the angles between their paths and the magnetic field direction. Cosmic rays are mostly composed of charged particles and are, therefore, bent by the earth's magnetic field at the end of their long journey from outer space to the earth.

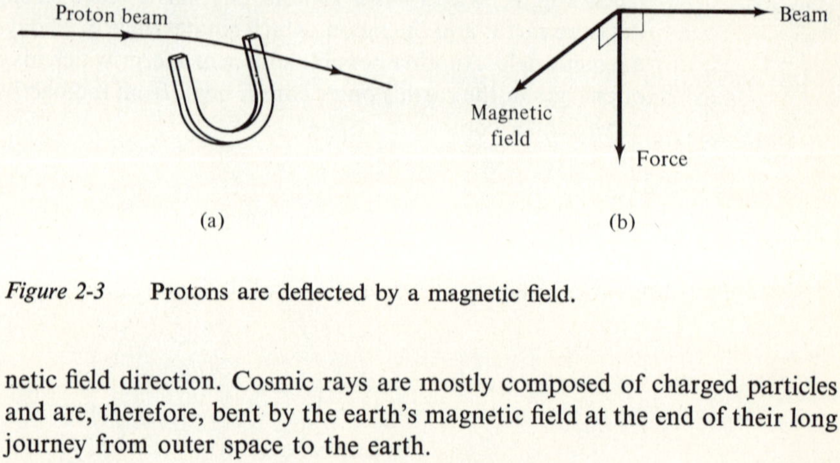

Figure 2-3 Protons are deflected by a magnetic field.

The hydrosphere and the water cycle

Hydrosphere The hydrosphere accounts for about 71% of the total surface area of the earth, but only 0.024% of the total mass of the earth. This, of course, is because the interior of the earth is very massive and contains little water.

The most common chemical compound in living matter is water. It constitutes about 75% by weight of living organisms. Life as we know it here on earth cannot survive without water. Of the other eight planets, only Mars may have water vapor in its atmosphere and water on its surface. This is one of the reasons why some scientists feel that there is a reasonable chance of finding life of some kind on Mars. The temperature range on the surface of Mars is also favorable for the existence of life.

Water cycle* The earth's supply of water is constant and is used over and over again. Most of the water (98%) is in liquid form, in the oceans, lakes, and streams. Of the remaining 2%, some is frozen in polar ice and glaciers, some is in the soil, some is in the atmosphere in the form of water vapor, and some is in the bodies of living organisms.

The water cycle (Fig. 2-4) is driven by solar energy. Energy from the sun evaporates water from the oceans, lakes, and streams, from the moist soil surfaces, from the leaves of plants (transpiration), and from the bodies of other living organisms. The water molecules are carried up into the atmosphere by air currents and eventually fall again as rain.

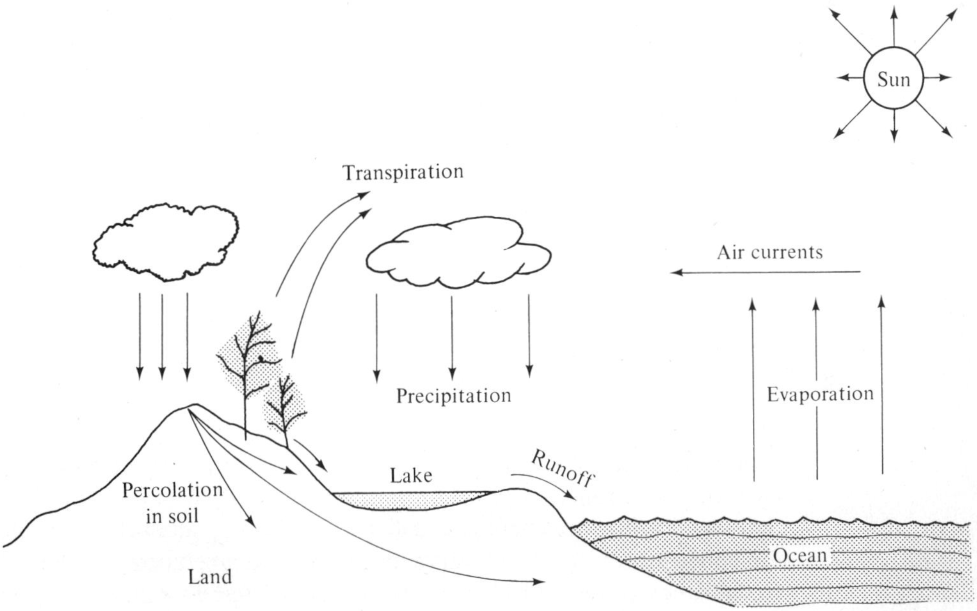

Figure 2-4 The water cycle

*The following three paragraphs have been adapted from *Invitation to Biology* by Helena Curtis, Worth Publishers, New York, 1972, page 424.

Some of the water that falls on the land percolates down through the soil until it reaches a zone of saturation. In the zone of saturation, all pores and cracks in the rock are filled with water. The upper surface of this zone of saturation is known as the water table. Below the zone of saturation is solid rock, through which the water cannot penetrate. The deep ground water, moving slowly, eventually emerges in lakes and streams and even in the ocean floor as seepages and springs, thereby completing the water cycle.

Composition of the atmosphere

Air is a mechanical mixture of gases, not a chemical compound. It is highly compressible, such that its lower layers are more dense than those above. Although the atmosphere is composed of a number of gases, below 50 miles five of them (nitrogen, oxygen, argon, carbon dioxide, and water vapor) make up 99.997% of it by volume. Rocket observations show that most of the atmospheric gases are mixed in constant proportions up to at least 30 miles. Table 2-1 gives some idea of the average composition of dry air, in addition to which there are significant quantities of water vapor and aerosols. Aerosols are particles of smoke, dust, and sea salt suspended in air.

Table 2-1 AVERAGE COMPOSITION OF THE DRY ATMOSPHERE BELOW 15 MILES

Component	*Symbol*	*Volume % (dry air)*
Nitrogen	N_2	78.08
Oxygen	O_2	20.94
Argon	Ar	0.93
Carbon dioxide	CO_2	0.03
Neon	Ne	0.0018
Helium	He	0.0005
Ozone	O_3	0.00006
Hydrogen	H	0.00005

Variations of composition When the composition of the atmosphere is looked at as a function of height, latitude, and season some variations are noticed. Water vapor is not found above 6 to 7 miles, and below that altitude the concentration of water vapor depends on temperature. In general, the water vapor content of the atmosphere is greatest in summer and in low latitudes. The concentration of water vapor in the air increases from an average of 0.56% at 60°N (the latitude of Stockholm) to 2.62% at the equator. This increase is due to the greater evaporation of water from the earth's surface at lower latitudes (more solar energy reaches the surface) and to the ability of air to hold more water vapor at higher temperatures.

Ozone (O_3), which protects life on earth by absorbing ultraviolet radiation, is concentrated mainly between 10 and 20 miles altitude. The upper layers of the atmosphere are irradiated by ultraviolet (u.v.) radiation from the sun which causes the breakup of oxygen molecules in the layer between about 50 to 60 miles.

$$O_2 \xrightarrow{\text{u.v.}} O + O \tag{2-1}$$

These separated atoms may then combine with other oxygen molecules to create ozone

$$O_2 + O + M \longrightarrow O_3 + M \tag{2-2}$$

where M represents a third atom or molecule which must be involved in the collision between the oxygen molecule and oxygen atom for this reaction to take place. This third body is not chemically altered in this reaction, but acts as a catalyst.

Such three-body collisions are rare at 50 to 60 miles because of the very low density of the atmosphere at that altitude, while below about 20 miles there is little u.v. radiation since it has already been absorbed at higher altitudes. Therefore, ozone is mainly formed between 20 and 40 miles where collisions of oxygen atoms and molecules are more likely. The atmospheric circulation transports ozone downwards so that most of it is found below 20 miles. Ozone is unstable, and it may be destroyed either by collisions with monoatomic oxygen to recreate oxygen (that is, $O_3 + O \rightarrow O_2 + O_2$) or by the action of radiation on it. Ozone concentration also varies with latitude and season. It is higher at northern latitudes and in the spring.

Atmospheric layers

The atmosphere can be conveniently divided into several well-defined horizontal layers, mainly on the basis of temperature. The evidence for this structure comes from radar, balloons, radio-wave investigations, and rocket and satellite flights. Figure 2-5 is a graph of atmospheric temperature as a function of altitude above sea level. A significant feature of this graph is that the temperature does not continue to decrease with increasing altitude, but actually reverses this trend in two locations. There are also variations in this temperature profile with the seasons and with location.

Troposphere The lowest layer of the atmosphere is called the *troposphere*, the zone where weather phenomena and atmospheric turbulence are marked. The troposphere contains 80% of the total mass of the atmosphere and virtually all of the water vapor and aerosols (particles suspended in the air). Throughout this layer there is a general decrease of temperature with height, at a mean rate of about 6.5°C/km (3.6°F/1,000 ft). The whole zone is capped

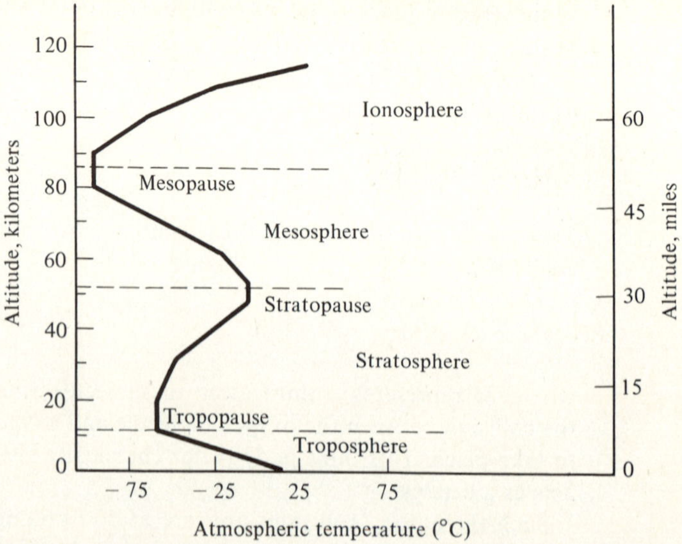

Figure 2-5 Temperature distribution of the atmosphere (1 kilometer = 0.62 mile.)

in most places by a temperature inversion level (that is, a layer of relatively warm air above cold air). In some places at the top of the troposphere there is a zone where temperature does not change with height. The troposphere thus remains to a large extent self-contained because the inversion acts as a lid that effectively limits mixing with upper layers of the atmosphere. This inversion level or weather ceiling is called the *tropopause*. Its altitude is not constant, either in location or time; at the equator the tropopause has an elevation of about 16 km (10 miles), and at the poles its elevation is only about 8 km (5 miles).

Stratosphere The *statosphere*, which is the second major atmospheric layer, extends upwards from the tropopause to about 50 km (30 miles). It contains most of the atmospheric ozone and almost no water vapor. The absorption of the ultraviolet component of the sun's radiation heats up the stratosphere (see Fig. 2-5). The ozone layer protects life on earth by absorbing most of the incoming u.v. radiation from the sun. Ultaviolet radiation is more energetic than visible light and can produce harmful effects in living organisms.

In the lower part of the stratosphere, where temperature is essentially constant with large changes in elevation (Fig. 2-5), there is very little turbulence and mixing of air. It has been proposed that supersonic transports should fly in this region of the atmosphere and thus gain the added advantage of good weather.

Mesosphere Above the warm stratopause, temperatures decrease to a minimum of about −90°C around 80 km. This layer is commonly known as the *mesosphere*. Pressure is very low in the mesosphere.

The upper atmosphere Above the mesopause (80 km) atmospheric densities are extremely low. However, the tenuous atmosphere still exerts a drag on space vehicles above 250 km. The lower portion of the upper atmosphere is composed mainly of nitrogen (N_2) and oxygen in molecular (O_2) and atomic (O) forms, whereas above 200 km atomic oxygen predominates over nitrogen. Temperature rises with height, due to the absorption of u.v. radiation by atomic oxygen. However, because of the rarefied air, solid objects such as artificial satellites do not reach such high temperatures as are indicated in Fig. 2-5.

Ionosphere We have already mentioned that solar ultraviolet radiation can break the bonds holding oxygen atoms together [Fig. 2-6(a)]. Extremely

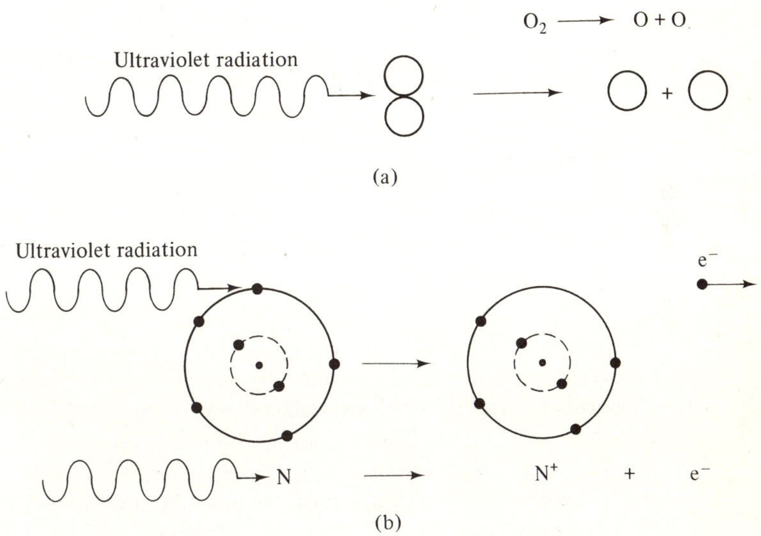

Figure 2-6 (a) Ultra violet light can break the chemical bond between two oxygen atoms. (b) Ultraviolet light can ionize an atom of nitrogen.

high-energy solar ultraviolet radiation can separate an electron from either an atom or molecule of gas, thus creating two charged particles from one neutral particle [Fig. 2-6(b)]. The two charged particles are the negative electron and the remaining positively charged atom or molecule, which is called an ion. This process, whereby radiation breaks up a neutral atom into two charged fragments, is called ionization.

Because small numbers of ionized particles occur at all altitudes higher than about 80 km (48 miles), the atmosphere above that altitude is called the *ionosphere.* It is estimated that at 300 km, the altitude of maximum ion concentration, only two out of each thousand atoms are ionized. These relatively few ionized atoms are responsible for the atmosphere's electrical properties.

In 1901 Guglielmo Marconi, an Italian inventor, showed that radio waves transmitted in England could be received in Canada, one quarter of the distance around the earth. A short time later, physicists in the United States and England independently suggested that the radio waves were being reflected back and forth between the earth's surface and a layer of charged particles in the atmosphere (Fig. 2-7), thus enabling communication between two cities which were not in a direct line of sight.

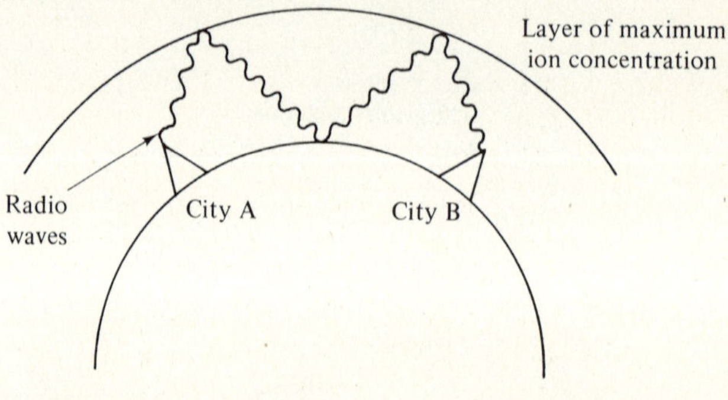

Figure 2-7 Radio waves can be sent between two cities which are not in a direct line of sight by reflecting them from the ionosphere and the earth's surface.

The distribution of charged particles in the ionosphere depends on fluctuations in incoming solar energy. There is a regular day-night difference in energy input, and there are also solar flares which are explosive emissions of energy lasting from a few minutes to a few hours. During flare periods, the

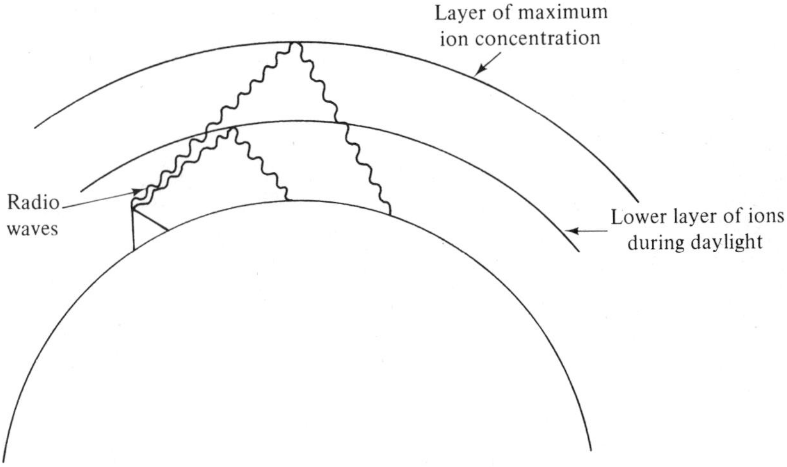

Figure 2-8 Radio waves travel farther when the lower layer of charged particles decays at sunset.

sun gives off additional amounts of high-energy ultraviolet radiation which increases the number of charged particles at all altitudes. Certain types of long-range radio communication become difficult or impossible during such periods (called ionospheric storms).

The ease of radio transmission varies with the day-night cycle on earth. During daylight hours, increased solar energy forms an ionized layer beneath the 300-km layer of maximum ion concentration. This lower layer quickly decays at sunset, whereas the maximum concentration layer persists through the night. At night the radio waves travel further because they are reflected from much higher ionized layers (Fig. 2-8).

Magnetosphere In addition to having much higher proportions of charged particles, the outermost zones of the atmosphere differ from lower levels in another fundamental way. The lower density allows the ions and electrons of these higher layers to move great distances without colliding with other particles and losing their electrical charges. Therefore, there is a strong interaction between these charged particles and the earth's magnetic field. (Remember, charged particles experience a force in a magnetic field.) Because of this strong interaction, the outer ionosphere above about 500 km (300 miles) is commonly referred to as the *magnetosphere.*

The charged particles in the magnetosphere are concentrated in two zones

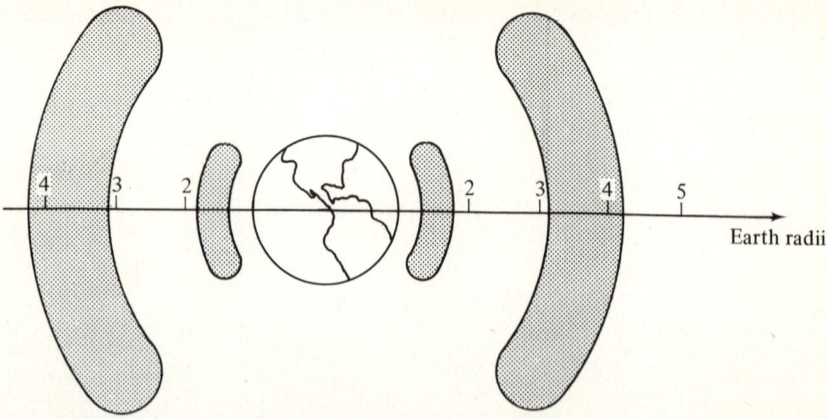

Figure 2-9 A cross section of the Van Allen radiation belts

at about 4,000 km and 20,000 km, and are known as the Van Allen radiation belts. These doughnut-shaped rings (Fig. 2-9) were discovered by James Van Allen from data gathered by the Explorer satellite in 1958. The inner ring is at about 4,000 km from the earth, and the outer ring, which is larger and wider, is about 20,000 km from the earth. Most of the charged particles, protons and electrons, trapped in the outer belt probably come from the sun. Storms on the sun send charged particles out in great quantities and many of those that pass near the earth are trapped in its magnetic field. The trapped particles follow corkscrew paths around the magnetic field lines of the earth in the upper atmosphere. Disturbances in the earth's magnetic field (which can be brought about by the arrival of particles from a solar flare) often allow the ions to dip into the lower atmosphere, causing it to glow like a fluorescent lamp. This phenomenon is known as the aurora borealis (northern lights; Fig. 2-10). Molecules of nitrogen and oxygen, and atoms of oxygen which have collided with the ions are raised to excited states. When they return to their normal states, light is emitted of an energy less than that of the exciting radiation (the ions). This phenomenon is known as fluorescence.

The particles trapped in the inner belt probably originated from the earth's atmosphere. This belt is now masked by electrons that were produced by high-altitude hydrogen bomb explosions in 1962. Those scientists in the A.E.C. who planned the atmospheric detonations did not expect the electrons to remain trapped in the Van Allen radiation belt for very long. As of this writing, however, the lower belt is still contaminated with electrons from these nuclear explosions.

Figure 2-10 An aurora photographed in Alaska. (Courtesy National Oceanic and Atmospheric Administration)

At a height of 80,000 km or so, the earth's atmosphere probably merges into that of the sun, but even the appropriate definitions of atmosphere and temperature are uncertain in these regions. The composition of the earth's atmosphere as a function of altitude is shown in Fig. 2-11. For example, at 500 km about 12% of the atmosphere is nitrogen, 86% is atomic oxygen, and 2% is helium. Above 2,000 km the atmosphere contains only He and H. At even higher altitudes there is only hydrogen.

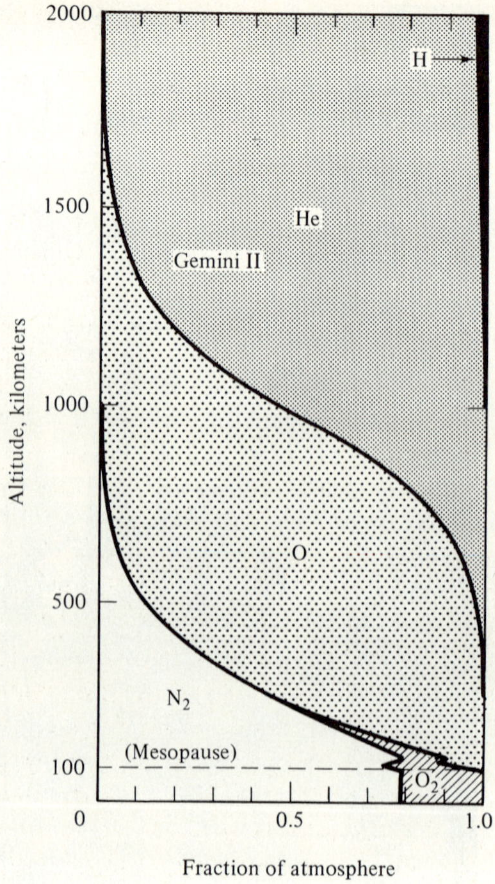

Figure 2-11 (From *Environmental Physiology Syllabus,* Nello Pace, Design Enterprises, 1972, Berkeley, CA.)

Suggested reading

1 *The Planet Earth,* a Scientific American book. New York: Simon & Schuster, Inc., 1957.

2 BARRY, R. G., and R. J. CHORLEY, *Atmosphere, Weather and Climate.* New York: Holt, Rinehart and Winston Inc., 1970.

3 MCALESTER, A.L., *The Earth, An Introduction to the Geological and Geophysical Sciences.* Englewood Cliffs, NJ: Prentice-Hall, Inc., 1973.

4 PACE, NELLO, *Environmental Physiology Syllabus.* Berkeley: Design Enterprises, 1972.

5 CLARK, S.P., *Structure of the Earth.* Englewood Cliffs, NJ: Prentice-Hall, Inc., 1971.

6 SKINNER, B.J., *Earth Resources.* Englewood Cliffs, NJ: Prentice-Hall, Inc., 1969.

7 FAGAN, JOHN J., *The Earth Environment.* Englewood Cliffs, NJ: Prentice-Hall, Inc., 1974.

Questions

1 If a man could reach the center of the earth, how many kilometers would he have to travel from the surface?

2 What assumptions are made in the determination of the direction of the earth's magnetic field as a function of time? How would the assumptions be affected if it were discovered that an earthquake took place in the region where the magnetite was found?

3 Jupiter rotates on its axis in a shorter time than it takes the earth to do so and has a stronger magnetic field than the earth. What other factors besides Jupiter's shorter day might account for its stronger magnetic field?

4 In the troposphere, which gaseous components of the earth's atmosphere vary most in percent composition?

5 (*a*) Where is most of the ozone found in the atmosphere?
(*b*) What effect does it have on the temperature in the stratosphere?
(*c*) What role does ozone play as far as life on earth is concerned?

6 All other things being equal, how much colder would you expect it to be at the top of an 8,000-foot mountain as compared to the temperature at its base (at sea level)?

7 List two important features of the troposphere and stratosphere.

8 (*a*) Why do you think that the height of the ionosphere varies with the time of day?
(*b*) What property of the ionosphere permits long-distance wireless communication?

9 How does the aurora borealis come about?

10 What is the approximate composition of the atmosphere at an altitude of 1,000 km?

Part Two

Climate and man

Chapter 3
Radiation and climate

Introduction

Climate may be defined as the prevailing weather in a particular region. Most people are aware of the strong influence of weather and climate on their lives. These influences range from the psychological to the physiological. In this chapter we are concerned with the physiological effects of climate on man. Man possesses a heat regulatory system which can maintain an almost constant internal temperature of 98.6°F as long as the temperature of the surrounding air does not fall too low or rise too high.

Without the aid of special clothing and shelters, primitive man was restricted to living over a small fraction of the earth's surface (Fig. 3.1) where the temperature did not get too high ($>$ 100°F) or too low ($<$ 65°F) for long periods. Remains of man's precursors have been recovered in African and Javanese deposits that are within the climactically favorable boundaries, and it is assumed that man spread from one or both of these locations.

With clothing, shelters, and heating, man was able to settle in most parts of the world. However, only in this century have the glaciated regions near the two poles and the icecap of Greenland become survivable. With air conditioning and imported water, even the deserts have become habitable by man. It is not likely that the deserts or polar regions of the earth will become very populated in our lifetime because tremendous expenditures of energy (and money) are required to keep people comfortable in the climates found in these regions.

Man has not adapted to his environment by genetic change; he has changed his environment to suit his genetic makeup. We discuss some of the con-

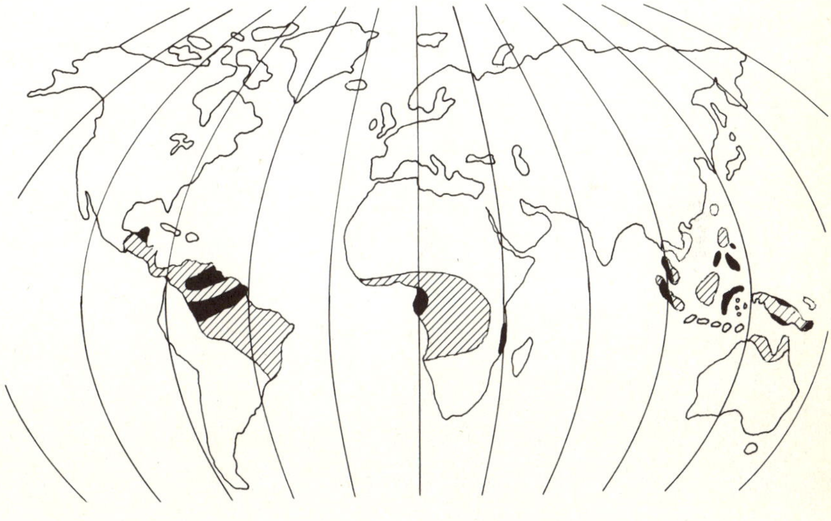

Figure 3-1 Areas with climatic conditions favorable to primitive survival: Dark areas are climatically best suited; hatched areas are suitable most of the time. (Figure 9 from the book *Weather and Health* by Helmut E. Landsberg. Copyright © 1969 by Doubleday & Co., Inc. Reproduced by permission of the Publisher.)

sequences of this alteration of the environment in the chapters on air pollution and energy and power. Before discussing man's thermoregulatory system we describe the properties of electromagnetic radiation in general and of solar radiation in particular, which is electromagnetic in nature.

Electromagnetic radiation Radiation is the emission and propagation (movement) of particles or waves. Energy is transferred from place to place by these particles or waves. For example, a heated filament in a light bulb emits radiation in the form of visible light and infrared radiation. Surprisingly, only about 10% of the radiation from a typical light bulb is in the form of visible light. Energy is transferred from the light bulb to nearby objects by both the visible light and the invisible infrared radiation. Much of the visible light is reflected by objects, allowing us to see them when the reflected light enters our eyes. The infrared radiation is absorbed by nearby objects, increasing the kinetic energy of the molecules of the objects and, therefore, their temperature. This absorption of infrared radiation makes us feel warmer when we sit near a light bulb of high wattage. Another source of radiation is radioactive material. An isotope of strontium, Sr 90, spontaneously emits electrons. These particles can transfer their energy to any material they happen to pass through, causing ionization of that material. If too much strontium 90 finds its way into our bodies, we have a radiation hazard. More about this in Chapter 6.

The term electromagnetic is used to describe all radiation which travels (in a vacuum) at the speed of light in a vacuum. Theoretically, a vacuum exists where no molecules or atoms of any substance are present. However, there is no such place; even in outer space there is still about 1 molecule in a volume of 1 cubic centimeter. Therefore, when we talk about the speed of light in a vacuum, we mean the speed with which light travels through outer space or through a man-made "vacuum." Actually, it turns out that the speed of light in air is almost the same as it is in a vacuum, so whenever we talk about the speed of light, we use the value 3×10^8 m/sec, or 186,000 miles/sec in either a vacuum or in air.

Example 3-1: How long does it take light from the sun to reach the earth?

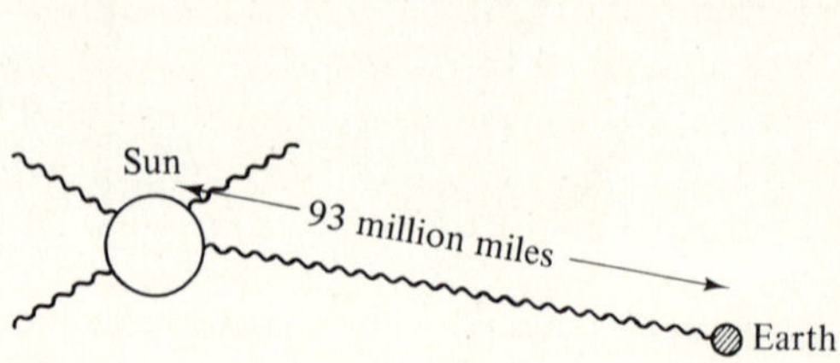

v equals the speed of light = 186,000 miles/sec.
d equals the distance from the sun to the earth = 93,000,000 miles.

t equals the time it takes for the radiation to reach the earth, which is equal to the distance the light must travel divided by its speed.

$$t = \frac{d}{v} = \frac{93{,}000{,}000 \text{ miles}}{186{,}000 \text{ miles/sec}} = 500 \text{ sec}$$

$$t = \frac{500 \text{ sec}}{60 \text{ sec/min}} = 8.33 \text{ min}$$

Visible light is the type of electromagnetic radiation most familiar to us; however, most of the electromagnetic spectrum (Fig. 3.2) is made up of invisible radiation. The only difference between these types of electromagnetic radiation, which make up what is called the electromagnetic spectrum, is their frequency or wavelength. In the following discussion of waves, the frequency and wavelength of a wave are defined.

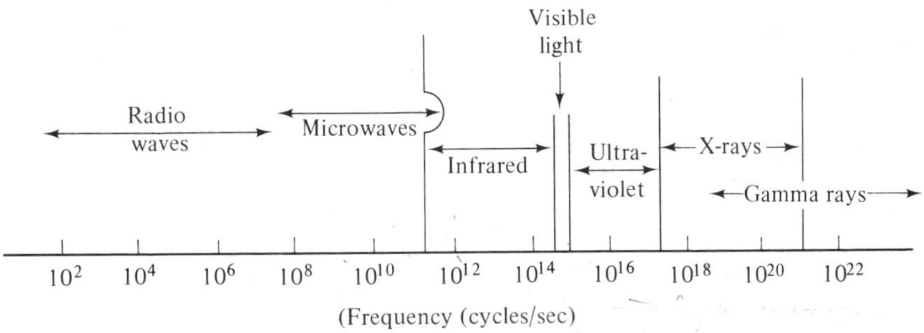

Figure 3-2 The electromagnetic spectrum extends from radio waves to gamma rays in a continuous fashion. The names of the sections are of a historical origin.

Wave model of radiation Energy is transferred from one location to another by radiation. This radiation is in the form of either particles or waves. Particulate radiation is easy to visualize; for example, a stream of bullets is one possible image which we can use. Waves, especially electromagnetic waves, are more difficult to understand. Some simple examples of waves are: the circular water waves we see when we drop a stone in water [Fig. 3-3(a)], and the waves which we set up by plucking a string or rope that is under tension [Fig. 3-3(b)].

In both waves pictured in Fig. 3-3, a disturbance is transmitted through some medium (water or string), but no matter is actually transmitted in the direction in which the disturbance moves. This is a general property of all wave motion, regardless of the type of medium through which a wave is

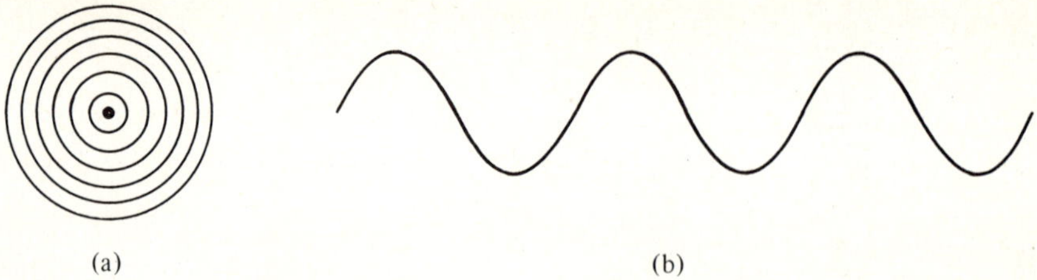

Figure 3-3 (a) Waves produced by a stone dropped in water; (b) Waves in a vibrating string

transmitted. In the case of the water waves, the water molecules move up and down when the disturbance passes them by, but they do not travel along with the disturbance. This can be seen by floating a cork in the vicinity of the place where a stone is to be dropped in the water. As the wave travels out from the point at which the stone is dropped, the cork moves only up and down in a direction perpendicular to the wave motion. Although it is harder to visualize a vibrating string, the molecules of the string also move up and down, but not in the direction of propagation of the disturbance.

We can learn some more about the general properties of waves by considering what happens when a rope is moved up and down at one end while the other end is fixed (Fig. 3-4). (We assume that the rope is very long so we won't have to worry about the reflection of the waves when they reach the other end.) The initial disturbance created by the man moves to the right at a speed, v, which depends on the tension in the rope. If the man moves his hand up and down regularly at the rate of 2 vibrations per second, then the number of complete waves which pass any point on the rope will also be

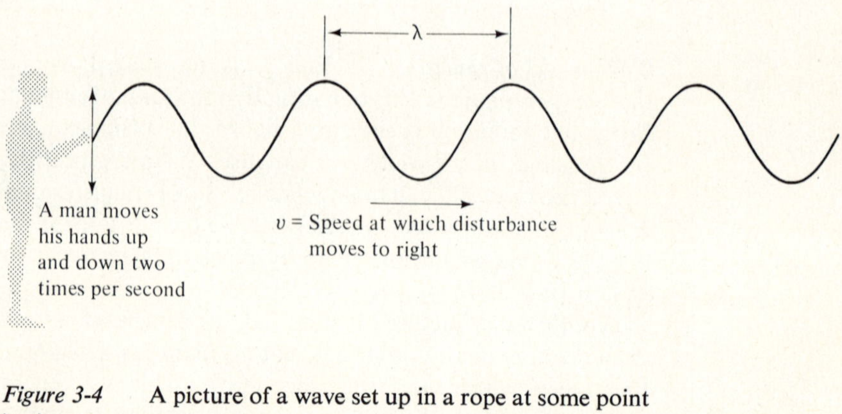

Figure 3-4 A picture of a wave set up in a rope at some point in time

2 per second. This is called the frequency of the wave. A complete wave contains both a trough and a crest such as $\sim$.

f = frequency = number of complete waves passing any point per unit time

The distance from peak to peak or trough to trough on the wave (the length of one complete cycle) is called the *wavelength* and is given the symbol lambda, λ. In the case of water waves, the wavelength is the distance between crests of the wave. There is a simple relationship between the frequency, f, and the wavelength, λ, of a wave. We can determine this relationship by considering what we observe if we are at a point to the right of the man. Let's assume that the wavelength, λ, is 1 meter and that the frequency, f, is 2 vibrations per second or 2 cycles per second. Each second we will see 2 complete waves pass us by. Each complete wave has a length of 1 meter, so the disturbance must be moving through the rope at a speed of:

$$v = 2 \text{ cycles/sec} \times 1 \text{ meter/cycle} = 2 \text{ meters/sec}$$

This rate at which the disturbance is propagated is called the wave speed, v, and in general is given by this equation:

$$v = f\lambda \tag{3-1}$$

In this example, $f = 2$ cycles/sec, $\lambda = 1$ meter, and $v = 2$ meters/sec. A unit called the hertz is often used instead of cycles/sec for frequency.

The wave speed, v, is not the speed at which the water or rope molecules vibrate about their equilibrium positions, but the speed at which the disturbance moves through the water or rope. We can see from Eq. (3-1) that if the frequency increases by a factor of two (while the wave speed remains constant), then the wavelength must decrease by a factor of two, and vice versa. Equation (3-1) can be used for any type of wave. If two of the quantities are known, then the third can be determined.

Example 3-2: When Eq. (3-1) is used for electromagnetic waves in a vacuum or in air, we can say that v equals the speed of light in a vacuum (which is usually given the symbol c). If the frequency of some electromagnetic radiation is known to be 5×10^{14} cycles/sec, then the wavelength can be calculated as follows:

$$v = f\lambda = 3 \times 10^8 \text{ m/sec} = c$$

$$\text{Rearranging: } \lambda = \frac{v}{f} = \frac{3 \times 10^8 \text{ m/sec}}{5 \times 10^{14} \text{ c/sec}} = 0.6 \times 10^{-6} \text{ meter}$$

Radiation of this wavelength is in the yellow region of the visible light spectrum.

When waves pass through a medium such as air, water, or rope, it is relatively easy to visualize what is vibrating at the frequency of the wave. That is, the air, water, and rope molecules, respectively, are vibrating about their equilibrium positions. Electromagnetic waves, however, are able to propagate through empty space, making the question "What is vibrating?" more difficult to answer.* This property of electromagnetic waves makes them so hard to visualize that some scientists in the late 1800s believed there was a transparent medium, called ether, through which visible light propagated. Experiments designed to detect the existence of this ether were unsuccessful. Finally, it was found that this concept of an ether was unnecessary and it was eventually dropped. Scientists are now more comfortable with the idea that electromagnetic radiation can travel through a vacuum.

Photon model In its interactions with matter, electromagnetic radiation is not always best described as a wave. In these interactions it sometimes behaves as a wave and sometimes as a stream of particles called photons or quanta. This so-called wave-particle duality has caused a great debate among physicists as to the true nature of light and other electromagnetic radiation. Even now no single model satisfactorily describes all interactions of radiation with matter. Therefore, it is also useful to discuss the photon nature of electromagnetic radiation.

In the wave picture of radiation, the energy of the wave is spread out along the wave and the radiation is emitted and propagated in a continuous manner. In the particle model of electromagnetic radiation, the photons or quanta carry the energy of the radiation in discrete bundles of energy given by this equation:

$$E = hf \tag{3-2}$$

where f is the frequency of the radiation, h is a constant equal to 6.63×10^{-34} joule sec, and E is the photon energy in joules. Only the values of E given by Eq. (3-2) are possible. The energy of a photon depends only on the frequency of the radiation and on the value of h, which is called Planck's constant after a German physicist, Max Planck.

*The electric and magnetic fields change with time at the frequency of the electromagnetic wave. We can picture an electric or magnetic field as a region where electric or magnetic forces exist.

Example 3-3: What is the energy carried by one photon of yellow light of frequency equal to 5×10^{14} cycles/sec?

$$E = hf = 6.6 \times 10^{-34} \text{ joule sec} \times 5 \times 10^{14} \text{ cycles/sec} = 33 \times 10^{-20} \text{ joules}$$

A joule is an extremely large amount of energy for a microscopic particle such as a photon to carry, but it is not large at all in terms of macroscopic interactions. We can see that this is the case by looking at the amount of energy it takes to heat up water. The temperature of 1 gram of water can be raised 1 degree Celsius by the addition of 4.186 joules. This amount of energy is also known as a calorie:

1 calorie = energy necessary to raise the temperature of 1 gram of water by 1 degree Celsius

We can gain a better feeling for the size of a joule by noting that it takes as much as 150,000 joules to heat up 1 pint (454 grams) of water from room temperature (20°C) to the boiling point of water (100°C). This amount of energy is about 10^{24} (1 followed by 24 zeros) times the energy carried by a photon of yellow light.

Table 3-1 THE METRIC SYSTEM

Quantity	*Units*
length	meters
mass	kilograms
time	seconds
velocity	meters per second (m/sec)
energy	joules
force	newtons

The first three quantities listed in Table 3-1 are the fundamental physical quantities from which all other physical quantities are derived. Except for the last entry (force), the other quantities have already been introduced. There are other systems of units but the metric system is used throughout this book because scientific measurements are customarily expressed in this system. It is only a matter of years before the United States converts to the metric system.

Black body radiation All bodies radiate electromagnetic radiation, whether they are hot or cold; the hotter a body, the greater its radiation. The source of this radiation is the vibrating electrons found in all materials. If the temperature is high enough (700°C), visible radiation is emitted. All bodies also receive radiation. This exchange of energy goes on continuously. Therefore, when a body remains at constant temperature it has not stopped radiating, but it is absorbing energy at the same rate that it radiates energy. When the temperature of a body is greater than that of its environment, it emits more radiation than it absorbs. A body which is a good radiator of energy is also a good absorber of energy. A black, rough surface, such as a coating of black paint, is an excellent radiator as well as an excellent absorber of radiant energy; a highly polished surface has opposite characteristics.

The way in which the temperature of a body affects the radiation coming from it was stated empirically by an Austrian physicist, Josef Stefan, in 1879 and theoretically derived five years later by Ludwig Boltzmann. The relationship which they developed applies to an idealized *blackbody*, that is a body which absorbs all of the radiation falling upon it and reflects none. Such a body is closely approximated by a hollow object in which there is a small opening. Radiation entering this opening undergoes repeated reflection inside the enclosure and very little escapes. When such a body is heated and used as a source, the radiation emitted from the interior is called "blackbody" radiation.

The amount of energy, E, radiated by a blackbody per unit time and per unit area is given by this equation:

$$E = \sigma T^4 \tag{3-3}$$

where T is the temperature of the body in degrees Kelvin (°K) and σ is a constant:

$$\sigma = 5.67 \times 10^{-8}\ \text{joule/sec}\cdot\text{m}^2\cdot{}^\circ\text{K}^4$$

The temperature of a body greatly affects its radiation; if the absolute temperature of a body is doubled, then its radiation is increased sixteenfold. The radiation emitted by the sun approximates that of a blackbody at a temperature of 6,000°K.

Quantum theory Equation (3-3) gives the total energy emitted by a blackbody as a function of its temperature, but it does not tell us how much radiation is emitted at different wavelengths. In the late 1800s the relationship between energy emitted and the wavelength of the radiation was experimentally determined. However, the existing theories concerning radiation

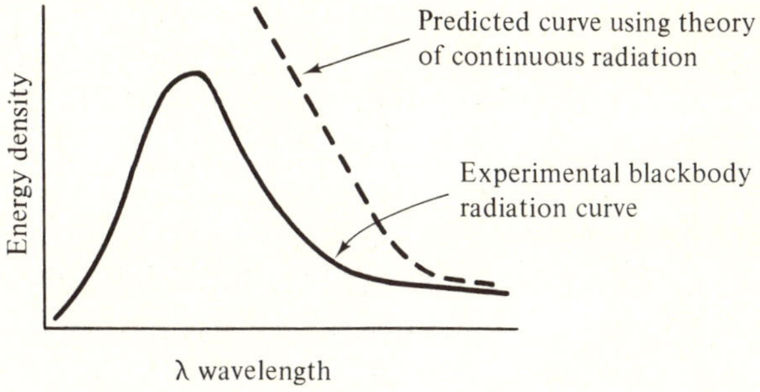

Figure 3-5 Experimental blackbody radiation curve and predicted curve, using theory of continuous radiation.

were not capable of explaining these experimental curves (Fig. 3-5). Max Planck, who experimented with blackbody radiation, was unable to explain his experimental results with the theory of continuous radiation (wave theory), but found that he could explain them by proposing a radically new theory which assumed that radiation is emitted in tiny but definite bundles of energy [given by Eq. (3-2)], each of which is now called a quantum.

This new quantum theory had little or no effect on the study of objects large enough to be observed, but it was extremely useful in explaining many phenomena of atomic physics. Some examples are the emission of light at particular wavelengths from gases (Fig. 3-6) and the photoelectric effect where electrons are ejected from some material by radiation which strikes its surface. Einstein extended Planck's quantum theory to explain the photoelectric effect, and received a Nobel prize in physics for his efforts.

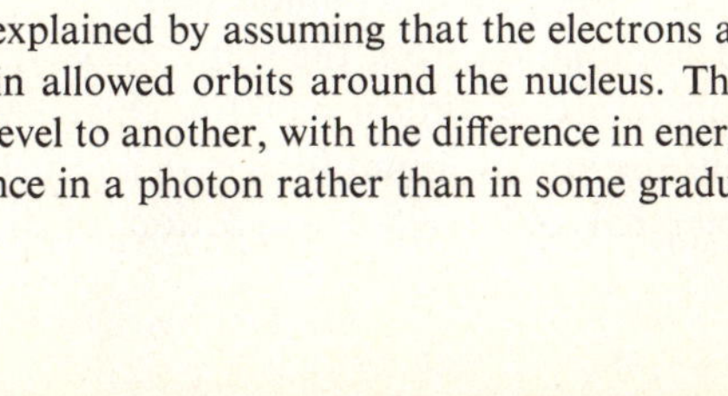

Figure 3-6 The emission of discrete radiation from excited atoms in a gas was explained by assuming that the electrons are found only in certain allowed orbits around the nucleus. They can jump from one level to another, with the difference in energy being given off at once in a photon rather than in some gradual manner.

Solar radiation The prime source of energy input to the earth is the sun,which is continually radiating energy into space by converting part of its mass into energy. This constant emission, termed insolation, is by far the largest of the three main energy inputs to the earth's surface. These main inputs are:

1 Solar radiation
2 Heat energy conveyed to the surface from the warmer interior
3 Tidal energy derived from gravitational energy in the earth-moon-sun system.

Less than one-tenth of 1% of the incident solar energy is captured by the chlorophyll of plant leaves, where it becomes the essential energy supply of the photosynthetic process and, eventually, of the plant and animal kingdom. Photosynthesis fixes carbon in the leaf and stores solar energy in the form of carbohydrate (Fig. 3-7).

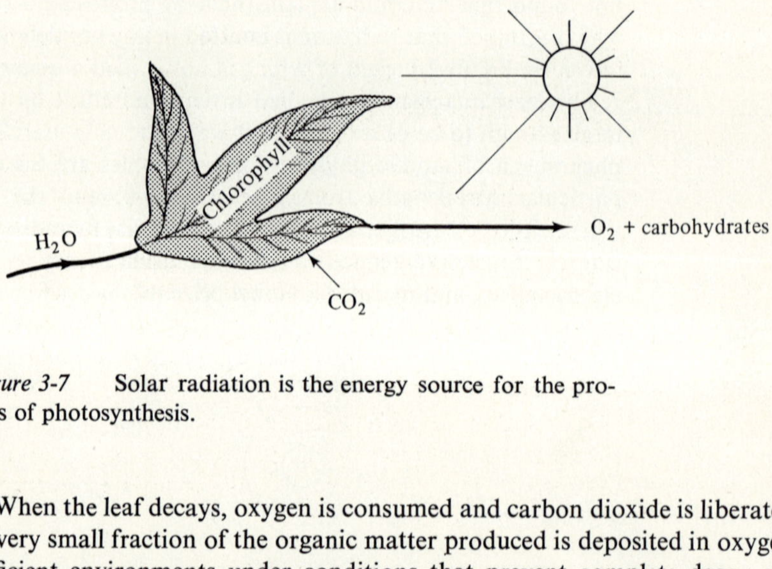

Figure 3-7 Solar radiation is the energy source for the process of photosynthesis.

When the leaf decays, oxygen is consumed and carbon dioxide is liberated. A very small fraction of the organic matter produced is deposited in oxygen-deficient environments under conditions that prevent complete decay and loss of energy. This is the origin of the fossil fuels: coal, oil shale, petroleum, and natural gas which are rich in energy stored up chemically from the sunshine of the past 600 million years.

The radiation from the sun approaches that of a blackbody which is at a temperature of almost 6,000°K (Fig. 3-8). The greatest output of radiation from the sun is found in the visible part of the electromagnetic spectrum. Note that only this tiny portion of the electromagnetic spectrum is re-

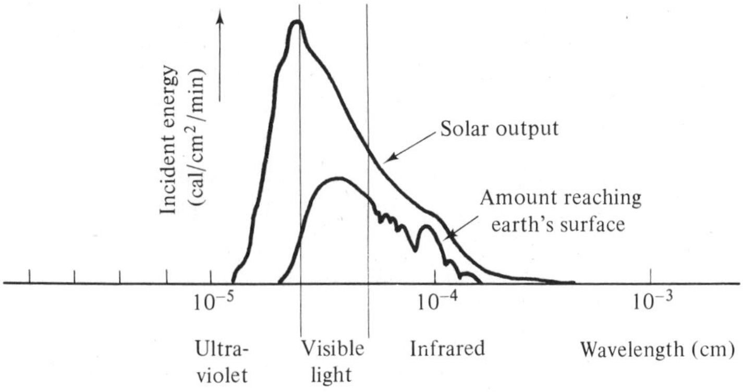

Figure 3-8 Sunlight is filtered by the atmosphere. Most of the ultraviolet, as well as some of the visible and infrared, radiation is absorbed by atmospheric gases and does not reach the earth's surface.

sponsible for vision, for the rhythmic day-night regulation of many biological activities, and for photosynthesis on which all life depends.

There are at least two reasons for this dependency on the visible part of the spectrum. First, the greatest output of radiation is in this region of the spectrum. Second, visible light photons are of just the right energy to only excite molecules; that is, to raise electrons from one energy level to another and, therefore, to produce biological changes. Radiation of longer wavelength (infrared, radio waves) increases the motion of molecules (their temperature goes up), but does not excite them. Shorter wavelength radiation, such as ultraviolet light and X-rays, both more energetic than visible light, can excite molecules, but it also can ionize molecules in the body and disrupt their normal structure and function.

This relationship between life on earth and the visible part of the solar spectrum can be looked at in two ways. We can say that life is well suited to the natural environment, or that the natural environment is well suited to the development of life.

Solar radiation and climate

Many factors enter into the determination of weather and climate in any particular area of the earth. High in importance is the amount of solar radiation received at that location. Topography and nearness to bodies of water are very important too. A combination of latitude, cloud cover, and time of year mostly determines the amount of solar radiation reaching any region of the earth on any particular day. Much more radiation is received at low latitudes that at high latitudes, and this excess energy is transported from the lower latitudes towards the poles; about 80% of the energy being transported by the atmosphere and 20% by the oceans. The atmospheric heat transfer occurs in the form of warm air masses and in the form of latent heat (water vapor which later condenses). This imbalance in solar heating of the earth's surface is the driving force responsible for the global wind patterns. The observed wind patterns are also influenced by the fact that they are seen from a rotating earth (Fig. 3-9). In turn, the global wind patterns, along with topographic features, play a large role in the determination of weather and climate at different localities.

For example, compare the climate of San Francisco and Washington, DC, which are situated at about the same latitude ($\simeq$ 38°N). At the latitude of the United States, the prevailing winds are out of the west. Therefore, air reaching San Francisco and much of the U. S. Pacific coast first passes over the Pacific Ocean. The ocean, which gains and loses heat more slowly than the land, gains heat from the warmer air in the summer and loses heat to the cooler air in the winter. Thus, temperatures are less extreme along the California coast than at Washington, DC, where arriving air passes over the continent which has no moderating influence on the weather. The effect of topography on climate can be seen by noting that summer temperatures are

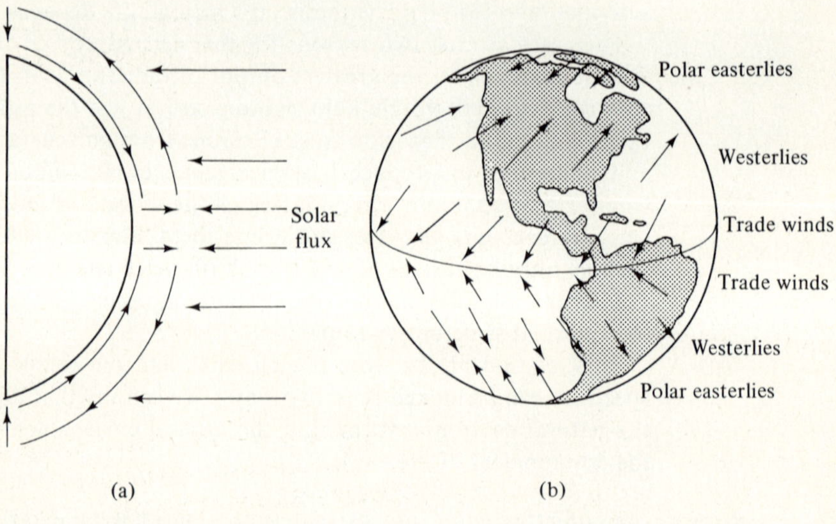

Figure 3-9 (a) More solar energy is absorbed near the equator than near the poles. Warm air rises over the equator and moves toward the poles, while cold air from the poles moves toward the equator. (b) The generalized wind directions are modified because of the earth's rotation on its axis, and appear as pictured above. Winds from the west predominate in middle latitudes, and winds from the east predominate in equatorial and polar regions.

much higher about 10 miles from the California coast than they are at the California coast. The Coast Range, a series of 2,000-foot hills which run north and south along the California coast, prevents cool Pacific air from reaching interior valleys, where 100-degree temperatures are not uncommon in summer. From this example we can see that determining the climate in a region is a complex process which depends on factors other than the amount of solar energy reaching that particular region.

Reflection and absorption of solar energy Nearly 40% of the incoming solar energy is immediately reflected back into space by the atmosphere, by clouds and by the earth's surface, leaving about 60% to heat the earth and its atmosphere. About 15% of the incoming radiation is absorbed in the atmosphere by ozone and water vapor. Ozone absorbs ultraviolet radiation while water vapor absorbs infrared radiation. The remaining 45% is absorbed by the earth's surface which eventually transfers most of this heat to the atmosphere (Fig. 3-10).

The earth, like any other body, radiates electromagnetic energy in all directions (Fig. 3-11). As the temperature of a body increases, the energy (or frequency) of the average photon emitted also increases. As a metal bar

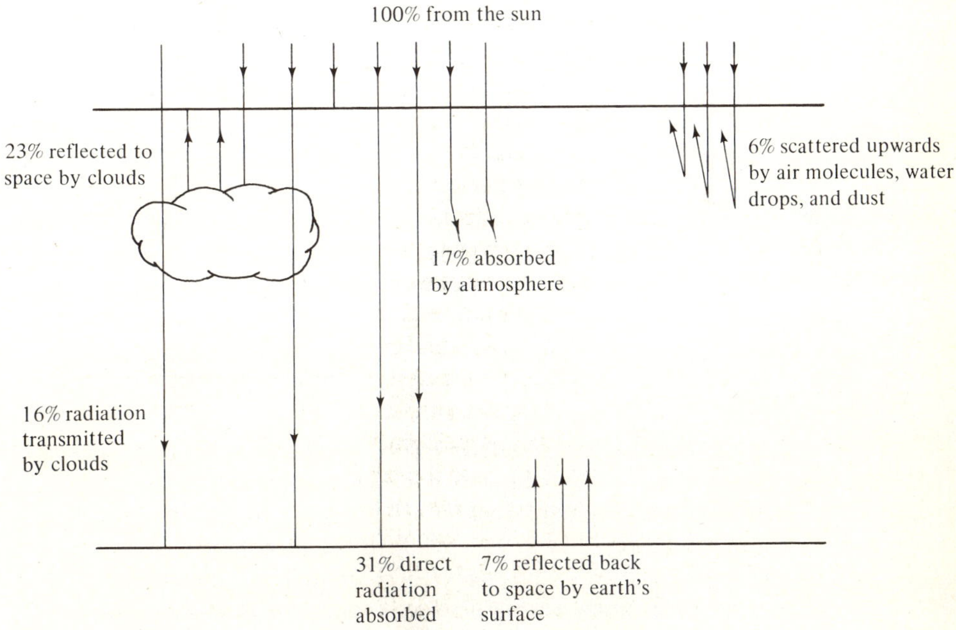

Figure 3-10 The earth's heat budget

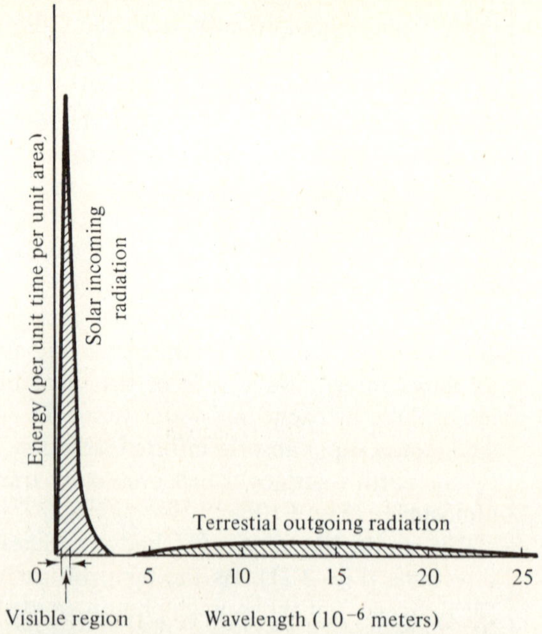

Figure 3-11 There is an obvious difference between incoming solar and outgoing terrestrial radiation. The ordinate scales of the two curves are different. The terrestrial radiation is of even lower magnitude than it appears to be.

is heated, its color will go from red to white as more photons of higher frequency are emitted. White light is a combination of all visible light frequencies (or colors). Since the temperature of the earth is much lower than that of the sun, its radiation is in the infrared part of the spectrum rather than the ultraviolet or the visible. Quite a bit of the earth's reradiated energy is absorbed by the water vapor, carbon dioxide and ozone in the atmosphere; the rest escapes into outer space. If the atmosphere did not retain energy, then the temperature of the earth's surface would drop to the point where most life would be impossible.

Most of the earth's outgoing absorbable infrared radiation is absorbed by water vapor in the troposphere. When there is night cloud cover, water droplets in the clouds absorb some of the earth's outgoing infrared radiation which would have been radiated to outer space otherwise. Some of this absorbed radiation is reradiated by the clouds back towards the ground, keeping the earth warmer than it would be on a clear night (Fig. 3-12).

Carbon dioxide shares with water vapor and water droplets the property of absorbing infrared radiation and transmitting visible light. This is why

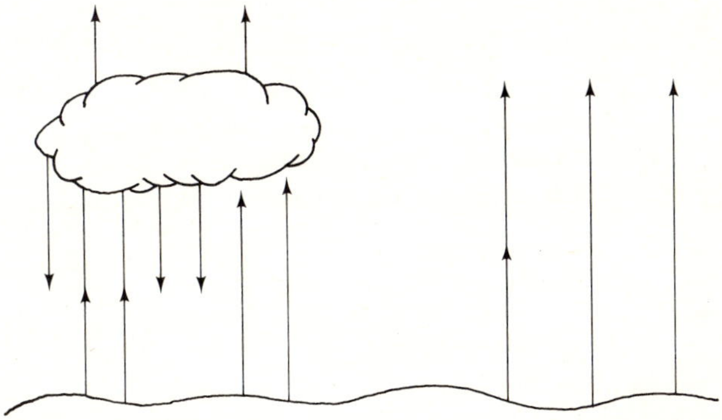

Figure 3-12 Clear nights tend to be colder than cloudy nights.

many scientists are concerned about the steady increase of carbon dioxide in the atmosphere. They fear an increase in world-wide temperatures due to the trapping of radiation (by CO_2) which would otherwise escape from the earth into outer space. The increase in CO_2 is mostly due to the increased burning of fossil fuels. (There is more about this problem in Chapter 9.)

Environmental variables

Introduction In order to attempt to predict whether or not an individual will feel comfortable under varying climatic conditions we need information about several factors. First, we must know the type of activity the individual is engaged in and what type of clothes he is wearing. There are several environmental variables which must be measured, such as relative humidity, pressure, temperature, radiant heat,* and wind velocity. An additional complicating factor is that each person has a different response to the same set of environmental variables, which means that we can talk only about the average response of some group of individuals to varying environmental conditions.

Temperature and radiant heat The concept of temperature has already been discussed and the most common method of temperature measurement, the mercury thermometer, was described in Chapter 1. The emission of radiant energy by all bodies must also be considered to determine the rela-

*Radiant heat: energy which is carried by radiation and transferred to our bodies, such as from the sun.

tive comfort of individuals under varying climatic conditions. There is obviously a greater heat input to our bodies when we are in sunshine in comparison to when we are in shade. This is due to the absorption by our bodies of radiant energy from the sun. A hot stove is also a major source of radiant energy. Depending on other factors (such as temperature), the radiant heat may or may not tend to make us feel comfortable. Two other environmental variables which we now discuss are wind speed and relative humidity.

Relative humidity In Chapter 1 we mentioned that air contains water vapor and that the amount of water vapor varies with location and time. The amount of water vapor which some fixed volume of air can hold without condensation taking place depends on the temperature of that volume of air. Condensation is a process whereby a substance changes from a vapor to a liquid, as when water vapor becomes liquid water. The amount of water vapor in any volume of air is usually given in terms of the relative humidity, which is the amount of water vapor in the volume of air divided by the maximum amount of water vapor that the same volume of air can hold at the same temperature. If a volume of air can no longer accept any water vapor without condensation occurring, then the relative humidity is 100%.

$$\text{relative humidity } \% = \frac{\text{amount of water vapor in some volume of air at temp. } t}{\text{maximum amount of water vapor that same volume of air can hold at temp. } t} \times 100$$

The amount of water vapor a volume of air can hold without condensation occurring increases as air temperature increases. Thus, warm air can hold more water vapor per unit volume than cold air. In warm air the water vapor molecules are moving more quickly than in colder air, and it is less likely that water vapor molecules will stick together during collisions with each other to form water droplets. One way of bringing about condensation is by lifting a volume of moist air to a higher altitude where the temperature is lower. Lifting of moist air due to an obstruction, such as a mountain, often leads to a higher rainfall on the windward side (side which the wind hits first) of a mountain than on the lee side. What happens is that most of the water vapor condenses before the moist air gets to the other side (lee side) of the mountain.

Measurement of humidity The moisture content of the air can be measured with an instrument called a *psychrometer*. This consists of two thermometers, one of which has a fiber covering over the mercury well (Fig. 3-13). Two temperatures are taken: the first, with the ordinary thermometer, is called the dry-bulb temperature; the second, with the fiber covering wet, is called the wet-bulb temperature. Evaporation of moisture on the fiber cools the wet-bulb thermometer to a temperature below that of the dry-bulb temperature (as long as the relative humidity is less than 100%).

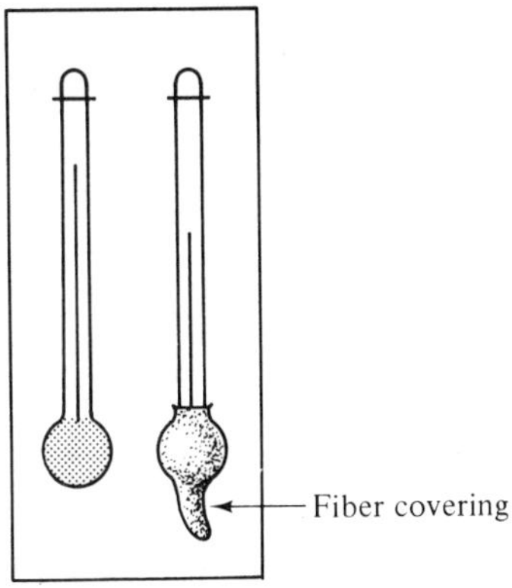

Figure 3-13 A psychrometer

The cooling effect of evaporation is discussed later, but note now that evaporation is an energy-removing process and, therefore, a cooling process. The psychrometer must be moved through the air so that the air in the vicinity of the wet-bulb thermometer does not get saturated with water vapor. If it were saturated with water vapor, then there could be no further evaporation of water from the fiber covering, and the wet- and dry-bulb temperatures would be the same. The lower the relative humidity, the greater the amount of water which will evaporate from the fiber covering into the surrounding air, and the greater the cooling effect on the thermometer. So if the air is dry, the wet- and dry-bulb temperatures will be quite different.

Once the wet- and dry-bulb temperatures have been obtained the relative humidity can be calculated from a psychrometric chart (Fig. 3-14). As the chart shows, if the wet- and dry-bulb temperatures are equal, then the relative humidity must be 100%. The value of the atmospheric pressure is included in Fig. 3-14 since the rate at which water evaporates from the wet-bulb thermometer depends on the atmospheric pressure.

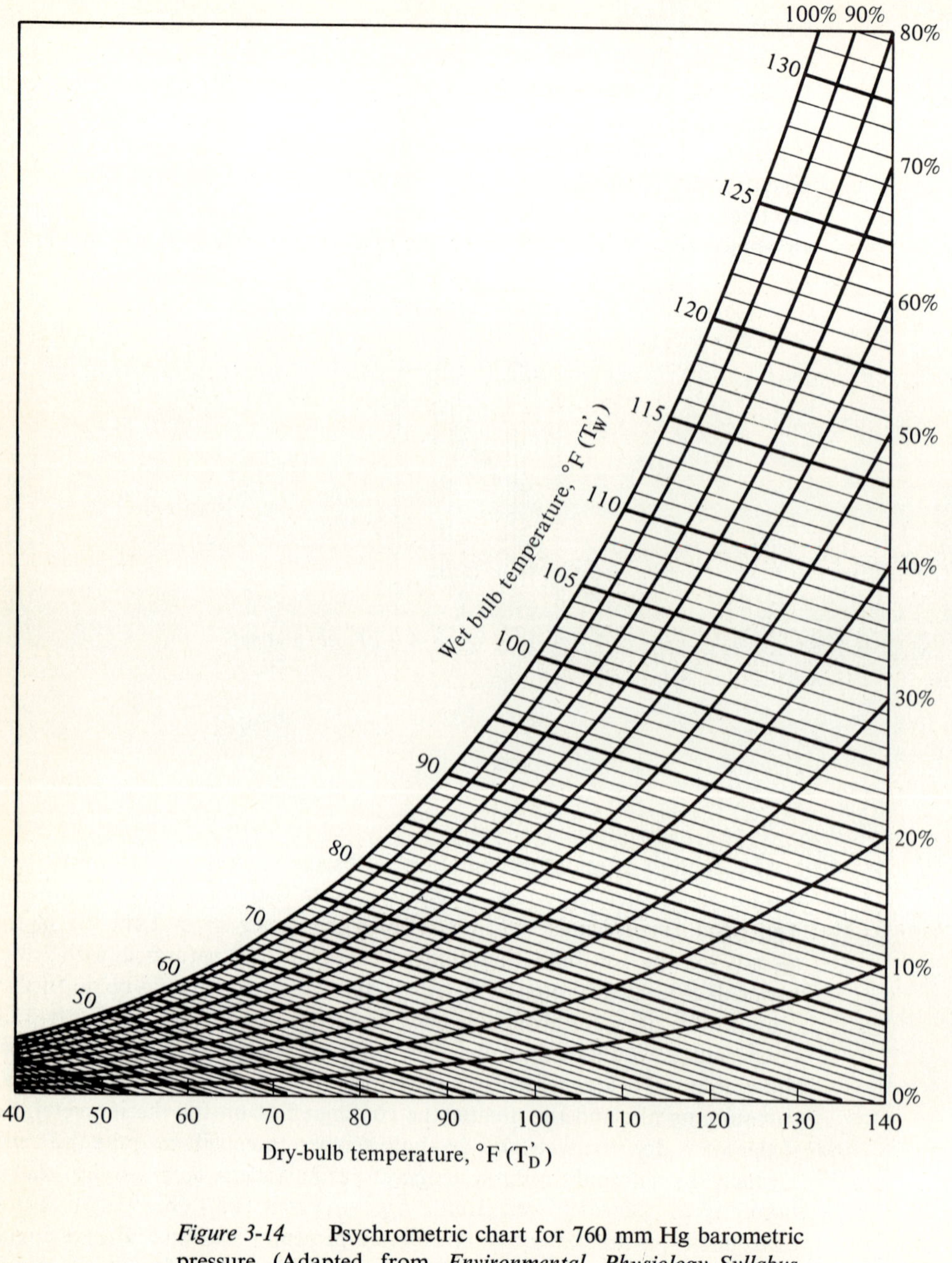

Figure 3-14 Psychrometric chart for 760 mm Hg barometric pressure (Adapted from *Environmental Physiology Syllabus*, Nello Pace, 3rd ed. Design Enterprises, 1972, Berkeley, CA.)

Example 3-4: The dry-bulb temperature is 70°F and the wet-bulb temperature is 60°F. What is the relative humidity under these conditions?

Find the point at which the 70°F dry-bulb and the 60°F wet-bulb temperature lines cross. This is between the 50% and 60% relative humidity curved lines. So we can estimate that the relative humidity is about 55% in this case.

Wind speed The speed of the wind near the earth's surface is measured by an instrument called an *anemometer*. It is composed of several hollow hemispheres that are attached by spokes to a central pivot (Fig. 3-15). The wind spins the cups around as it would a pinwheel, at a speed that is proportional to the wind speed. The wind speed is registered on a meter with a dial which has been previously calibrated.

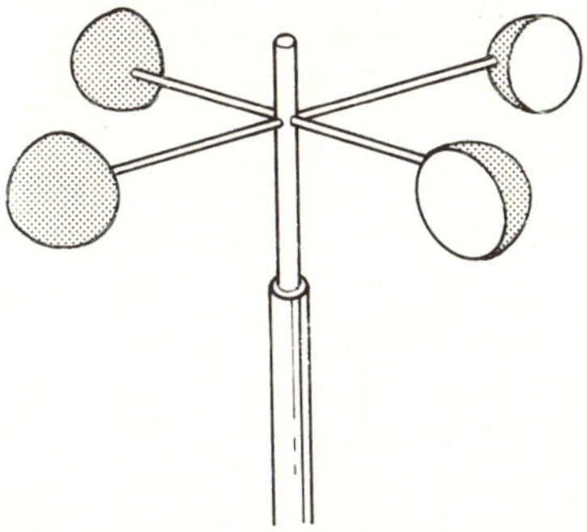

Figure 3-15 An anemometer

Balloons are used to determine the direction and the speed of winds above the earth's surface. These balloons are released into the air and then followed by radar or other means, allowing the speed and direction of the wind to be calculated. Wind speed and direction both vary with altitude above the ground. The upper winds tend to flow in the same direction most of the time, but the lower winds are usually more variable.

Atmospheric pressure We have all heard weathermen say that the barometer is falling and bad weather is expected. Those of us who pictured a *barometer*, which is an instrument for measuring air pressure, falling through space had the wrong idea. By air pressure or *atmospheric pressure*, we mean the force exerted per unit area at some particular location by the air above that location. We can picture this force as a push upon any object beneath some quantity of air. Gravity is responsible for this force and, therefore, for what we call atmospheric pressure. Without the force of gravity the earth would have no atmosphere.

$$\text{Pressure} = \frac{\text{Force}}{\text{Area}} \tag{3-4}$$

Law of gravitation The law of gravitation was proposed in 1687 by Isaac Newton, an English physicist. Popular legend claims that Newton "discovered" the idea of gravity, while lying under an apple tree, when an apple fell to the ground. Whether or not this story is true, Newton did not make his discovery in a vacuum, but was aided by the work of those scientists who came before him. In particular, there was Galileo (1564–1642) who studied the motion of moving bodies here on earth, and Tycho Brahe (1540–1601), a Danish astronomer who carefully observed the motion of the moon, stars, and planets.

It took the insight of a genius like Newton to realize that one relatively simple physical law could be used to explain both types of motion. Newton saw that the motion of the moon around the earth and the falling of an apple to the ground were both due to the force of attraction between these bodies and the earth. This was an important breakthrough, because until then it was believed that there was no connection between the motion of the planets and the motion of objects on or near the surface of the earth. Newton was able to show that gravity was responsible for both types of motion.

The statement of the universal law of gravitation is that all bodies attract each other with a force, F, that is proportional to the product of their masses, m_1 and m_2, and inversely proportional to the square of the distance, r, between them (Fig. 3-16):

$$F \propto \frac{m_1 m_2}{r^2} \tag{3-5}$$

The body of mass m_1 exerts a pull on the body of mass m_2 which is equal in magnitude, but opposite in direction, to the pull exerted by m_2 on m_1.

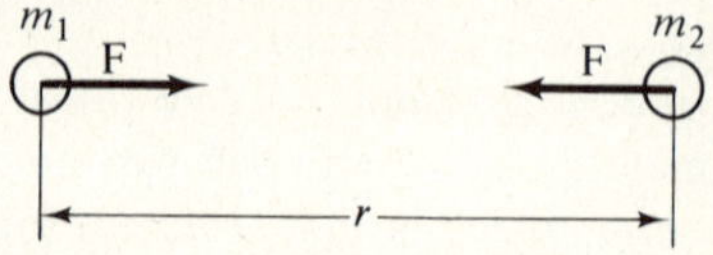

Figure 3-16 Gravitational forces are always attractive and depend on the masses of the bodies involved and the distance between them.

From Eq. (3-5) we can see that if the mass of one of the bodies is doubled, then the gravitational attraction between them is also doubled; or if the distance, r, between them is doubled, then the gravitational force of attraction is one-quarter of what it was before. As opposed to electrical forces, which can be either attractive or repulsive, gravitational forces are always attractive.

From the law of gravitation we can conclude that the earth exerts a force on the atmosphere above it. This force is exactly equal to the weight of the atmosphere since the *weight* of any object is equal to the force of gravitational attraction between it and the earth. The concepts of mass and weight are sometimes confusing. Hopefully, the following example explains the difference between these two quantities.

Consider what happens to an object when it moves from the earth's surface to some altitude above the earth. The mass of the object, which is equal to the sum of the masses of its constituent atoms, does not change with location.* This is not so for the weight of the object, which is equal to the gravitational force between it and the earth. If we look at Eq. (3-5), we see that as the distance, r, between the center of the earth and an object of mass m increases, the force of attraction between them decreases (Fig. 3-17). Therefore, the weight of the object decreases as it moves further from the earth.

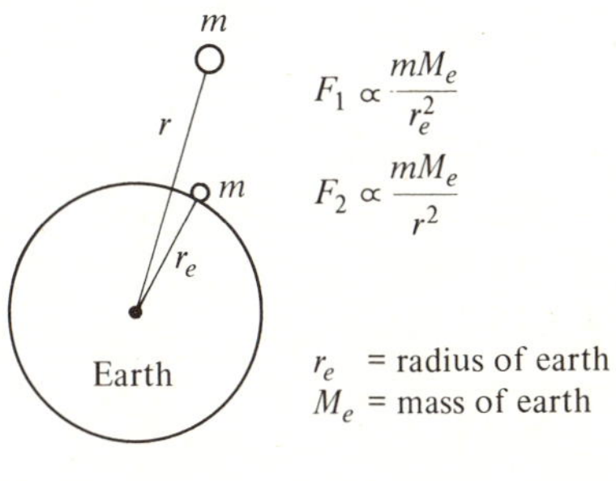

Figure 3-17 Since r is greater than r_e, F_1 is greater than F_2, and the object weighs less further from the earth.

Now we can describe the way in which a barometer measures the air pressure to which we are all subject. A barometer measures the height of a column of mercury that the atmosphere is able to support in a vertical glass tube (Fig. 3-18).

The closed upper end of the tube has a vacuum space and its lower end is immersed in a pool of mercury. By exerting pressure downwards on the

*For objects traveling close to the speed of light, their mass increases with speed.

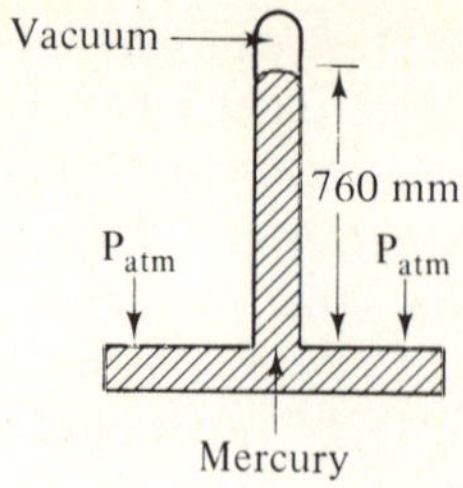

Figure 3-18 A mercury barometer

surface of the mercury in the open dish, the atmosphere is able to keep a mercury column in the glass tube at a height of about 760 millimeters (a millimeter is equal to 1/1,000 of a meter) or 30 inches. When equilibrium is reached (that is, when the mercury is stationary), the downward force acting on the column of mercury, which is its weight, W, is equal to the upward force, F_{atm}, exerted on the mercury column by the air above the dish.

In equilibrium, $W = F_{atm} = PA$, by Eq. (3-4), where A is the cross-sectional area of the glass tube. So

$$P_{atm} = W/A$$

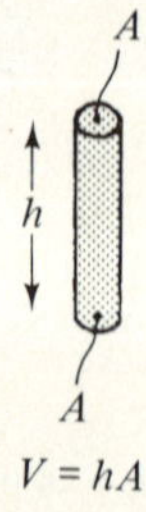

In discussing pressure we will use a more familiar system of units than the metric system. $W = \rho V$, where ρ is the density of mercury in pounds per cubic inch, and V is the volume of mercury in the column in cubic inches: $V = hA$, where h is the height of the mercury column

$$P_{atm} = \rho Ah/A = \rho h$$

$$P_{atm} = \rho h \qquad (3\text{-}6)$$

Example 3-5: (*a*) If the average atmospheric pressure at sea level is 14.7 pounds per square inch, to what height will a column of mercury rise when the atmospheric pressure is at its average level?

$$P_{\text{atm}} = \rho h \quad \text{where } \rho = 0.49 \text{ pounds/in.}^3$$

$$\text{Rearranging, } h = \frac{P_{\text{atm}}}{\rho} = \frac{14.7 \text{ pounds/in.}^2}{0.49 \text{ pound/in.}^3} = 30 \text{ inches}$$

(*b*) How many millimeters is that equal to?

$$30 \text{ in.} = 30 \text{ in.} \times 25.4 \text{ mm/in.} = 762 \text{ mm} \qquad 2.54 \text{ cm} = 1 \text{ in.}$$
$$10 \text{ mm} = 1 \text{ cm}$$

The height of the mercury column depends only on its density and on the existing air pressure. It is independent of the cross-sectional area of the glass tube.

If the atmospheric pressure rises, then the weight of the air above the pool of mercury also increases, and this forces the mercury higher into the glass tube. On the other hand, a falling barometer is due to decreasing air pressure and a lower weight for the same air column. Low-pressure systems often

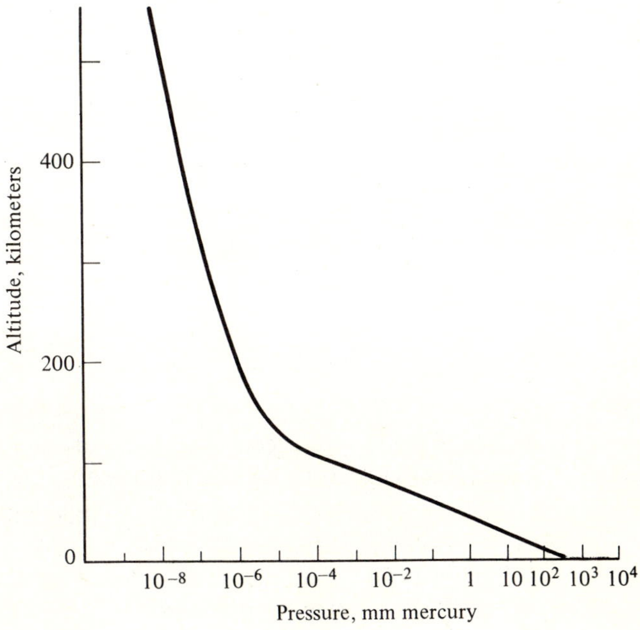

Figure 3-19 Pressure versus altitude

bring clouds and rain. In weather reports the atmospheric pressure is usually expressed in terms of the height of a mercury column, in either inches or millimeters. As we have already seen, this height in inches can be converted to a pressure in terms of pounds per square inch by using Eq. (3-6).

As we climb to higher altitudes, less of the atmosphere is above us and, therefore, the atmospheric pressure decreases. For example, at about 18,000 feet the atmospheric pressure is about 7.35 pounds per square inch or 380 mm of mercury, which is half its value at sea level. In terms of weight, half of the atmosphere is below this altitude. The highest permanent human habitations are a little below 18,000 feet in elevation. Figure 3-19 shows how the pressure changes with altitude. It falls off very rapidly and is only about 50 mm of mercury at an altitude of 7 miles or 12 kilometers. One kilometer is equal to about 0.62 mile. Jet planes flying at 6 or 7 miles altitude must carry their own supply of oxygen because of the lower atmospheric pressure.

Suggested reading

1 BARRY, R.G., and R.J. CHORLEY, *Atmosphere, Weather and Climate.* New York: Holt, Rinehart and Winston Inc., 1970.

2 BLAIR, THOMAS A., and ROBERT C. FITE, *Weather Elements.* Englewood Cliffs, NJ: Prentice-Hall, Inc., 1961.

3 LANDSBERG, HELMUT E., *Weather and Health.* New York: Doubleday & Co. Inc., 1969.

4 OLISHIFSKI, JULIAN B., and FRANK E. MCELROY, eds., *Fundamentals of Industrial Hygiene.* Chicago: National Safety Council, 1971.

5 PACE, NELLO, *Environmental Physiology Syllabus.* Berkeley: Design Enterprises, 1972.

6 SILVERBERG, ROBERT, *The Challenge of Climate.* New York: Meredith Press, 1969.

7 All of the introductory physics texts listed at the end of Chapter 1 have sections on electromagnetic radiation.

Questions

1 Where do you think more energy would be required for survival and comfort, the desert or the polar regions of the earth? List some energy needs in each region and justify your answer.

2 List the following electromagnetic radiation types in order of increasing photon energy: X-rays, radio waves, and visible light.

3 (*a*) What is the wavelength of radiation with a frequency of 2×10^{15} cycles per second?

(*b*) What do you think is vibrating at such a high frequency?

(*c*) Suppose this radiation strikes an ordinary glass window pane. Will it penetrate the window?

4 Suppose an X-ray machine emits X-rays with a frequency of 3×10^{18} cycles per second. What energy will such photons carry?

5 Express your height and weight in the metric system of units.

6 Explain what is implied when we state that the nude human body acts like a blackbody in the infrared region of the electromagnetic spectrum.

7 All objects at a temperature greater than zero degrees Kelvin radiate energy. Radiation from the human body is all in the infrared region of the spectrum. If a person's skin temperature is 27°C and his body surface area is 2 square meters, how much energy (in joules) will he radiate each second? Use Eq. (3-3) and the conversion equation: $T_{°K} = 273 + T_{°C}$.

8 (*a*) What is the smallest unit of electromagnetic radiation?
(*b*) Does visible light always behave as though it is composed of such units?

9 Explain why man ultimately depends on photosynthesis for his food whether he eats plants, meat, or fish?

10 Is the prevailing wind direction in your locality what you would expect from Fig. 3-9(a)? If not, why do you think it is different?

11 Does the earth behave like a blackbody for solar radiation? Explain your answer.

12 Why are you more likely to find dew or frost on a car after a clear, cold night rather than a cloudy, cold night?

13 As a volume of air descends on the lee side of a mountain, is condensation more likely or less likely to occur than at the top? Assume that the absolute amount of moisture in this volume of air remains constant.

14 Suppose a psychrometer is used to obtain a dry-bulb temperature of 60°F and a wet-bulb temperature of 50°F. What is the relative humidity?

15 What are the units of pressure in the metric system? Do you think we will use these units to express pressure when the U. S. converts to the metric system? Explain.

16 (*a*) According to the definition of weight given in this book, is it possible for an object to be weightless? Discuss.
(*b*) What do you think is meant when people say that the astronauts in orbit around the earth or the moon are weightless?

17 Your weight, W, on any planet or other celestial body is given by $W = mMG/r^2$, where m is your mass, M is the mass of the planet or other body, G is a constant, and r is the radius of the planet. The mass of the moon is about 1/80 the mass of the earth, and the radius of the moon is about 1/4 the radius of the earth. Use this information to find out how much you would weigh on the moon.

18 (*a*) Imagine that water is used in a barometer instead of mercury. If the atmospheric pressure were 14.7 pounds/in.^{2}, how high would a column of water rise due to atmospheric pressure? From this example you can see why water is not ordinarily used in barometers. The density of water is 1 g/cm^3 or 0.036 lb/in.^{3}

(*b*) Suppose the atmospheric pressure dropped by 10%. What height would the water column reach then?

19 (*a*) Satellites orbiting the earth at an altitude of 120 miles are still slowed down by the air in their path. What is the air pressure at that altitude?

(*b*) There are still quite a few molecules in 1 cm^3 of air at this altitude. If there are 5×10^{20} molecules/cm^2 at sea level, how many molecules per cubic centimeter are there 120 miles above the earth? Air pressure is proportional to the density of molecules.

Chapter 4

Body heat balance and energy exchange

Metabolic heat production

The food we eat is broken down by chemical reactions (called metabolic reactions) in the cells of our bodies. Most of the energy stored in the food is converted into thermal energy, with only a small percentage being used for the performance of work such as that required for contraction of our muscles. The energy stored in the food we eat (whether of animal or plant origin) can be traced back to the energy contained in solar radiation. Photosynthesis [Eq. (4-1)], which proceeds via visible light, combines molecules of carbon dioxide and water to form carbohydrates and oxygen. Compounds such as starch and sugar, which contain only carbon, oxygen, and hydrogen, are known as carbohydrates. From carbohydrates, animals and plants can synthesize proteins and other molecules for the building of new cells. Therefore, all life depends on the energy contained in sunlight.

$$6CO_2 + 6H_2O \xrightarrow[\text{chlorophyll}]{\text{sunlight}} C_6H_{12}O_6 + 6O_2 \qquad (4\text{-}1)$$

The energy stored in such foods as glucose is called chemical energy and most of it resides in the chemical bonds between atoms of carbon and hydrogen. A chemical bond is a union between two or more atoms. Electrical forces are responsible for the attraction between the atoms in a chemical bond, and the force of attraction (or strength of the bond) depends on the nature of the atoms making up the bond.

The solar energy which has been stored in glucose as chemical energy can be released by the process of oxidation, which is the opposite of photosynthesis [Eq. (4-2)].

$$C_6H_{12}O_6 + 6O_2 \longrightarrow 6CO_2 + 6H_2O + \text{energy} \qquad (4\text{-}2)$$

Instead of adding energy (sunlight) to make the reaction go, in this case, energy is released. Oxidation is the combination of a substance with oxygen (as in the above reaction), or a reaction where the atoms in an element lose electrons. The above reaction is a typical energy-producing reaction in our body, as are the oxidation of proteins or fats. Carbohydrates and fats contain only carbon (C), hydrogen (H), and oxygen (O) while protein contains nitrogen and usually some other elements such as sulfur and iron in addition to C, H, and O. Equation (4-2) can be read as follows: 1 molecule of glucose combines with 6 molecules of oxygen to form 6 molecules of carbon dioxide and 6 molecules of water, with energy being liberated in the reaction. There are intermediate steps in the oxidation process, but they are not important for the purposes of our discussion.

Specific heat The addition or subtraction of heat is measured by its effect upon the temperature of a substance. When the energy content of foods or the body's heat production is being measured, the effect upon a known amount of water is used as a standard. The unit of energy which is used is the calorie, defined earlier as the amount of heat necessary to raise the temperature of 1 gram of water by 1 degree Celsius (to be more exact from 14.5°C to 15.5°C). One kilocalorie (kcal) is equal to 1,000 calories. Nutritionists use an energy unit called the large *calorie*, which is equal to 1 kilocalorie, when discussing the energy derived from various foods. In the oxidation reaction in Eq. (4-2), 4.1 kilocalories are released for every gram of glucose that is oxidized in our bodies. In general, we can use the values 4.1 kcal per gram for oxidation of carbohydrates and protein in our bodies and 9.3 kcal per gram for fat which is oxidized in our bodies.

Example 4-1: An 8-ounce glass of whole milk has a mass of 244 grams. Whole milk contains 3.5% protein, 3.9% fat, and 4.9% carbohydrate. What is the fuel value in calories of 1 glass of milk, assuming that all of the protein, fat, and carbohydrate are oxidized?

First, we must determine the amount of protein, fat, and carbohydrate contained in a glass of whole milk:

0.035 (244 grams) = 8.55 grams of protein
0.049 (244 grams) = 12.0 grams of carbohydrate
0.039 (244 grams) = 0.52 grams of fat

The fuel value of the protein plus carbohydrate is equal to:

$$(8.55 + 12.0) \text{ grams } (4.1 \text{ kcal/gram}) = 84.3 \text{ kcal}$$

The fuel value of the fat is equal to 9.52 grams (9.3 kcal/gram) = 88.5 kcal. Therefore, the fuel value of 1 glass of whole milk is equal to 172.8 kcal or 172.8 calories.

When an amount of heat, Q, is added to a substance of mass, m, we can calculate what the temperature increase, Δt, will be if we know the specific heat, c, of the substance, which is the amount of energy in calories necessary to raise the temperature of 1 gram of that substance by 1 degree Celsius [Eq. (4-3)]:

$$Q = mc\Delta t \tag{4-3}$$

The specific heat of water has already been given as being equal to

1 calorie/gram°C

Water has one of the higher specific heats of common substances. Therefore, it takes a lot more energy to raise the temperature of water than that of most other substances, and a lot more energy is released when the temperature of water drops. The high specific heat of water as compared to land surfaces is responsible for the moderating influence of large bodies of water on the climate. In the summertime a body of land will heat up more rapidly than an adjacent body of water (of equal mass), and in the wintertime a body of land will cool off more rapidly than an adjacent body of water. This property of water, being able to react more slowly to temperature changes than land surfaces, tends to moderate the climate in certain land areas near large bodies of water. The prevailing wind direction and warm water currents also play a major role in the determination of the effect of a large body of water on the climate. For example, along the Pacific coast of the U.S. and the northwest coast of Europe the prevailing winds are from the west so air masses pass over the ocean before reaching the land. Therefore, there can be an exchange of energy between the air and the ocean, which tends to moderate the climate. On the eastern coast of the U.S. the prevailing winds are also from the west, so arriving air masses pass over the continental U.S. rather than the ocean before reaching the coastal locations. Therefore in this case there is no moderating influence from the nearby Atlantic Ocean. It should be mentioned that Eq. (4-3) also applies to cases where heat is lost by a substance and its temperature decreases.

Calorimetry The body's heat production can be measured experimentally by having its heat transferred to a known quantity of water which surrounds the person being studied, and then measuring the increase in water temperature. This method is known as direct calorimetry and is difficult to carry out in practice. The subject must be placed in a thermally insulated room (one where heat does not enter or leave; Fig. 4-1), and any increase in the temperature of the air or other surroundings must be measured in addition to

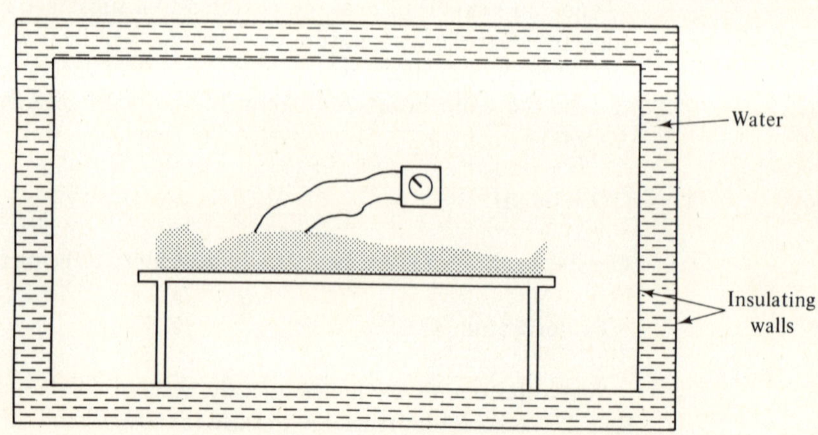

Figure 4-1 The body's heat production can be measured through direct calorimetry.

the temperature change of the surrounding water. Any evaporation of water from the subject's skin or respiratory system must also be measured since evaporating water takes energy from the body surface.

The total heat production of an individual depends on many factors such as sex, age, activity, and environmental temperature. If the subject is fasting and resting, his total rate of heat production is called his basal metabolism. This is the least amount of energy necessary to maintain his vital functions when at complete rest. For adults, the basal metabolic rate is usually somewhere between 1,200 and 2,000 kcal per day.

Example 4-2: A man remains at rest in a well-insulated room for 1 hour. If there is no temperature change in the air but the surrounding water, which has a mass of 80,000 grams (about 20 gallons), changes temperature by 1°C, what is the man's rate of heat production? Assume that there is no evaporation of water from his skin or respiratory membranes.

Heat transferred to water $= Q = mc\,\Delta t$

$m =$ mass of water $=$ 80,000 grams

$c =$ specific heat of water $=$ 1 cal/gram°C

$\Delta t =$ temperature increase of water $=$ 1°C

$Q =$ 80,000 grams (1 cal/gram°C) 1°C $=$ 80,000 cal

$Q =$ 80 kcal/hour equals man's rate of heat production

The energy released within the body can also be calculated by an indirect method. In this method the rate of elimination of the end-products of the oxidations is measured. These are carbon dioxide, water, and nitrogen-containing compounds (chiefly urea) derived from protein. The small amount of water produced is difficult to identify and a more convenient measurement is used in its place, that of oxygen consumption. The composition and the amount of each type of compound in the metabolic mixture can be determined by measuring the intake of oxygen, output of carbon dioxide, and excretion of nitrogen.

Both the direct and indirect methods of calorimetry have been used to determine the rate of energy expenditure for individuals during various activities. As Table 4-1 shows, an individual's metabolism can vary by very large factors in passing from one activity to another, although the higher levels of activity can be kept up for only a short time. The temperature-regulating system of the body is able to maintain an almost constant body temperature even though the body's heat output may vary over a large range.

Food and Exercise Calorimetry is also used to measure the amount of energy released when foods are oxidized. An 8-ounce glass of whole milk is

Table 4-1 Energy Expenditure Rate and Oxygen Requirement for 70 kg Man Engaged in Various Activities

Activity	*Energy expenditure (kcal/hr)*	*Oxygen requirement (liters/min)*
Sleeping	70	0.24
Lying quietly	80	0.27
Sitting	100	0.34
Standing	110	0.38
Driving a car	140	0.48
Office work	145	0.50
Housekeeping	150	0.52
Slow level walking, 3 km/hr	170	0.58
Piloting an airplane	175	0.60
Laboratory work	190	0.65
Level bicycle riding, 10 km/hr	210	0.72
Carpenter work	230	0.79
Laundress work	270	0.93
Brisk level walking, 6 km/hr	300	1.03
Walking up 5% grade, 5 km/hr	360	1.24
Baseball pitcher	390	1.34
Pick and shovel work	400	1.37
Walking down stairs, normal	415	1.43
Making beds	420	1.44
Swimming crawl stroke, 2 km/hr	440	1.51
Chopping wood	450	1.55
Level snowshoeing, 4 km/hr	520	1.79
Level skiing, 5 km/hr	550	1.89
Walking up 10% grade, 5.6 km/hr	580	1.99
Lumbering	600	2.06
Fencing	630	2.16
Rowing, 5.6 km/hr	660	2.27
Level running, 9 km/hr	720	2.47
Level walking in knee-deep snow	760	2.61
Level skating, 20 km/hr	780	2.68
Level running, 11 km/hr	870	2.99
Running up 8.6% grade, 11 km/hr	950	3.26
Football	1,000	3.44
Walking up stairs, normal	1,100	3.78
Level running, 18 km/hr	1,300	4.47
Swimming crawl stroke, 3.5 km/hr	1,600	5.50
Swimming breast stroke, 4.3 km/hr	3,600	12.37
Level running, 25 km/hr	3,900	13.40
Level running, 30 km/hr	9,400	32.30

(From *Environmental Physiology Syllabus*, Nello Pace, Design Enterprises, 1972, Berkeley, CA)

known to release about 170 large calories of energy when oxidized. However, in terms of the definition of a calorie already given, the above would read 170 kilocalories. This should be kept in mind when comparing caloric values of foods (as given in ordinary calorie charts) to metabolic heat production, which is expressed in true calories, not kilocalories.

Many people feel that they can work off some of their body fat by exercising. We can see that this is very difficult to do from the following calculation.

Example 4-3:

(a) How much energy is expended by a 70-kg man who runs 2 miles a day at a speed of 9 km/hr?

(b) How much weight would this man lose if this excess energy were supplied by stored body fat?

In Table 4-1 we see that a 70-kg man requires an energy expenditure of 720 kcal/hr to run at a speed of 9 km/hr (or 5.5 miles/hr), and 100 kcal/hr if he is just sitting. So we can assume that while running he is using energy at a faster rate than normal, an excess of about 600 kcal/hr. Answers:

(a) The time it takes to run 2 miles at 5.5 miles/hr is 2 miles/5.5 miles/hr = 0.364 hr. The excess energy expended during this time is

$$(600 \text{ kcal/hr})(0.364 \text{ hr}) = 218.4 \text{ kcal}$$

So we can say that this man expends an extra 218 kcal/day.

(b) When 1 gram of fat is oxidized in the body, 9.3 kcal of energy is released. This is true for fat stored in the body and fat contained in food. Assume that the man's excess energy comes from oxidizing stored body fat. Then the reduction in body fat is given by:

$$\frac{218.4 \text{ kcal/day}}{9.3 \text{ kcal/g fat}} = 23.4 \text{ g fat/day}$$

It would take about 19 or 20 days of running 2 miles a day to lose 1 pound of body fat, assuming that the man's caloric intake did not increase and that he continued all other activities as usual. Not many of us will be motivated to undertake such a program of exercise solely to lose weight. The main motivation of jogging is the expectation of an improved physical condition and a feeling of well-being. Caloric intake can be decreased by as much as 330 kcal of energy by giving up one rich dessert such as apple pie. A possible side benefit from exercise such as jogging is that it may result in a higher basal metabolism. If this happens, then more energy that is derived from daily food intake will go towards maintaining vital body functions and less towards energy storage in the form of body fat.

Body heat balance

Introduction The body temperature is maintained at a relatively constant level because of the balance which exists between heat production, heat loss, work output, and energy storage [Eq. (4-4)]:

$$\text{Metabolic heat production} = \text{heat loss} + \text{work output} \pm \text{energy storage} \tag{4-4}$$

The energy storage has a negative sign (reserves being used) when heat loss plus work output is larger than the heat production. If the metabolic heat production is greater than the heat loss plus work output, then the extra energy is stored within the body. That is, not all of the food we eat will be oxidized in this case. The first term on the right hand side of Eq. (4-4) is actually composed of several other terms, some of which may represent heat losses and some of which may represent heat gains. These methods of energy transfer will be discussed after we determine what would happen to the body temperature if heat loss were not possible.

We will calculate the expected temperature rise for a normal young adult male if he were not able to lose the heat created by his metabolism. Let us assume a basal metabolic rate of 80 kcal/hr for this person who weighs about 155 pounds or has a mass of 70 kilograms (2.2 kg = 1 pound).

If we add an amount of heat, Q, to a body which cannot dissipate it to the environment, then the temperature rise, Δt, is given by Eq. (4-3) where the various terms have already been defined:

$$Q = mc\Delta t$$

we know that $Q = 80 \times 10^3$ cal/hr $= 8 \times 10^4$ cal/hr
$m = 70 \times 10^3$ grams
$\Delta t =$ temperature change

However, we do not know the value of the average specific heat, c, of a human body. We do know that our bodies are composed of about 65% water, which has a specific heat of 1, and that the rest of the body has a specific heat of less than 1. From this information we can show that the body's specific heat is greater than 0.65, but less than 1. The actual value of the body's specific heat is:

$$c_{\text{body}} \cong 0.83 \text{ cal/g°C}$$

Rearranging,

$$\Delta t = \frac{Q}{mc} = \frac{8 \times 10^4 \text{ cal/hr}}{7 \times 10^4 \text{ g (0.83 cal/g°C)}}$$

$$\Delta t = 1.38\text{°C/hour} \quad \text{or} \quad 2.5\text{°F/hour}$$

We could not survive for more than a few hours if all channels of heat loss from our bodies were blocked. Now we are going to discuss the ways in which energy is transferred between our bodies and the environment.

Energy transfer

Conduction Heat tends to flow from warm objects to cooler ones. One way in which this comes about is by transfer of thermal energy between two objects in contact with each other. This process is known as conduction, and is responsible for the feeling of warmth we get when we hold the handle of a frying pan which is over a fire. The fire causes the molecules at one end

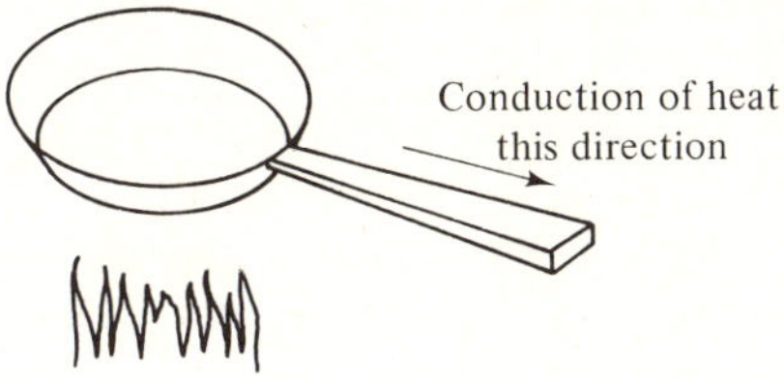

of the frying pan to move more rapidly than at the other end. These molecules and free electrons collide with neighboring molecules and electrons, causing them to move more rapidly, too. This process continues on down the handle until, finally, the molecules and electrons near our hand are moving more rapidly than the molecules making up our fingers. Collisions between the molecules in the frying pan handle and the molecules making up our fingers transfer thermal energy to our fingers and we feel warmer.

Most of us are probably aware of the fact that some materials conduct heat better than others. If the frying pan handle were made of plastic or wood it would take much longer for the heat to reach our fingers. Wood and plastic, which are poor conductors of heat, are called insulators. Air is a very poor conductor of heat and, therefore, a very good insulator. Because of the poor heat conductivity of air, double window panes with an air space in between them will keep a house warmer in winter than two window panes back to back.

Heat can also be *lost* by conduction. If you place your hand on a cold window pane your skin will become colder rather quickly. In this case, there is a flow of heat from the warmer hand to the cooler glass. In all cases of energy transfer by conduction, the amount of energy transferred between two objects is proportional to the temperature difference between them.

Usually the air temperature is lower than our skin temperature, and our body loses heat to the environment by conduction. Most of the heat is lost to the surrounding air. Loss of heat by conduction is minimized by the insulation of clothing, especially clothing with lots of air spaces through it. The effectiveness of fur as an insulator allows Eskimos to tolerate very extreme cold. In northwest Greenland the natives dress with one layer of fur facing inward and another layer facing outward. This tends to trap air which is warmed by the body in between the two layers of clothing. When the air temperature is greater than our skin temperature (skin temperature is normally around 91°F), our body has to lose its heat by other methods with no contribution from conduction.

Convection Conduction and convection are very intimately related. Conduction of heat from the body to the air increases the temperature of the air adjacent to the skin. If the heated air did not move away from the skin, conduction would quickly cease as the temperature of the air near the skin would reach the same value as the skin temperature. A small amount of convection almost always occurs around the body because the heated air near the skin rises, allowing new unheated air to take its place. Hot air expands, becoming less dense than cooler air, and rises. When the body is exposed to wind, the layer of air near the skin is replaced by new air more rapidly than if there were no wind, and heat loss by convection rises. Both conduction and convection together account for only a small fraction of the body's total heat loss. For example, a nude person seated in a comfortable room, without much air movement, loses about 12% of his heat by conduction and convection. When clothing is worn, heat loss by conduction and convection is reduced.

As we have seen, the term *convection* is applied to the transfer of heat from one place to another by the actual motion of hot material. Two examples of this process are a hot-air furnace and a hot-water heating system. In the case of the hot-air furnace, heat is transferred to the air inside the furnace. This heated air, which is less dense than cooler air, rises and is replaced by cooler air which descends to take its place. The cooler air is heated and this process repeats itself (Fig. 4-2).

Convection currents are very common in the atmosphere. The daytime sea breeze and the nightime land breeze can both be explained by convection (Fig. 4-3). In the early afternoon the land heats up more rapidly than the sea and there is a convection of heat upwards from the land with cooler air from over the ocean moving in to replace it. At night the land cools off more quickly than the sea and the air above the sea is warmer. It rises and is replaced by cooler air from the land. The result is a land breeze.

Radiation All objects, whether hot or cold, emit electromagnetic radiation. The hotter the object, the more energetic the average photon will be and the greater the total energy radiated. Heat can be transferred from the surface of one object to that of another by radiation. There does not have to be any contact between the two objects in order to transfer energy between

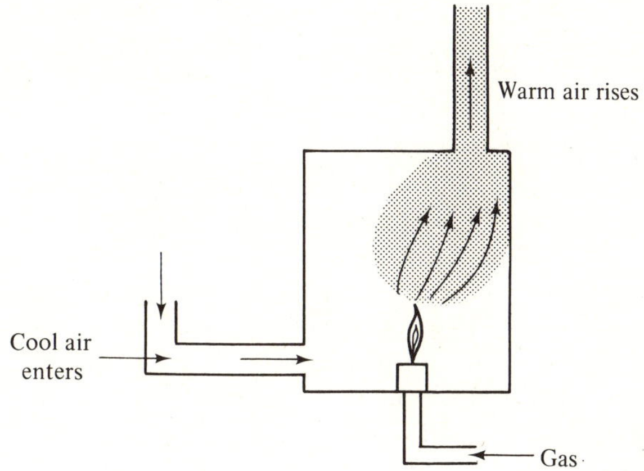

Figure 4-2 Hot-air furnace

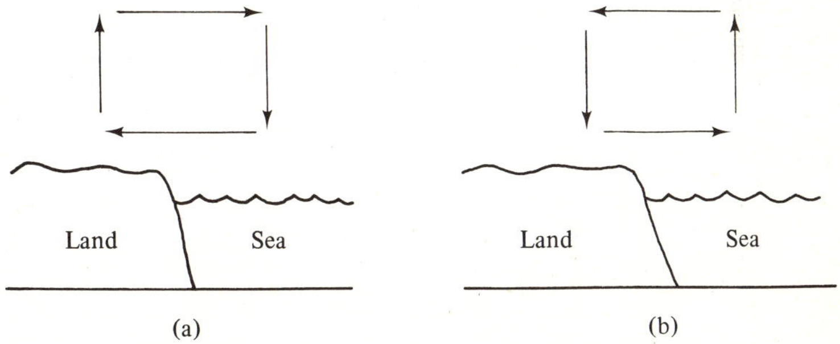

Figure 4-3 (a) Early afternoon sea breeze; (b) Late evening land breeze

them. Our own bodies are no exception to this phenomenon; they absorb and emit radiation all the time.

Solar radiation travels through the near emptiness of space before it is partially absorbed by our bodies. Human skin acts like a blackbody radiator for infrared radiation irrespective of the actual skin color. Therefore, a nude body will absorb all the infrared radiation incident upon it (for example from a stove). Solar radiation is mostly in the visible region of the electromagnetic spectrum, and absorption of visible light depends on skin color. Approximately 35% of visible light is reflected from white skin, but only a small amount is reflected from dark skin. Therefore, in sunlight, dark skin does absorb more heat than white skin. Wearing highly reflective clothing will cut down on the amount of heat gained from solar radiation.

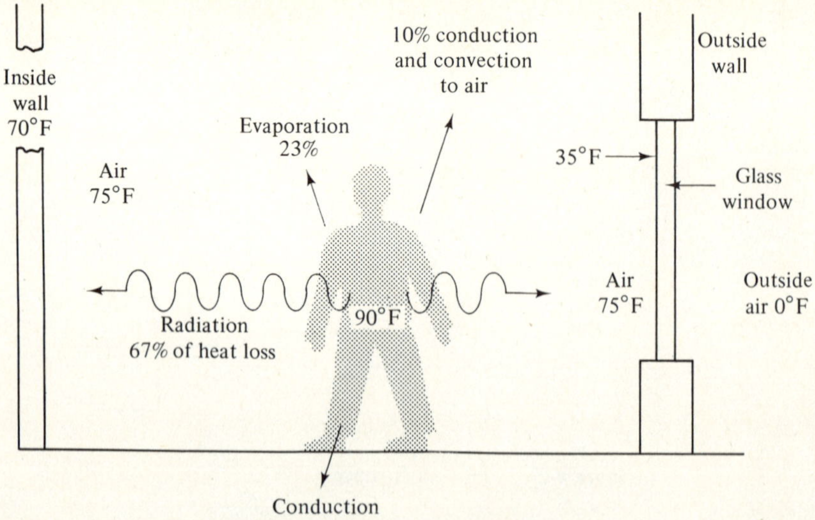

Figure 4-4 Avenues of heat loss from man, with an air temperature of 75°F. (From *Introduction to Environmental Physiology*, G. E. Folk, Jr., Lea & Febiger, Philadelphia: 2nd ed. 1973)

When we are indoors most of the heat loss from our bodies is due to radiation (Fig. 4-4).

In order to illustrate how there might be a net heat loss from the body, consider what happens when we sit near an uncovered window on a cold day. Even when there is no draft around this window, we will feel colder than we do in another part of the same room. The amount of heat lost through radiation has not changed for us since our skin temperature has not changed. However, the amount of heat gained from radiation emitted by surrounding objects has decreased since we are receiving much less radiation from the cold surface of the window pane than we would if we sat near a wall which was at room temperature. Therefore, the difference between the energy we radiate to our surroundings and the energy received from our surroundings increases and there is a net energy loss through radiation, making us feel colder.

The amount of heat lost or gained through radiation is a complex function of the position and temperature of surrounding objects. If we are near an uncovered window, an operating oven, or out in the sun, the situation is even more complex. If we are outdoors in the sunlight, radiation received from the sun must be taken into account, and it can result in as much as 200 kcal/hr being gained by a nude person. This is equivalent to the amount of heat generated when riding a bicycle at 6 miles per hour on level ground (Table 4-1). If the air temperature is high (around 90°F), then a human body is not able to rid itself of all this excess thermal energy by the methods already discussed. Some of the heat will be removed by the evaporation of sweat.

Evaporation The process whereby a substance changes from its liquid state to its vapor state is known as *evaporation*. An open glass of water is continually undergoing evaporation; that is, some of the water molecules are leaving the surface of the liquid and reaching the air as individual molecules of water vapor. Some water vapor molecules will return to the glass of water because of collisions with those molecules which make up the air above the glass of water. If the glass is uncovered, more molecules will leave than will return, and there will be a decrease in the amount of water in the glass. This phenomenon is known as evaporation. If the glass were covered, an equilibrium situation would result after a short time since as many molecules would be returning to the glass as would be leaving it.

We have not yet explained why some molecules of water leave the liquid. The molecules of water in the liquid state are constantly moving about and colliding with each other. If a water molecule near the surface is struck a little harder than usual, it may gain enough energy to escape from the pull of nearby water molecules and leave the liquid. Only the more energetic molecules break away from the liquid, leaving the less energetic and slower ones behind. Therefore, the average energy of the water molecules left behind is less than the average energy before evaporation, and the temperature of the liquid drops. Remember, temperature is proportional to the average kinetic energy of the molecules. Since returning water vapor molecules generate heat when they collide with water molecules at the surface, in order for a liquid to cool by evaporation there must be more molecules leaving the surface than returning to it.

When water evaporates from the human body surface, or any other surface, 0.58 kcal of heat is lost for each gram of water that evaporates. Whether the air temperature is high or low, there is always some evaporation of water (called insensible perspiration) from the skin and lungs. This evaporation—about 1,000 grams per day—cannot be controlled for the purpose of temperature regulation. However, evaporative loss of heat can be controlled by regulating the rate of sweating.

As long as the body temperature is greater than that of the surroundings, heat is lost by radiation and conduction; but when the temperature of the surroundings is greater than that of the skin, the body gains heat by these processes. In this case, the only way the body can get rid of heat is by evaporation.

Example 4-4: Suppose that the air temperature reaches 100°F and you are driving a car. How much sweat would you have to evaporate from your body each hour in order to dissipate your metabolic heat? Assume that your insensible perspiration is negligible in this case.

If you weigh about 155 pounds, then you will use energy at the rate of about 140 kcal/hr:

0.58 kcal of heat is removed for each gram of water that evaporates from the body.

$$\text{Evaporation rate} = \frac{140 \text{ kcal/hr}}{0.58 \text{ kcal/g}} = 242 \text{ g/hr} = 0.53 \text{ lb}$$

$$\cong 8.5 \text{ ounces water}$$

So the equivalent of an 8-ounce glass of water is lost each hour through sweating, under the above conditions.

Effects of humidity and air movement on evaporation Many of us have experienced the uncomfortable feeling of being wet with sweat on a hot and humid day. If the surrounding air already holds almost as much water vapor as it can at the prevailing temperature, (relative humidity near 100%), then very little of the water secreted by our sweat glands can evaporate from our skin. Instead, the sweat remains in the liquid state and we do not lose very much heat by evaporation. Man cannot survive for very long at high temperatures ($>$ 95°F) if the relative humidity is around 100%. This situation, where evaporation is severely limited, is similar to that of the glass of water with a cover over it. If the humidity is 100%, as many water molecules leave our skin as return to it and there is no net loss of water molecules by evaporation (hence, no cooling).

Lack of air movement prevents effective evaporation in the same way that effective cooling by conduction of heat to the air is prevented when there is little convection. The surrounding air becomes saturated with water vapor and further evaporation cannot occur. Convection currents cause air that has become saturated with moisture to move away from the skin while unsaturated air replaces it. On a hot, still day a fan can change the situation by blowing moist air away from the skin. But a fan will not be very useful when the

Table 4-2 Modes of Heat Loss

Room temperature	*Radiation*	*Conduction: convection*	*Evaporation*
Comfortable 77°F	67%	10%	23%
Warm 86°F	41%	33%	26%
Hot 95°F	4%	6%	90%

(From *Introduction to Environmental Physiology*, G. E. Folk, Jr., Lea & Febiger, Philadelphia: 2nd ed. 1973.)

relative humidity is high, since in that case the fan blows air which is almost saturated with water vapor towards the skin.

The relative contributions of the various methods of heat loss at several particular temperatures are given in Table 4-2 for a nude subject when there is a constant low air movement.

How the body regulates its various methods of heat loss for the purpose of maintaining a constant internal temperature is our next topic of discussion.

Temperature regulation

Skin and core temperature When we speak of body temperature we usually mean the temperature of the interior, called the core temperature. This core temperature is usually different from the skin temperature (typically around 90°F), which changes with the temperature of the surroundings. No single core temperature can be considered to be normal since measurements on many so-called normal people have shown a range of "normal" temperatures from around 97°F to over 99°F. The average normal temperature is considered to be 98.6°F when measured orally, and approximately 1°F higher when measured rectally. We need to consider an average temperature since an individual's body temperature varies with the time of day.

Since the body's temperature regulatory system is not 100% effective, our body temperature varies somewhat with exercise and with the temperature of the surroundings. When excessive heat is produced in the body by strenuous exercise, the rectal temperature can rise to as high as 104°F, and when the body is exposed to extreme cold, the rectal temperature can often fall to as low as 96°F.

As stated earlier, if the body had no mechanism for heat loss, its core temperature would increase very rapidly to the point where the individual could not survive. In very cold surroundings it is often necessary for the body to increase heat production in order to maintain a balance between heat production and heat loss. When the rate of heat production is exactly equal to the rate of heat loss, the person is said to be in thermal equilibrium. In the following pages we discuss the temperature regulatory system of the body which maintains this equilibrium.

The hypothalamus The hypothalamus (Fig. 4-5) is that part of the brain which regulates body temperature. It is often called the body's thermostat since it functions almost as a thermostat does in a home-heating system. When the temperature of the tissues in this part of the brain rises, temperature-sensitive nerve cells emit signals that increase heat loss while at the same time decreasing heat production. As a result, the body temperature falls back towards normal. On the other hand, when the body temperature falls below normal, these thermosensitive cells transmit signals which increase the rate

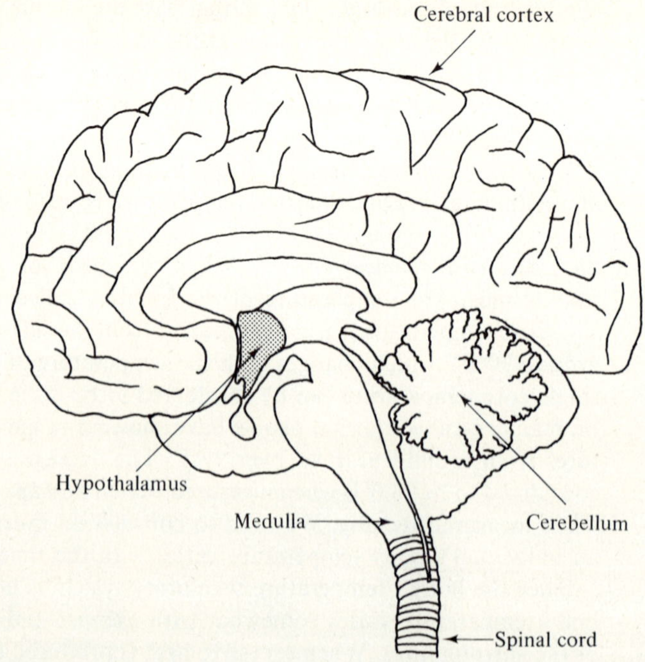

Figure 4-5 Location of the hypothalamus in the brain

of heat production, thus increasing the body temperature up towards normal.

Figure 4-6(a) illustrates the effect of hypothalamic temperature on heat production and heat loss. The solid curve shows that almost precisely at 37°C (98.4°F) sweating begins and then increases rapidly with temperature. If the body temperature falls below 37.1°C (dashed curve), the various methods for increasing heat production become activated, especially the increase in muscular activity which culminates in shivering. Both of these curves, but especially the one for heat loss, show the body responding to internal temperature changes only when certain critical temperatures are reached. These critical temperatures are almost always stated as exact numbers, as if there were no uncertainty in their values. However, there is always some experimental error associated with the measurement of a physical quantity. So we now digress from our discussion of temperature regulation in order to discuss in some detail the important subject of experimental errors.

Experimental Errors In order to determine the shape of the curves in Fig. 4-6, measurements were made of the rates of heat production and loss

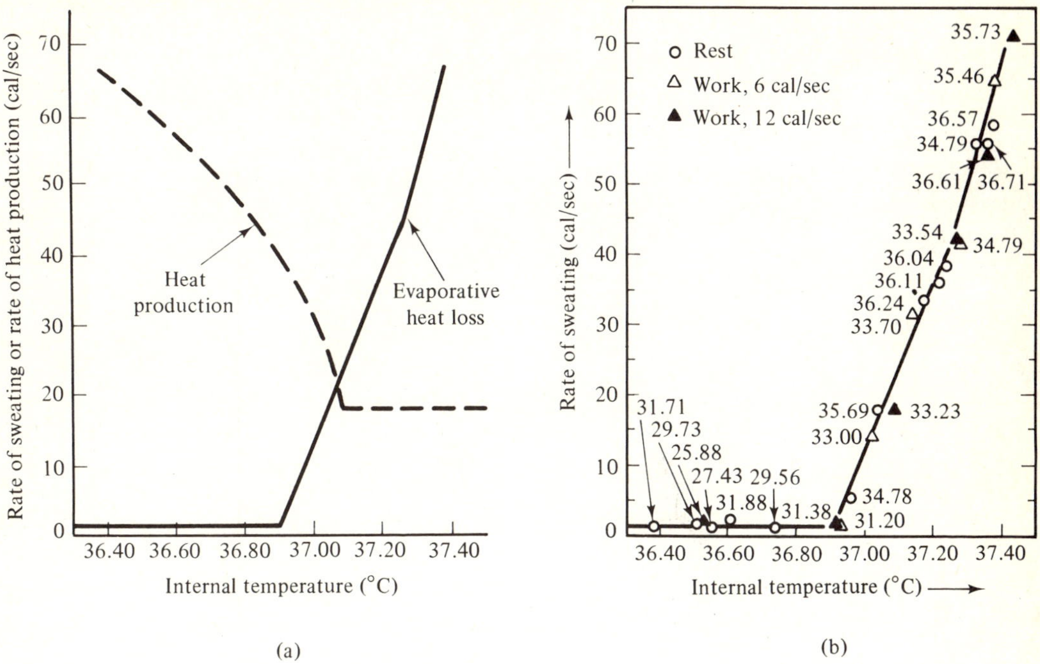

Figure 4-6 (a) The effect of internal temperature on heat production and heat loss; (b) Data from which solid graph in (a) was drawn. The internal temperature was measured at the tympanic membrane in the ear. The temperatures next to each data point are the skin temperatures corresponding to that particular measurement. No correlation was seen between sweating rate and skin temperature. (From *Temperature*, edited by James D. Hardy. © 1963. Reprinted by permission of Van Nostrand Reinhold Company.)

while the internal temperature of some individual was varied. Each measurement of the internal temperature and of the heat production and loss at that temperature has some experimental error or uncertainty associated with it.

As an example of how experimental error may arise, consider what happens when we measure the length of an object with a ruler. Suppose we place a ruler alongside a block of wood whose length we need to measure (Fig. 4-7). Since the length of the block is not such that it lines up exactly with one of the lines on the ruler, we can say only that the block has a length somewhere between $1\frac{3}{4}$ inches and 2 inches. In such cases we usually estimate the length. In this particular example we estimate the length of the block as $1\frac{7}{8}$ inches.

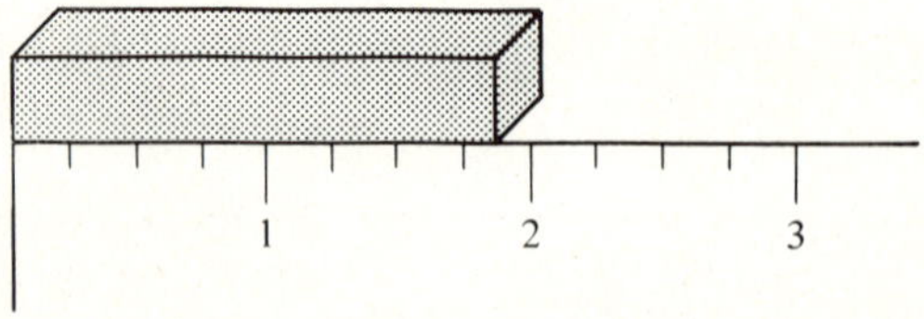

Figure 4-7 Measuring the length of a block of wood with a ruler

However, in order to indicate that we have made an estimate we should express the result along with the uncertainty in the measurement, which is $\pm\frac{1}{8}$ of an inch in this case. That is, the length is somewhere between $1\frac{3}{4}$ inches ($1\frac{7}{8} - \frac{1}{8}$) and 2 inches ($1\frac{7}{8} + \frac{1}{8}$). The experimental error in this measurement has been stated as plus or minus $\frac{1}{8}$ inch, which means that the estimate of length may be either too large or small by as much as $\frac{1}{8}$ inch. The length of the block should be stated as $1\frac{7}{8}$ inches $\pm\frac{1}{8}$ inch. Many other simple measurements, such as weighing produce on a supermarket scale, require estimating and, even though we don't normally think about it, these measurements contain experimental errors. Often the determination of experimental errors in a measurement is more difficult than in the example given, but they are present nevertheless.

Returning to the data in Fig. 4-6(b), each measurement of temperature has an experimental error of ±0.02°C associated with it. This amount of error is appropriate for the particular instrument used to measure the temperature in these experiments. Figure 4-6(b) shows the actual data from which Fig. 4-6(a) was obtained. Here we see that each data point is represented by either a circle or a triangle. The diameter of each circle is equal to the experimental error in the temperature measurement. Since we can draw more than one upward sloping straight line through the circles, there is really some uncertainty in the determination of the critical temperature. Therefore, the critical temperature for the onset of sweating should be written as 36.90°C ±0.02°C, and the critical temperature for the onset of shivering should be written as 36.90°C ± 0.02°C.

All of the experimental data and all of the physical constants which are used in this book (or any other book) have experimental errors associated with them. We will not discuss these errors, but it should be kept in mind that they exist and that physical quantities are never determined exactly.

Mechanisms of increased heat loss when the body becomes overheated Hot climates can be classified as hot dry or warm humid. The hot dry climates are usually desert regions which are scattered throughout the world. In the U.S.

Figure 4-8 (a) In hot-arid zone, buildings are close together and have thick walls and small windows; (b) In hot-humid zone, houses are far apart, have raised roofs, and are not completely walled in.

Figure 4-8 (*Cont'd.*) (c) Also hot-arid area; (d) Also hot-humid area. (From Victor Olgyay, *Design with Climate: Bioclimatic Approach to Architectural Regionalism* (Copyright © 1963 by Princeton University Press : Figs. 14 and 15, p. 5, and Fig. 22, p. 8, Fig. 23, p. 9. Reprinted by permission of Princeton University Press.)

the desert regions are mainly found in the Southwest. The warm humid climates are typically represented by the tropical rain forests lying within latitudes that are 10 or 20 degrees from the equator.

In both hot climate types man experiences heat stress. In the tropics his problem is how to effect efficient evaporation of sweat under conditions of high humidity, whereas in the desert his sweat evaporation is no problem. The difficulty there is to maintain sweat production at a sufficiently high rate to cool the body. In Fig. 4-8 we can see some of the ways in which shelters have been designed to the climate in these two different types of hot regions.

Vasodilation The term *vasodilation*, which means an increase in the diameter of blood vessels, is one of two main ways the body uses to rid itself of excess heat. The other is sweating. Normally, the muscles surrounding blood vessels in the skin are contracted in such a way as to constrict the space through which blood can flow. Overheating of the hypothalamus causes nerve impulses to be sent to the muscles surrounding the blood vessels in the skin. This results in relaxation of the muscles, an increase in the diameter of the blood vessels, and, therefore, an increase in blood flow to the skin. This increase in blood flow from the interior of the body to its outer surfaces often shows up as a flushing of the face. As the blood flow from the body core to the skin increases, the skin temperature rises and more heat is lost to the surroundings via radiation and conduction.

Also in the skin are temperature sensors which are sensitive to temperature increases as small as 0.002°F per second. If the skin temperature increases at or greater than this rate, then impulses are automatically sent from receptors in the skin to the spinal cord and back to the muscles in the skin. These impulses cause vasodilation with results as described above. Another important function of the skin receptors is to inform the individual that his skin temperature is rising, allowing him to seek shelter or a change of clothing in order to lower the heat load on his body.

Sweating When the body becomes overheated, large quantities of sweat, which is a dilute solution of salt (sodium chloride; NaCl), are secreted onto the surface of the skin by the sweat glands to provide evaporative cooling of the body. Impulses from the hypothalamus cause sweating and are distributed to the skin everywhere on the body. In cold weather, the rate of sweat production is essentially zero, but in very hot weather the maximum rate of sweat production can be as much as 4 quarts per hour, which is equivalent to the loss of 8 pounds of body weight per hour. However, this high rate of sweating cannot be maintained indefinitely.

The sweat glands are tubular structures (Fig. 4-9) consisting of two parts: a deep, coiled portion that secretes the sweat and a duct portion passing outward through the dermis of the skin. The salt concentration in sweat, which is normally less than 1%, is important when there is exposure to extreme heat. Large quantities of salt are lost to the skin during the sweating process, and the salt lost must be replaced in the diet. Therefore, men living in a desert must have access to salt in addition to water.

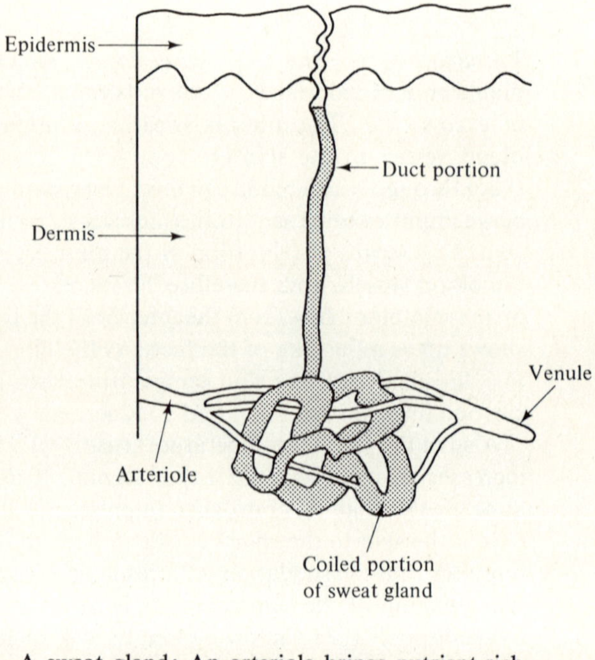

Figure 4-9 A sweat gland: An arteriole brings nutrient-rich blood to a sweat gland and a venule carries the remains back to the heart.

A person who is constantly exposed to hot weather, or heat, sweats more and more each day until a maximum rate of sweating of about 10 quarts per day is reached after about 5 or 6 weeks. This increased effectiveness of the sweating mechanism is caused by a direct increase in the output capability of the sweat glands themselves. A physiological change such as this one, which occurs within the lifetime of an individual, is called *acclimitization*. It allows the individual to adjust to a particular environment in such a way as to increase his comfort. When sweating increases, there is usually a decrease in the concentration of salt in the sweat, which allows a greater conservation of salt.

Example 4-5: (*a*) In a hot environment 8–11 quarts of sweat can be lost each day. How much salt is lost per day at this rate of sweating?

Assume that one-half of 1% of the weight of sweat is salt. Then if 10 quarts of sweat are lost each day, 10 quarts $\times$ 2 pounds/quart = 20 pounds of body weight are lost each day:

$$\text{amount of NaCl lost} = 0.005\ (20\ \text{pounds}) = 0.1\ \text{pound NaCl}$$

$$0.1\ \text{pound}\ (454\ \text{grams/pound}) \cong 45\ \text{grams NaCl per day}$$

(*b*) What percent of the total amount of salt in the body does this represent?

In Chapter 1, Table 1-1, we learned that 0.13% of the total body weight is Na and 0.15% is Cl. If all of this Na and Cl went into forming NaCl in the body, then the total amount of NaCl in the body of a 70-kg man would be:

$$0.0028(70\ \text{kg}) = 0.196\ \text{kg} = 196\ \text{g}$$

This is the maximum amount of salt that a 70-kg man might contain; however, the actual amount of salt in a 70-kg man is more likely to be about 170 grams.

$$\%\ \text{of body salt lost each day} = 45\ \text{grams}/170\ \text{grams} \times 100 = 26.4\%$$

Cold environments About 20% of the land area of the earth (50% of Canada and the USSR) is located in zones where the mean annual temperature is below 30°F. In these regions the underlying soil remains permanently frozen (permafrost) while a thin layer temporarily thaws during summer. Frostbite is a fairly common occurrence in these regions. Even in the U.S., where indoor temperatures of 70°F are common in the winter, some individuals die of cold exposure each year. It is quite possible that your physiological system for combating cold may one day suddenly be called upon. So we now discuss the body's methods for adjusting to a cold environment.

Vasoconstriction When the temperature of the hypothalamus drops below 37.1°C (98.4°F), the diameter of the blood vessels in the body decreases, slowing the transfer of heat from the interior of the body to the skin. This lowers the skin temperature and thus decreases heat loss to the environment by radiation and conduction. With maximum vasoconstriction the only heat that can leave the body is that which can be conducted directly through the insulator layers of the skin. This direct conductivity is a less efficient method of heat transfer from the interior to the skin than when the blood transfers heat to the skin.

Shivering Shivering results when the hypothalamus reaches a temperature below 37.1°C. When we begin to shiver, the rhythmic contraction of muscles throughout the body increases the rate of heat production to as much as four times normal ($\cong$ 400 kcal/hr). Heat production is increased as greater amounts of carbohydrates are oxidized to provide energy for muscle contraction. By this means the temperature of the muscles rises to approach that of the core, and the core temperature itself may rise by as much as 1°F.

The rate of metabolism can be increased by as much as 20% to 30% by the release of a chemical called norepinephrine at fat deposits. This release takes place wherever the body temperature falls low enough and does not seem to be controlled by the hypothalamus. Norepinephrine decomposes fat. Then fatty acid is released in the blood plasma, increasing heat production. This mechanism for additional heat production is called nonshivering thermogenesis or nonshivering heat generation.

Much more is known about the body's reaction to hot and cold environments than has been discussed here. There are several references on this subject at the end of this chapter for those who are interested in learning more.

Suggested reading

1 LANDSBERG, HELMUT E., *Weather and Health*. New York: Doubleday & Co. Inc., 1969.

2 BLAIR, THOMAS A., and ROBERT C. FITE, *Weather Elements*. Englewood Cliffs, NJ: Prentice-Hall, Inc., 1961.

3 CURTIS, HELENE, *Introduction to Biology*. New York: Worth Publishing Co., 1972.

4 FOLK G.E., Jr., *Introduction to Environmental Physiology*. Philadelphia: Lea & Febiger, 1966.

5 GUYTON, ARTHUR, *Textbook of Medical Physiology*. Philadelphia: W.B. Saunders & Co., 1966.

6 RUCH, THEODORE, and JOHN FULTON, eds., *Medical Physics and Biophysics*, 18th ed. Philadelphia: W.B. Saunders & Co., 1960.

7 BENZINGER, T.H., "The Human Thermostat," *Scientific American*, **204** (Jan. 1961), 134.

8 "Temperature; its measurement and control in science and industry, Part 3" James D. Hardy, ed.; *Biology and Medicine*. New York: Reinhold Publishing Corp, 1963.

9 ADOLF, E.F., ed., *Physiology of Man in the Desert*. New York: Interscience Publishing Inc., 1947.

10 OLGYAY, VICTOR, *Design with Climate*. Princeton, NJ: Princeton University Press, 1963.

11 GIVONI, BARUCH, *Man, Climate and Architecture*. New York: Elsevier Publ. Co., American Inc., 1969.

Questions

1 If atoms are electrically neutral, how can there be any force of attraction between them to form chemical bonds?

2 One source of water for our bodies is the water which is contained in liquids and solids which we ingest. Can you think of another source of water input, one which is related to the oxidation of food?

3 Peanuts are one of the most concentrated sources of protein. They contain 26.9% protein, 44.2% fat, and 23.6% carbohydrates. Suppose you eat 100 grams (about a $\frac{1}{4}$ pound) of peanuts, and all of the nutrients are oxidized. How much energy will be released?

4 (*a*) Assume that all the heat generated in the oxidation of an 8-ounce glass of milk goes into warming the body of a 70-kg man. Calculate the expected temperature increase of his body by using Eq. (4-3). Take the specific heat of this man's body to be the same as that of water.

(*b*) Assume that the energy generated in oxidizing the glass of milk does not heat up the man's body, but instead is all used to allow him to walk up a 5% grade at 3 miles per hour. How long would he be able to continue this activity by using only the energy gained from oxidizing the one glass of milk?

5 On page 43 it was stated that it takes as much as 150,000 joules to heat up one pint of water from room temperature to the boiling point of water. Verify this statement by using Eq. (4-3): $Q = mc\Delta t$.

6 Explain in detail why double window panes tend to keep a house warmer than a single window pane.

7 Why do your feet feel colder when you walk barefoot on a wooden floor as opposed to a carpeted floor?

8 Why is heat loss by conduction and convection lowered when clothing is worn?

9 Some convection of heat away from the body takes place even on a very still day. Explain how.

10 (*a*) Suppose that you have placed your hands a few inches above a toaster which is turned on. Explain in detail how energy is transferred from the warmer toaster to your hands.

(*b*) What happens if you move your hands away from the top of the toaster and near one of the sides of the toaster?

(*c*) Suppose that a vacuum is created between the toaster and your hands. When you place your hands above the toaster, does any energy transfer take place now? If it does, explain how.

11 Why does evaporation from a liquid increase when you heat it?

12 (*a*) Even if the air temperature is low, about 1,000 grams of insensible perspiration evaporates from the skin and lungs each day. Even though we are not aware of this evaporative cooling, it is always proceeding. How much heat is lost in this manner every day?

(*b*) Is it significant compared to the total metabolic rate?

13 The outside air temperature is 105°F and the relative humidity is 50% both inside and outside your house. If the inside air temperature is 95°F, would you be more or less comfortable if you opened all the windows, letting in warmer air but increasing the air movement inside? Explain by considering what happens to all the methods of heat loss.

14 Which heat loss mechanisms become activated when you put your foot into a bathtub full of very hot water? Is the hypothalamus involved? Why?

15 (*a*) You are in the desert during a hot summer day. Which heat loss mechanisms are important in maintaining thermal equilibrium? Are they controlled by the hypothalamus? Why?

(*b*) Suppose you remain in the desert for a few weeks. Would you expect any acclimitization? What kinds?

16 (*a*) If the temperature of the hypothalamus drops below 37.1°C, which heat conservation mechanisms become activated?

(*b*) Which heat production mechanisms are activated?

17 What special design features would you incorporate into a house you were going to build in the desert? (See Fig. 4-8.)

18 What special design features would you incorporate into a house you were going to build in the rain forest of South America? (See Fig. 4-8.)

19 If the initial internal temperature is 37°C, estimate the temperature increase required for a doubling of the sweating rate.

20 Estimate the temperature decrease required for a doubling of the heat production rate when the initial internal temperature is 37°C.

21 Which process has a greater response to an equal temperature change, heat production or heat loss?

Part Three

Radiation in the environment

Chapter 5

Interactions of radiation with matter and natural background radiation

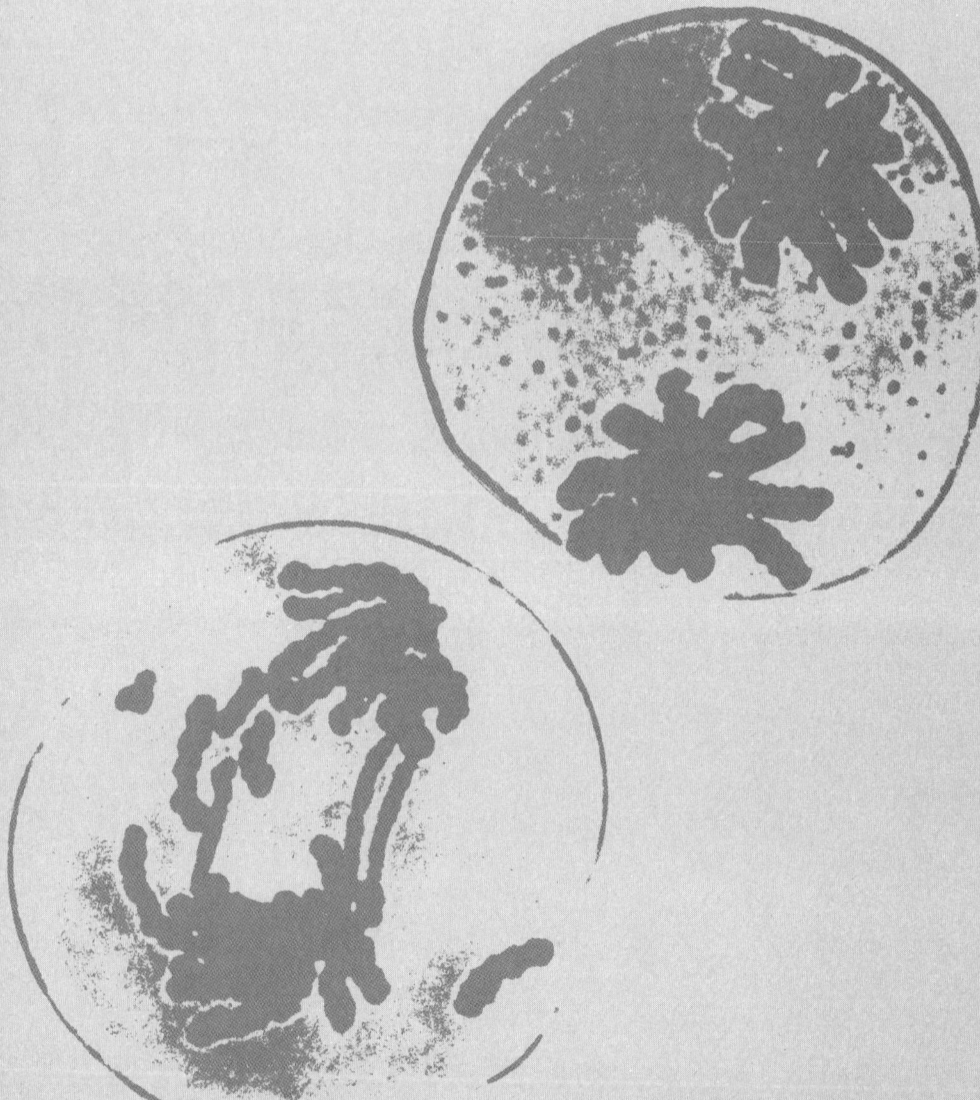

Introduction and history

Man has been exposed to radiation since his appearance on earth. In fact, radiation, which can cause mutations in living things, may have had an influence on man's evolution from a simple organism to a very complex organism. Radiation is often thought of as consisting of particles (such as electrons, protons, and neutrons) or waves (such as sound and electromagnetic), although particles can sometimes display wave properties and waves can sometimes display particle properties. All radiation which travels at the speed of light in a vacuum is called electromagnetic radiation; some examples are radio waves, infrared radiation, visible light, and X-rays.

In Part Three we explore the effects of radiation upon matter in general and living tissue in particular. The sources of man-made and natural radiation in our environment are discussed and the extra risk introduced by man-made radiation sources is evaluated in terms of the benefits received. The protection afforded the general population by the radiation protection guidelines is also discussed. Finally, some hazardous incidents involving radiation use are studied.

Our familiarity with radiation, especially high-energy radiation (in this text *high-energy radiation* means radiation of energy greater than that of ultraviolet light), is rather recent. The study of high-energy radiation began towards the end of the 19th century when two extremely important scientific discoveries took place just four months apart. Both discoveries were accidental in the sense that they were unexpected. However, in each case the scientist involved possessed the necessary insight and objectivity to take advantage of the unexpected.

In 1895 Wilhelm Roentgen, who was experimenting with electron beams in Germany, discovered that a paper screen coated with a phosphorescent* substance glowed when placed near a discharge tube.† The glow still appeared even when blocks of wood or thousand-page books were placed in front of the paper screen. Roentgen, who called this penetrating radiation X-rays, deduced that the radiation originated at the point where the electrons struck the glass tube. The discovery of X-rays, which took place during the Victorian age, quickly attracted public attention. The fact that X-rays permitted one to "see" through opaque objects was sensational and caused great concern lest, by their use, fully dressed people might be made to appear unclothed (Fig. 5-1). When such speculation ceased, there was wide appreciation of the value of X-rays in setting broken bones and they were quickly put to this use. Since X-rays penetrate flesh more readily than the denser bones, they produce a shadowgraph of the bones (Fig. 5-2).

In 1896 a French physicist, Henri Becquerel, was studying the phosphorescence of minerals exposed to the newly discovered X-rays. When he

*Phosphorescent materials are those which can absorb energy at one time and re-emit some of that energy in the form of light at a later time.

†A discharge tube is a tube containing low-pressure air or some other gas which is used to generate a beam of negatively charged particles of small mass called electrons.

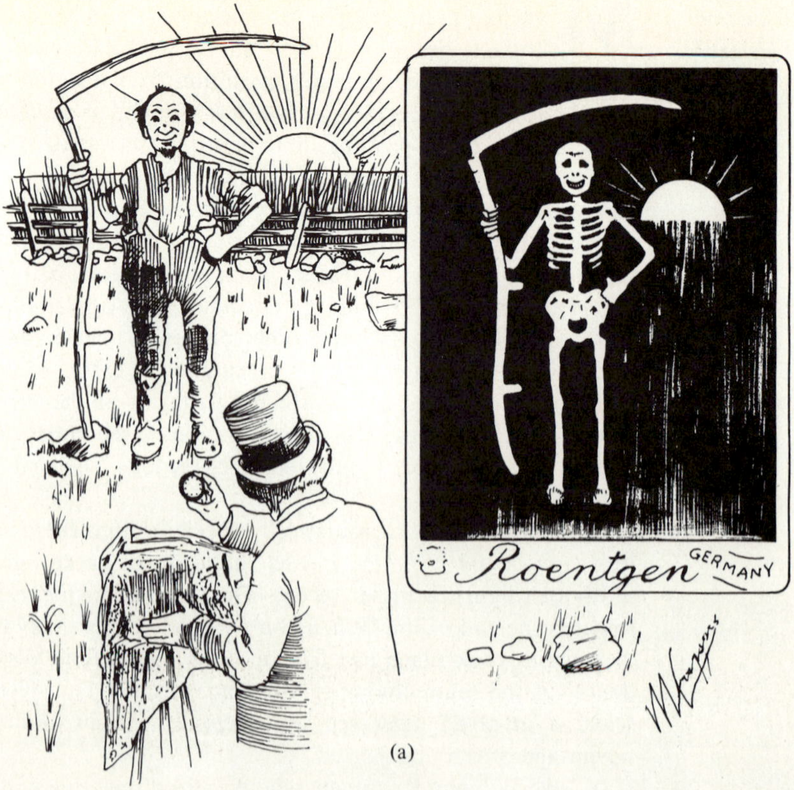

(a)

(b)

Figure 5-1 (a) Cartoon from *Life* (Feb. 1896). The new Roentgen photography: "Look pleasant, please." (b) From cartoon in *La Nature*, Paris (May 1896). (From *Wilhelm Conrad Röntgen and the Early History of the Roentgen Rays*, Otto Glasser, Charles C. Thomas, Publisher, 1934 Springfield, IL)

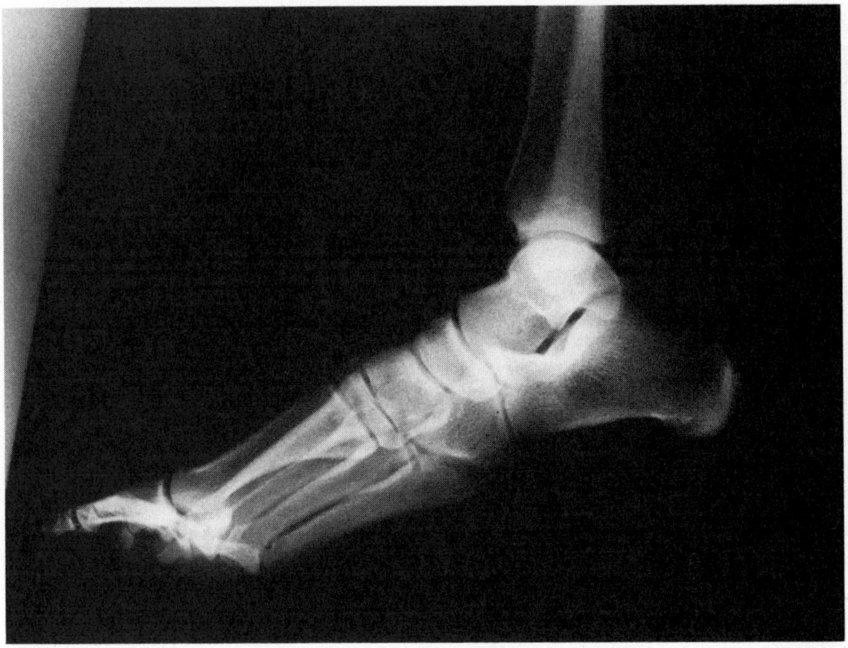

Figure 5-2 In this X-ray of a foot we can also see the two leg bones: the slender fibula, and weight-bearing tibia.

came to the uranium compounds, he discovered by chance that even unexposed uranium compounds emitted penetrating radiation which ruined a photographic plate covered with black paper. This property is shared by X-rays, but Becquerel's "rays" were continuously emitted even in the absence of a source of energy. This property of uranium, spontaneous radiation, is called radioactivity. For a time it was thought that the laws of the conservation of energy (energy is neither created nor destroyed, but just changes form) was violated during radioactive decay. It was later determined that energy stored in the nuclei of radioactive atoms (nuclear energy) was being converted into energy of radiation. By 1900, the Curies (Marie and her husband, Pierre) had isolated two other radioactive elements (polonium and radium), and other workers in the field had found that there were three types of radiation emitted by these radioactive elements: helium nuclei (or alpha particles), electrons (or beta particles), and gamma rays (to be discussed).

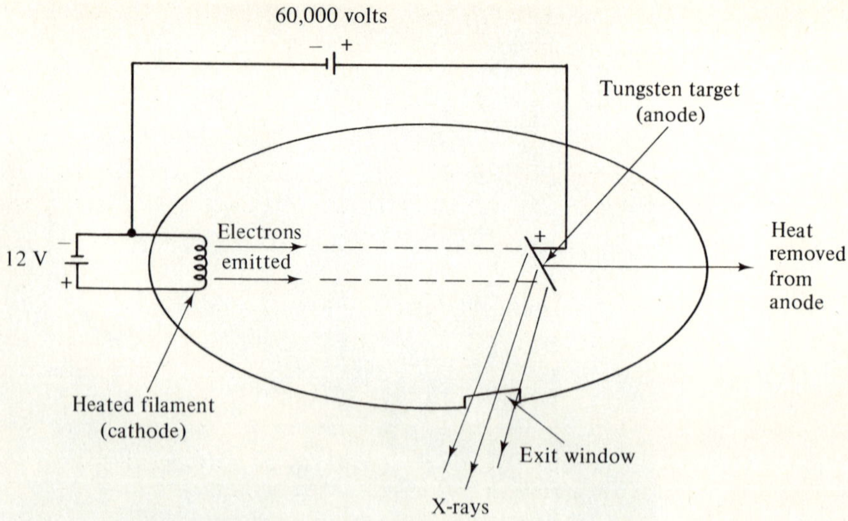

Figure 5-3 A simplified schematic of a modern X-ray tube. A thin piece of wire is heated by an electric current (just as is the filament in a light bulb) and emits electrons which are accelerated towards the positively charged tungsten anode. When these electrons are decelerated by the tungsten, radiation is emitted by both the electrons and the tungsten atoms. Only a small fraction of the electrons' kinetic energy is converted to X-ray energy; the rest goes into heating the anode. Tungsten is usually chosen for the anode because of its high melting point (3,650°K).

Types of radiation

The radiation reaching the earth is in the form of both charged particles and electromagnetic radiation. The charged-particle component is mostly cosmic rays whose origin is unknown. This type of radiation, consisting mainly of protons, is discussed in detail later in this chapter. The principal source of energy reaching the earth is electromagnetic radiation from the sun.

Electromagnetic radiation In Chapter 3 both wave and photon models of electromagnetic radiation were discussed. When we calculated the energy of a single photon of yellow light in joules we found the resulting number (33×10^{-20} joule) to be very small and unwieldy. It is much more convenient to use a different unit to measure energy when we discuss the interaction of radiation with matter.

As was indicated in Chapter 3, a joule is much too large a unit of energy for expressing the energy of a photon or a particle. An energy unit named the electron-volt (eV) is the one that we will use.

$$1 \text{ eV} = 1.6 \times 10^{-19} \text{ joule}$$

An electron-volt is the energy an electron gains in moving from the negative to the positive terminal of a 1-volt battery. Two examples will show the advantage of working with electron-volts instead of joules.

Example 5-1: What is the energy of a photon of frequency 5×10^{14} cycles/sec?

$$E = hf = 6.6 \times 10^{-34} \text{ joule}\cdot\text{sec } (5 \times 10^{14} \text{ c/sec}) = 33 \times 10^{-20} \text{ joule}$$

Now we can convert from joules to electron-volts by dividing by the number of joules in 1 eV:

$$E = \frac{33 \times 10^{-20} \text{ joule}}{1.6 \times 10^{-19} \text{ joule/eV}} = 2.08 \text{ eV}$$

which is less than the amount of energy needed to ionize molecules.

Example 5-2: What is the energy carried by a gamma-ray photon of a frequency equal to 10^{20} cycles/sec?

$$E = 6.6 \times 10^{-34} \text{ joule}\cdot\text{sec}\cdot(10^{20} \text{ c/sec}) = 6.6 \times 10^{-14} \text{ joule}$$

To convert this answer into electron-volts:

$$E = \frac{6.6 \times 10^{-14} \text{ joule}}{1.6 \times 10^{-19} \text{ joule/eV}} = 4.17 \times 10^{5} \text{ eV}$$

The joules cancel each other and the electron-volt goes into the numerator.

For energies this large we often convert to millions of electron-volts (MeV): 10^6 eV = 1 MeV. Therefore, $E = 0.417 \times 10^6 \text{ eV} = 0.417$ MeV. This energy is far greater than the energy needed to ionize any molecule.

Even 1 MeV is not a great deal of energy in ordinary terms since more than 10^{13} MeV would be needed to raise the temperature of 1 gram of water by 1 degree Celsius. However, photons and particles occur in very large numbers, so even though the energy per particle may be small, the total energy can be quite large. In addition, most molecules can be ionized by the addition of several electron-volts of energy and 32.5 eV will cause the ionization of almost any molecule in living tissue. Ionization of a molecule or atom is the splitting of that atom or molecule into charged fragments called ions. Both X-ray and gamma-ray photons are energetic enough to ionize any molecules found in living tissue. This can be seen by applying the equation $E = hf$ to the frequency values of X-rays and gamma rays. Water, which is the most common molecule found in our bodies, has an ionization energy of 12.6 eV.

Ionization of any of the human body's atoms, which were not meant to be ionized, can create disruption in the body's well-ordered structure. When radiation moves through space it carries energy with it. This energy can be transferred from its source to other bodies as in the case of light reaching a photographic film or X-rays striking body tissue. The main property of radiation which we are concerned with here is its ability to transfer energy to human tissue and thus ionize it.

Particulate radiation In evaluating the effects of radiation on man, the types of radiation we are concerned with most are charged particles and electromagnetic radiation. Particles can be considered as tiny charged or uncharged balls. The charged particles carry an electric charge just as we do when we walk across a carpet on a dry, cold day. Charged particles are attracted to other charged particles of opposite sign and repelled by particles of the same sign. This is an electrical force.

Some examples of positively charged particles are protons, positrons, and alpha particles; an example of a negatively charged particle is an electron. Alpha (α) particles are the nuclei of helium atoms (they carry twice the positive charge of a proton). Positrons are particles with the same mass as an electron, but with a charge equal in magnitude but opposite in sign to that of an electron.

Interactions of radiation with matter

The energy of a charged particle is used up very quickly in pushing and pulling the protons and electrons of the matter through which it is passing. Therefore, charged particles are not very penetrating. Neutral particles, such as neutrons, have no charge and consequently travel greater distances through matter before losing all their energy in direct collisions with other particles. Neutrons lose most of their energy in collisions with hydrogen nuclei when passing through living tissue (which is rich in H_2O).

Electromagnetic radiation, which we consider to be photon-like in most of its interactions with body tissue, also travels a greater distance through matter than charged particles of the same energy. This should not be surprising since photons are neutral particles. Photons of high energy (X-rays and gamma rays) transfer most of their energy to the electrons surrounding the nuclei of atoms. These electrons can either be raised to higher energy levels, which are further from the nucleus (excitation), or removed from the pull of the positively charged nucleus altogether (ionization). Charged particles have the same general effect upon matter as photons; that is, excitation and ionization.

Heating and other effects Electromagnetic radiation which is less energetic than visible light (such as infrared waves and microwaves) is converted into heat when absorbed by living tissue. The heat thus formed is sufficient to cause atoms and molecules to vibrate more rapidly, but this added vibration is not usually sufficient to pull molecules apart and, therefore, does not bring about chemical changes. Visible light brings about some chemical changes. It breaks up silver compounds and produces tiny black grains of metallic silver (the chemical basis of photography). Living tissue, however, is largely unaffected by visible light, but the retina of the eye is one obvious exception.

Another very important exception is the process of photosynthesis (which proceeds by means of blue light and red light) whereby molecules of carbon dioxide and water combine to form carbohydrate and oxygen. Ultraviolet light, which is more energetic than visible light, can bring about chemical changes in man more easily. It will redden the skin, stimulate the production of pigment, and break up certain steroid molecules to form vitamin D. There is evidence that persistent exposure to sunlight brings about a heightened tendency to skin cancer. Ultraviolet light is not very penetrating, however, and its effects are confined to the skin.

Ionization Electromagnetic radiation more energetic than ultraviolet light, such as X-rays and gamma rays, carries sufficient concentrations of energy to penetrate the skin and ionize the tissue it passes through. Electromagnetic radiation of high enough energy can release electrons from the atoms making up the tissue to leave behind positively charged nuclei (positive ions). The released electron (negative ion) moves through the surrounding tissue, creating more ions as it collides with other electrons. The greatest energy loss occurs at the end of a particle's path as it slows down. Ionization can also be caused by particulate radiation which strikes the tissue of the body with high enough energy. Ionizing radiation is capable of imparting so much energy to molecules as to cause them to vibrate themselves apart, producing not only ions but also high-energy, uncharged molecular fragments called free radicals. Free radicals are much more reactive than ions and the net damage in tissue is even greater than that caused by ions alone. The track of an electron or positron in body tissue is random and might look like Fig. 5-4(a).

Each dot represents a collision of the incident particle with an electron or proton of the tissue which results in the production of some ions and free radicals. This picture holds whether the electron was incident upon the tissue as the primary radiation or secondarily produced by photons (X-rays or

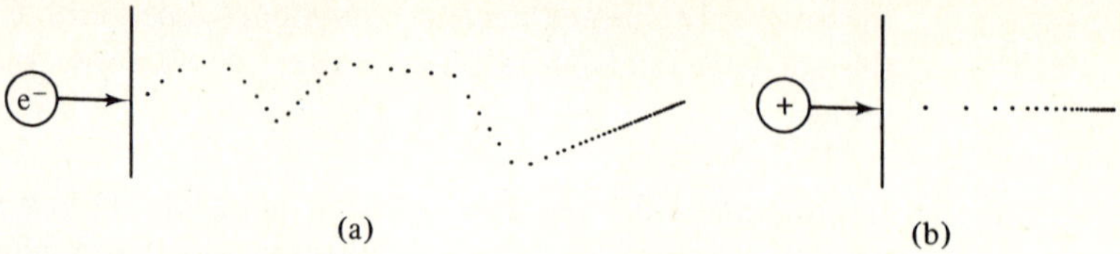

Figure 5-4 (a) An electron track in body tissue; (b) An alpha particle track in body tissue

gamma rays) colliding with electrons. The track of a proton or alpha particle, of energy equal to that of the electron pictured above, is much shorter and more dense [Fig. 5-4(b)]. Protons and alpha particles are much heavier than electrons and ricochet less easily.

Radiation damage Considerable evidence has accumulated to indicate that the cell, and particularly its nucleus, is the primary site of radiation damage. Disorganization of the body follows only after a sufficient number of cells have been so injured that they can no longer carry out their normal functions. Ionizing radiation may disrupt enough chemical bonds of a chromosome* to break it into fragments. Even if the chromosome remains intact, an individual gene† along its length may be badly damaged and a mutation‡ may be produced. Mutations can be produced by the direct collision of ionizing radiation with a chromosome or indirectly by the production of free radicals near the chromosome. If the free radical strikes the neighboring chromosome, it may carry enough energy to break chemical bonds and cause a mutation. Figure 5-5 shows the effect of ionizing radiation on chromosomes.

Much more radiation is needed to kill a cell outright than to damage the nucleus. Radiation damage to a cell's nucleus can cause the cell to grow uncontrollably as a cancer cell [Fig. 5-7(a)]. If the damage is more severe, the cell might lose its ability to divide properly and will die after a period of confused growth of giant cells [Fig. 5-7(b)]. If damage is very severe, the cell will die more quickly. If enough cells become deranged, the tissue, then the

***Chromosome:* A DNA (deoxyribonucleic acid)-containing body of the cell nuclei of plants and animals, responsible for the determination and transmission of hereditary characteristics.

†*Gene:* A functional hereditary unit that occupies a fixed location on a chromosome and is capable of mutation.

‡*Mutation:* Mutation can be used in two ways. First, it can mean a drastic change in the structure of a chromosome. Second, it can be equated with the production of an organism differing from the parental strain due to a change in the chromosome structure of one or both of the parents.

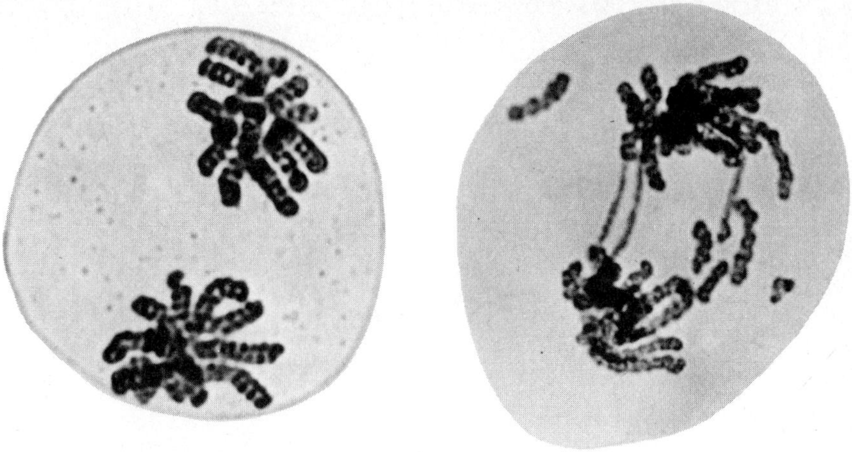

Figure 5-5 Effects of ionizing radiation on chromosomes: (a) A normal plant cell, showing chromosomes divided into two groups; (b) The same type of cell after X-ray exposure, showing broken fragments and bridges between groups—typical abnormalities induced by radiation. (Courtesy U.S. A.E.C)

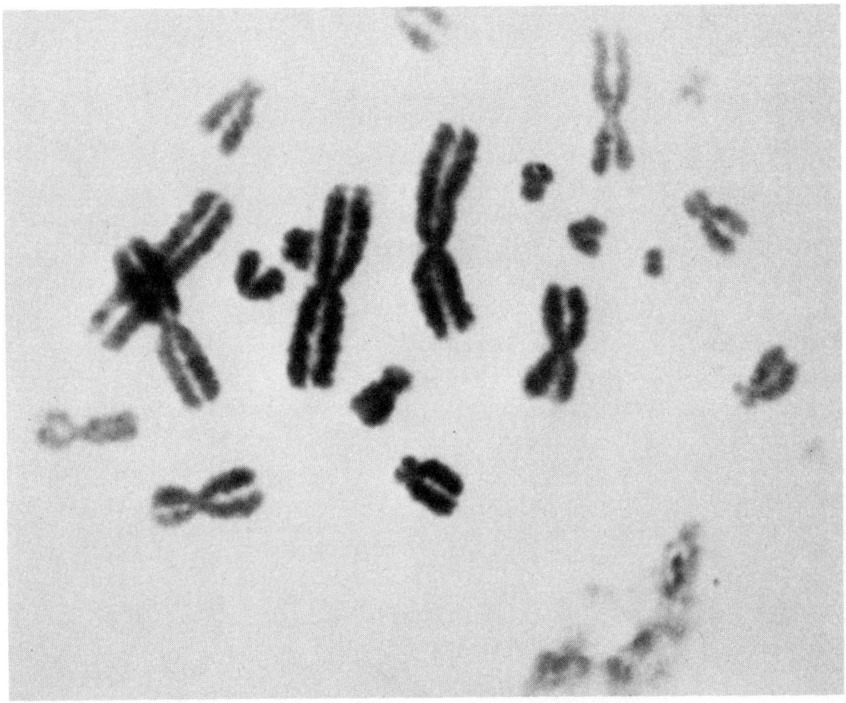

Figure 5-6 This normal cell nucleus, enlarged 550 times, has 23 pairs of chromosomes. (Courtesy Dr. C. K. Yu, Argonne National Laboratory)

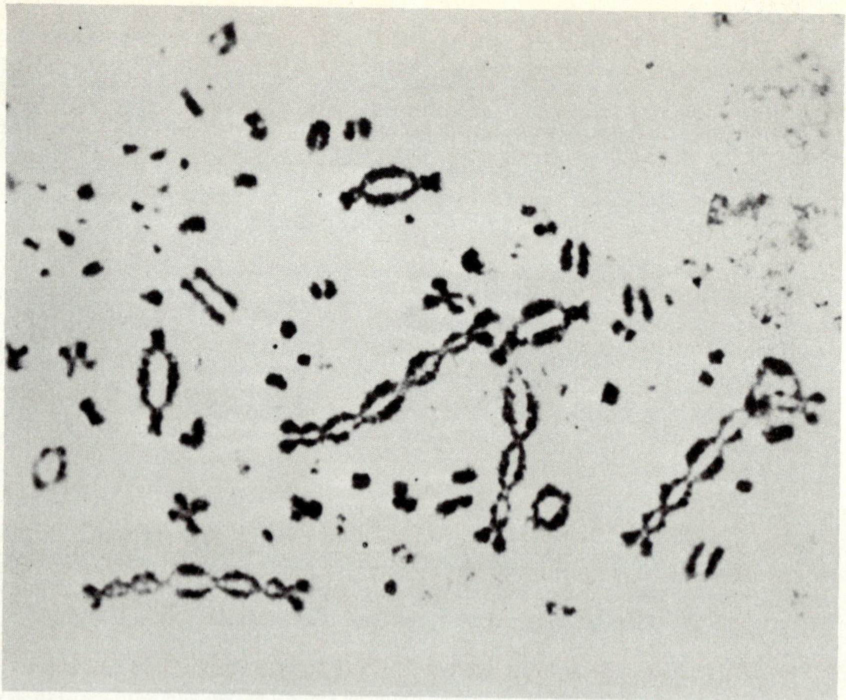

(a)

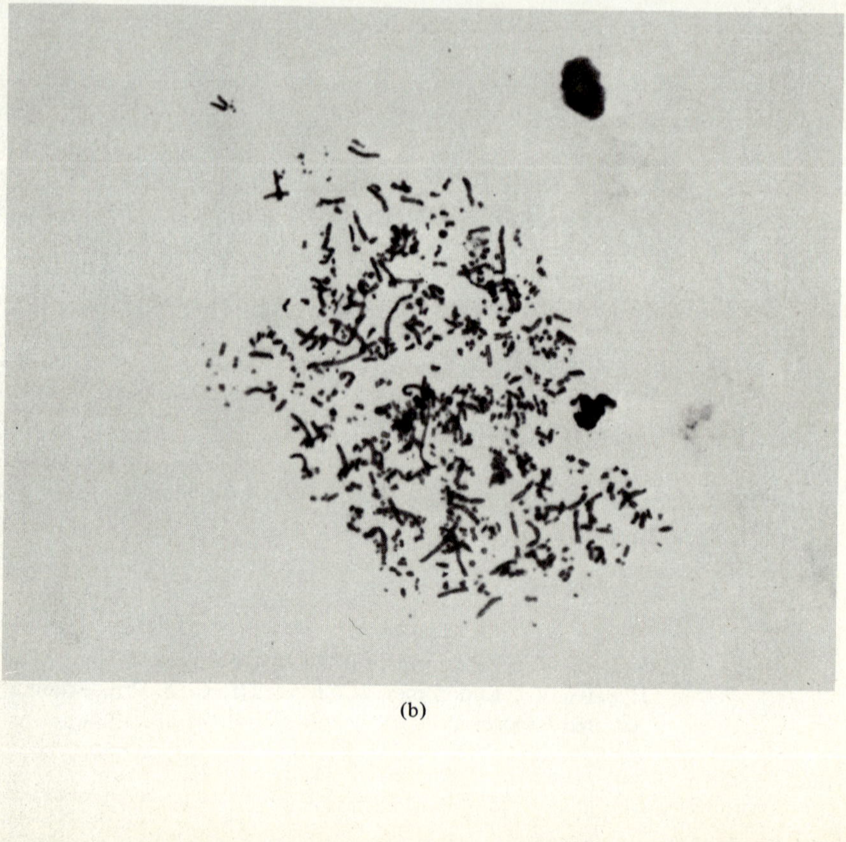

(b)

organ, and later the body become disordered, with the final outcome depending on the radiation dose and on the individual. Radiation damage ranges from the unimportant death of a few replaceable cells (much like a mild sunburn) to nearly immediate death with the induction of tumors and cancers, premature aging, and acute illness (radiation sickness) as intermediate possibilities.

The most serious threat to mankind from radiation may result from those mutations which are produced in sex cells (genetic mutations). The offspring of persons whose genes (in sex cells) have been mutated may possess new characteristics that might not otherwise exist. It is assumed that most mutations are bad in so far as they make it more difficult for an organism to live easily and comfortably. We can make this assumption since man is the product of millions of years of evolution, and a random change cannot be expected to improve his set of genes (although some mutations have been postulated as beneficial). The position that most mutations are undesirable and that any excess radiation above background should be avoided is often justified from the foregoing argument. In the rest of Part Three we learn the sources of natural background radiation and man-made radiation, and the possible effects that radiation may have on man's well-being.

Natural background radiation

Cosmic rays The usual use of the term *cosmic rays* is in reference to those charged particles that reach the earth from outer space. The sun is also a source of charged particles (especially during solar flares), but it is much weaker in total energy than the source of cosmic rays. The exact mechanism of cosmic ray production is unknown although many theories have been advanced to explain their origin. Almost 90% of the cosmic ray flux at the top of the atmosphere is protons. Helium nuclei contribute another 9% and the remaining 1% are heavier nuclei. Meteorites and moon rock samples contain a record of cosmic ray bombardment covering the last billion years of solar system history. Measurements of the abundances of nuclides* and of the cosmic-ray tracks contained within these samples have shown that the cosmic-ray sources have been emitting rather uniformily over this time period and

*Nuclide: Any given combination of protons and neutrons.

Figure 5-7 (a) This cell nucleus, enlarged 275 times, has chromosomes that have doubled, tripled, etc., abnormally because of irradiation. Such a cell could easily become cancerous. (b) This giant cell nucleus, enlarged 50 times, has grown abnormally due to irradiation. The cell now has more than 700 chromosomes and is about 10 times the size of a normal cell. (Courtesy Dr. C. K. Yu, Argonne National Laboratory) that the composition of the cosmic rays cannot have changed drastically over the last 50 million years. However, the cosmic-ray flux reaching the earth's surface has varied in the past, whenever the earth's magnetic field has changed since changed particles are deflected by magnetic fields.

The primary cosmic rays consist of extremely energetic particles, some as high as 10^{14} MeV in energy. As the primaries enter the atmosphere they interact with various nuclei to produce a variety of secondary radiation consisting of electrons, neutrons, protons, other charged particles, and gamma rays. The intensity of the primaries and secondaries is lessened as they pass downward through the atmosphere. Thus, the composition and intensity is a strong function of altitude. As the altitude is increased, there is less protection from the atmosphere; hence, the cosmic-ray dose (amount of radiation absorbed) increases. For example the dose at the elevation of Denver is more than double that at sea level. The cosmic-ray dose also varies with latitude (greatest at poles) since the strength of the earth's magnetic field varies with latitude.

Table 5-1 compares the cosmic-ray dose for several locations in the United States. Near sea level, the dose rate, which is given in millirems (mrem) per year, doubles with an increase in altitude of 5,000 feet. Figure 5-8 shows that the cosmic ray dose does not increase indefinitely. A broad maximum in the dose rate occurs at 12-15 miles altitude as a consequence of the various mechanisms for converting the primary cosmic rays. At flight altitudes for commercial jet aircraft, the dose rate is in the range 0.3–0.7 mrem/hr. During a transcontinental round-trip flight, the average passenger receives a 5-mrem dose.

Table 5-1 Approximate Cosmic-Ray Doses for Several American Cities

Location	*Nominal elevation (ft)*	*Dose rate (mrem/yr)*	*Difference from sea level (mrem/year)*
Albuquerque	4,958	60	32
Atlanta	1,050	33	5
Baltimore	20	28	0
Chicago	579	31	3
Denver	5,280	67	39
Kansas City	750	32	4
Oklahoma City	1,207	34	6
Phoenix	1,090	33	5
Salt Lake City	4,260	54	26
San Francisco	65	28.5	0.5

(From Sailor, V. L., "Population Exposure to Radiation: Natural and Man-Made," U.S. A.E.C.) [Note: At the end of 1974, the U.S. Atomic Energy Commission was abolished and two new agencies were created in its place—the Energy Research and Development Administration and the Nuclear Regulatory Commission.]

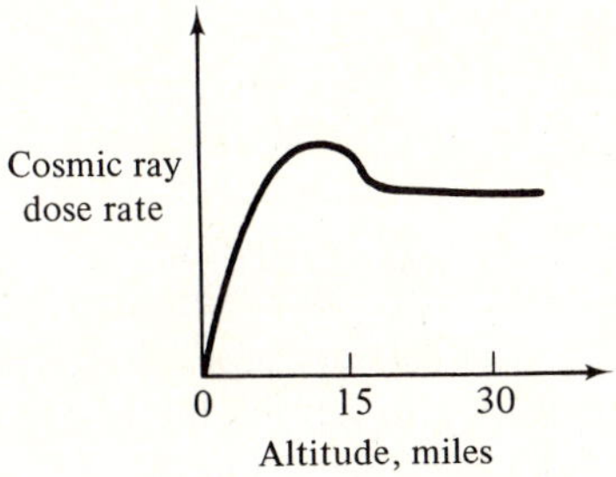

Figure 5-8 Cosmic-ray dose rate versus altitude

Radiation units and measurements

The intensities or strengths of radiation sources were initially measured by detecting the amount of ionization they produced in air. One roentgen is defined as the amount of X-ray or gamma-ray radiation necessary to produce 3×10^9 charge pairs in 1 cubic centimeter of dry air at 0°C and at 1 atmosphere of pressure. A charge pair consists of one electron and one positive ion (in the case of air, the positive ions are mostly nitrogen and oxygen).

There are many methods of measuring radiation intensities. A common one uses an ionization chamber (Fig. 5-9) which may be filled with air or some other gas such as argon. Radiation which can pass through the outer shield of the chamber ionizes the gas in the chamber, forming ion pairs. If the voltage is high enough (a few hundred volts) the ions will be attracted to the

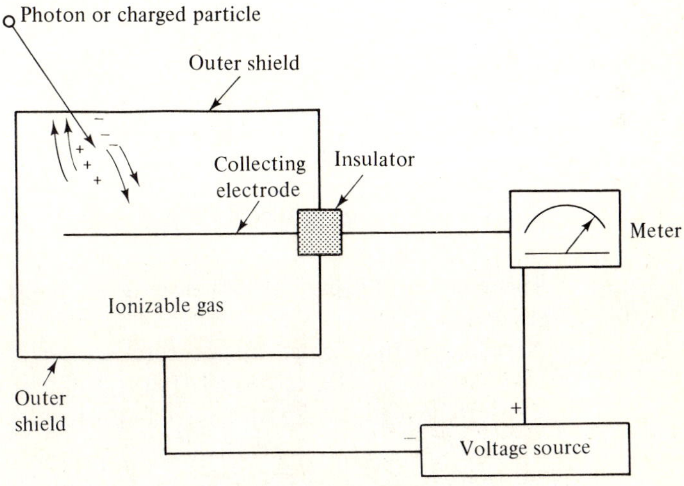

Figure 5-9 An ionization chamber

shield or to the collecting electrode, depending on their charge, before they can recombine. As a result, a pulse of current will flow for each radiation particle that interacts in this way, and these pulses will be measured by the meter. Each pulse will be proportional to the ionization energy delivered by the particle. If the voltage is slightly increased, the pulses get larger but still proportional to the energy left by the particle. (This device is called a proportional counter.) If the voltage is increased still further (to about 1,000 volts), then the pulses become still larger but all alike in size. Thus, the pulse size is no longer proportional to the ionization energy delivered by the particle. This is what happens in the well-known Geiger-Muller counter. This device is very sensitive to the passage of radiation particles, but it tells us very little about their energy. Therefore, it makes a crude dosimeter, but a sensitive detector of small amounts of radiation.

The energy absorbed by a given volume of air can be converted from roentgens/cm^3 to ergs*/cm^3 since the amount of energy necessary to ionize an oxygen or nitrogen molecule is known. Therefore, if we know the volume which is absorbing the radiation, the amount of energy absorbed can easily be computed. The amount of energy absorbed when an X-ray or gamma ray beam enters a medium other than air depends on the nature of the atoms making up that medium (since different elements have different absorption characteristics). Human body tissue acts pretty much like air in its radiation absorption characteristics if we consider the energy absorbed per unit mass rather than the energy absorbed per unit volume.

$$\text{in air:} \quad 1 \text{ roentgen} = 1.13 \times 10^{-1} \text{ erg/cm}^3 \tag{5-1}$$

if we convert to mass absorption,

$$\text{in air:} \quad 1 \text{ roentgen} = \frac{1.13 \times 10^{-1} \text{ erg/cm}^3}{\text{density of air}} = \frac{1.13 \times 10^{-1} \text{ erg/cm}^3}{1.2 \times 10^{-3} \text{ g/cm}^3}$$

$$= 93 \times 10^2 \text{ ergs/g}$$

$$\text{in tissue:} \quad 1 \text{ roentgen} \cong 93 \text{ ergs/g}$$

This amount of energy absorbed per gram holds approximately for body tissue.

A radiation unit which avoids specification of a particular absorbing material is the rad. One rad equals 100 ergs of absorbed energy per gram of any absorbing material. So we can see that 1 roentgen is approximately equal

*It is convenient to use another energy unit in this calculation. It takes 10^7 ergs to make one joule; 10^7 ergs = 1 joule

to 1 rad for X-rays and gamma rays:

$$\begin{aligned} 1 \text{ rad} &= 100 \text{ ergs/gram} \\ 1 \text{ roentgen} &\cong 1 \text{ rad} \end{aligned} \tag{5-2}$$

Since beams of different radiations cause very different biological damage even when the body absorbs the same amount of energy from each type, it is necessary to define certain units in addition to the roentgen and the rad. The absorbed dose of any ionizing radiation is the amount of energy imparted to matter per unit mass of irradiated material at the place of interest. This leads to the definition of the Relative Biological Effectiveness (RBE) of an ionizing radiation as the ratio of the absorbed dose in rads of X-rays of about 200 keV of energy which produces a specific biological effect, to the absorbed dose in rads of the ionizing radiation which produces the same effect.

Some RBE factors are listed in Table 5-2. The first line in the table tells us that X-rays, gamma rays, and electrons have the same biological effectiveness when they are absorbed by body tissue or any kind of tissue. It is not surprising that X-rays and gamma rays have the same RBE as electrons since they lose their energy in matter by transferring it to the electrons of the material they pass through. Going down the table, we find that neutrons which lose their energy via collisions with the nuclei of the material they are passing through cause more damage to tissue than X-rays, gamma rays, or electrons. (Fast neutrons are more effective than thermal or slow neutrons.) Finally, alpha particles (which are helium nuclei) and other heavier ions have the greatest RBE. An example at this point may help to clarify the meaning of the RBE.

A man receiving a dental X-ray absorbs a dose of 1 rad in the tissues of his cheeks. A week later this same man who works at a nuclear power plant is accidentally exposed to a beam of fast neutrons. Again he receives a dose of 1 rad in the same tissues. Will the biological effect be the same in both cases? The answer is no, since fast neutrons have an RBE of 10 compared to an RBE of 1 for X-rays. Therefore, it is as if 10 rads were absorbed by the man while at work compared to 1 rad in the dentist's chair. Indeed, this is how the

Table 5-2

Radiation	*RBE*
X-ray, electrons, gamma ray	1
Thermal neutrons	2–5
Fast neutrons	10
Alpha particles, high Energy ions of O, N, etc.	10–20

rem (radiation equivalent man), which is the unit of radiation usually used, is defined. The rem can be used for all types of radiation. The biological dose in rem is equal to the absorbed dose in rads times the RBE factor:

$$\text{rem} = \text{rad} \times \text{RBE} \tag{5-3}$$

Since 1 rem is a fairly large dose, we often find that mrem are used in expressing a dose. In Table 5-1 mrem/yr (1/1,000 rem/yr) was used as a dose rate or as the dose delivered in mrem per year at the listed locations. From the table we can see that a person who lives at sea level can expect to receive a dose of about 28 mrem each year. If he lives about a mile above sea level (Denver or Albuquerque), his cosmic-ray dose for the year will double to about 60 mrem. Thus far, no ill effects have been noticed from this excess radiation received by people living at high altitudes.

Radioactive decay

The artificial production of gold from mercury was one of the dreams of alchemists, but by 1895 it had generally been agreed that the elements are stable and unalterable. However, with the discovery of radioactivity by Becquerel it was realized that the elements can spontaneously transmute themselves. By 1919 artificial transmutation of nitrogen into oxygen had been accomplished by Rutherford who was working in England.

Most* radioactive disintegrations involve the emission of either an electron (beta decay) or a helium nucleus (alpha decay) from the nucleus of the disintegrating atom. Often gamma rays are also emitted (when a nucleus makes a transition from an excited state to a more stable one) as part of the decay process. Radioactive disintegrations occur in a random manner, which can be discussed only statistically. The unpredictability of individual distintegrations was the first experimental indication that the mechanistic determinism in the physical sciences of the 19th century was due for revision.

Law of radioactive disintegration If we start off with N_0 radioactive atoms at time $t = 0$ and if λ represents the probability that one of these atoms will disintegrate in unit time, then the number of atoms of the original kind still present at time t is given by the following equation:

$$N = N_0 e^{-\lambda t} \tag{5-4}$$

where e is the base of the natural logarithms ($e = 2.718$). This equation holds only for very large N_0. If N is plotted as a function of time, an exponential curve, called the radioactive decay curve, is obtained (Fig. 5-10).

*Positron emission and orbital electron capture are the two other possibilities.

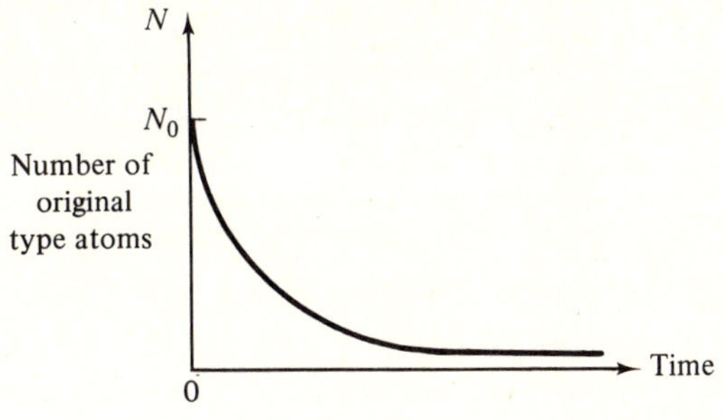

Figure 5-10 Number of original type atoms of a radioactive substance as a function of time

Since the decay is exponential there will never be zero atoms of the original type left ($N = 0$ only for $t = \infty$). Although we cannot get a meaningful answer to the question, "How long will a given radioactive sample last?" we can compute the time it will take before the sample is reduced to some fraction of its initial amount. If the time for half the sample to decay (called the half-life, $T_{1/2}$) is computed one finds that

$$T_{1/2} = 0.693/\lambda \qquad (5\text{-}5)$$

where λ is called the decay constant and can be determined experimentally.

Example 5-3: I^{131} is an isotope of iodine which is found in fallout from nuclear bombs. It emits electrons and gamma rays (.30 MeV energy) and has a half-life of 8 days. If we start with N_0 atoms of I^{131}, how long will it take before $N_0/4$ atoms are left?

$$T_{1/2} = 0.693/\lambda \quad \text{so} \quad \lambda = 0.693/T_{1/2} = 0.693/8 \text{ days} = 0.0865 \text{ day}^{-1}$$

$$N = N_0 e^{-\lambda t}$$

We want to know what t is when $N = N_0/4$.

$$N_0/4 = N_0 e^{-\lambda t} \quad \text{or} \quad \tfrac{1}{4} = e^{-\lambda t} = 0.25$$

Looking in a table of exponential functions, we find that $e^{-1.39} = 0.25$. So

$$\lambda t = 1.39 \quad \text{and} \quad t = 1.39/\lambda$$

$$t = 1.39/0.0865 \text{ day}^{-1} = 16 \text{ days}$$

This answer could have been arrived at more easily by recognizing that the half-life, which is the time it takes for half of the original number of atoms to decay, can be applied two times in succession. At $t = 0$, start with N_0 atoms. After one half-life, $N_0/2$ of original type are left. At $t = 8$ days, start with $N_0/2$ atoms. After another half-life of 8 days, $\frac{1}{2}(N_0/2) = N_0/4$ of original type of atoms are left.

Alpha and beta decay As stated before, nuclei usually decay in two ways: alpha decay and beta decay. In both processes: (1) the total electric charge before the disintegration equals the total electric charge after the disintegration, and (2) the sum of the mass numbers of the initial particles must equal the sum of the mass numbers of the final particles. Ra^{226}, which is one of the isotopes of the element radium, decays by emission of alpha particles. The new element which results from this alpha decay can be determined from the above two rules.

$$_{88}Ra^{226} \longrightarrow {}_2He^4 + {}_AX^M \qquad (5\text{-}6a)$$

X represents the unknown new element, M represents the mass number, and A the atomic number of the new element. We know from rules (1) and (2) that:

(1) $88 = 2 + A$
(2) $226 = 4 + M$

So $M = 222$ and $A = 86$. Therefore, the new element has atomic number 86 (radon).

The final equation can now be written:

$$_{88}Ra^{226} \longrightarrow {}_2He^4 + {}_{86}Rn^{222} \qquad (5\text{-}6b)$$

An example of beta decay is given by:

$$_{19}K^{40} \longrightarrow {}_{-1}e^0 + {}_{20}Ca^{40} \qquad (5\text{-}7)$$

In this type of decay an electron is emitted, and a neutron in the nucleus of the decaying atom (potassium in this case) is converted into a proton. Thus, beta decay can be described by this reaction:

$$_0n^1 \longrightarrow {}_1p^1 + {}_{-1}e^0 \qquad (5\text{-}8)$$

The new element which results from the decay of K^{40} must have an atomic number of 20 (calcium). In beta decay and alpha decay, the decay particles (electrons and alpha particles) are usually released with large amounts of kinetic energy and can, therefore, cause ionization in tissue through which they pass. The same is true for any gamma rays released during either of the decay processes.

External exposure Rocks and soil contain varying amounts of natural radioactivity, mostly due to members of the Th^{232} (thorium), U^{235} (uranium), and U^{238} radioactive decay series (Fig. 5-11). A typical annual dose from rocks and soil is 50 mrem. This is mostly due to the gamma-ray component of the decay, since beta and alpha particles are absorbed by relatively small thicknesses of material. For example a 3-MeV beta particle will be stopped by $\frac{1}{25}$ inch of aluminum, and a 3-MeV alpha particle by only 1 inch of air. The

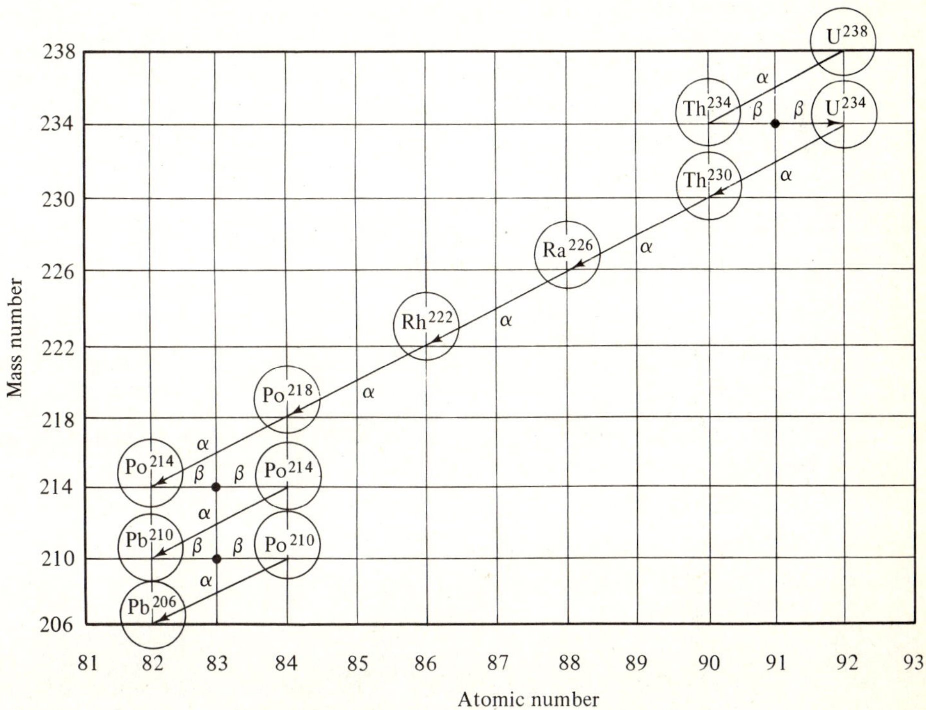

Figure 5-11 The radioactive isotope of uranium, U^{238}, decays to the stable isotope of lead, Pb^{206} through a series of alpha and beta decays. Each downward decay to the left is an alpha decay, and each rightward decay is a beta decay. Some nuclei can decay in both ways.

radioactive content of rock and soil varies widely throughout the U.S. and throughout the world. Relatively high radioactive content in rock is found in regions of Vermont, New Hampshire, Connecticut, Manhattan Island, and the Rocky Mountain states.

Several radioisotopes are present in the atmosphere as gases or attached to dust particles. Among these are the isotopes produced by cosmic rays, and the noble gases radon and thoron which are formed in the thorium and uranium decay series. The radioactive content of a volume of air varies with meteorological conditions, being enhanced during periods of inversion and reduced after a heavy rainfall.

Effect of shelter Many of us spend much time indoors where the walls and the roofs of the buildings reduce exposure to external radiation. However, often this reduction is offset by exposure from the natural radioactivity in the structural materials. A typical wooden house will reduce exposure to about 70% of the outdoor level. Brick, concrete, and stone buildings typically contribute exposures 50 to 100 mrem/yr greater than outdoor values at the same site; however, much larger indoor exposures have been observed (see Table 5-3).

Table 5-3 GAMMA EXPOSURE RATES INSIDE VARIOUS BUILDINGS, COSMIC RAY COMPONENT INCLUDED

		Exposure rate (mrem/year)	
Location	*Structure*	*Typical*	*Extreme*
East Germany		106	1,200
New York City	brick	79–118	
Grand Central Station	New York granite		350
United States	wood	60	
" "	concrete	130	
Aberdeen, Scotland		81	110
Cornwall, U.K.	granite	145	
Sweden	wood	48–57	
"	brick	99–112	
"	concrete	158–202	

(From Sailor, V. L., "Population Exposure to Radiation: Natural and Man-Made," U.S. A.E.C.)

Within a building, exposure rates can vary tremendously. In rooms with poor ventilation, radon and thoron content of the air can build up to very high levels. The lower floors of a multi-story building are better shielded against gamma rays produced by cosmic rays than the higher floors. At the present time there is no systematic effort to select building material on the basis of low radioactivity.

Internal exposure Every human body naturally contains radioactive materials. Since the radioactive content of vegatation and animal life depends on the soil and water, the human body burden of radioactivity depends on the source of foodstuffs and drinking water. For example, approximately one million people in Illinois and Iowa consume drinking water containing an unusually high content of Pb^{210}, Po^{210}, and Ra^{226}. The bones of these people contain as much as four times the normal amounts of radium and its decay products.

Most of the internal dose comes from K^{40} [Eq. (5-7)] which is a radioactive isotope of potassium and has an isotopic abundance of 0.0118%. It is both a beta and gamma ray emitter. Therefore, of the approximately 140 grams of potassium in the body, 0.0118% or 0.165 gram is of the radioactive isotope K^{40}. The dose rate from K^{40} is about 20 mrem/yr. Other radioisotopes in the body, such as C^{14}, Rn^{222}, and Ra^{226}, contribute another 3 mrem/yr to the internal dose. Since potassium does not concentrate in any one organ, the internal dose from K^{40} is spread throughout the whole body equally. This type of dose is called a whole-body dose.

By now it should be apparent that the radiation dose from natural sources is highly dependent on geographical location. There is even a temporal variation due to the fluctuation of cosmic-ray dose rates during the solar cycle ($\pm 5\%$). However, typical whole-body doses to an average member of the population have been established. For the year 1970 this dose was estimated to be 130 mrems. Table 5-4 summarizes the internal and external dose rates from natural sources. When the subtotals are added, taking an average value for the two cosmic-ray dose rates listed, we find a dose rate which is equal to only 121 mrem/yr. Therefore, a weighted average of the cosmic-ray

Table 5-4 TYPICAL WHOLE-BODY DOSES FROM NATURAL SOURCES

Source	***Dose***
Internal	(mrem/year)
Potassium—40 in human body	20
Other radionuclides in body (C^{14}, Rn^{222}, $Ra^{222, 228}$, etc.)	3
External out-of-doors	
X-rays from soil and rocks	50
Cosmic rays (Sea level, 50° Geomagnetic lat.)	28
Cosmic rays (Denver)	67

dose rates and the external dose rates must have been taken to arrive at a figure of 130 mrem/yr for a total dose rate.

Why are some atoms radioactive? We have spoken of stable and unstable atoms without trying to explain why some atoms are unstable. Actually, the first question that should be asked is, "What holds the nucleus together in the first place?" since it is composed of positively charged particles (protons) and uncharged particles (neutrons). If only an electrical force was acting within the nucleus, then it should fly apart due to the repulsive force between the protons. However, there is another force, called a nuclear force, which is important only when nucleons (protons and neutrons) are very close together. It is always an attractive force and it exists between neutrons and neutrons, neutrons and protons, and protons and protons.

When protons are near each other, as in small nuclei, the nuclear force can overcome the electrical repulsive force between the protons. In a large nucleus, distant protons may have a larger repulsive electrical force than attractive nuclear force between them. This tends to make that nucleus more unstable than the smaller one. Electrical forces weaken with the inverse square of the distance between charges, whereas nuclear forces weaken more quickly with distance, between the nucleons.

Neutrons tend to hold nuclei together since they only attract other nucleons, and many of about the first twenty elements achieve stability with approximately the same number of neutrons and protons. This is not true for the heavier elements where we find many more neutrons than protons in the nucleus. Even so, all elements with an atomic number greater than 82 are unstable. For example, Ra^{226} has 88 protons and $226 - 88 = 138$ neutrons, and it decays by alpha-particle emission.

This explanation of nuclear stability or instability is only qualitative in nature and it still is not capable of explaining the instability of a simple nucleus such as tritium. Tritium, which has 2 neutrons and 1 proton in its nucleus, is unstable with a half-life of about 10 years. There is no electrical repulsive force in the nucleus since there is only 1 proton. Obviously, other considerations are necessary to explain nuclear stability for all the elements. The simple model that we have described does explain why the ratio of neutrons to protons must increase with atomic number if there is to be any chance of relatively stable heavy atoms existing.

With the introduction of the nuclear force, we have now discussed three of the four types of forces known to exist in nature. These four forces are: gravitational, electric, nuclear, and weak interactions. They all influence the structure of matter, but on different scales. Gravitational forces explain the structure of the solar system and the universe, electric forces govern the structure of matter from the size of a nucleus to a redwood tree, nuclear forces are important for matter of the size of a nucleus on down, and a weak interaction is the shortest range force; it takes place within a particle such as a neutron.

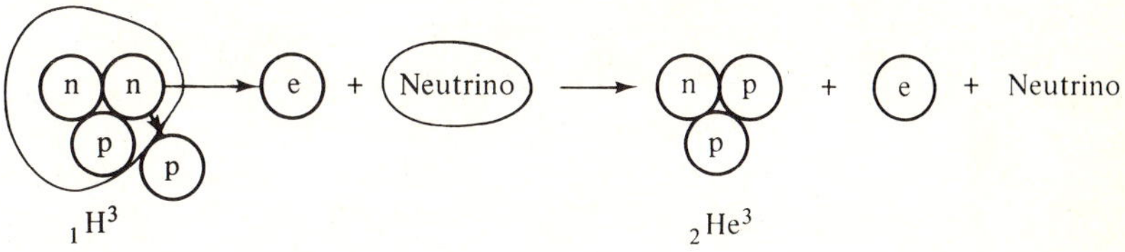

Figure 5-12 Beta decay of a tritium nucleus

Weak interactions keep nuclei with an "excess" of neutrons from being stable. For example, this force causes the tritium nucleus to be unstable. A neutron decays into a proton, electron, and neutrino through a weak interaction. A neutrino is a massless uncharged particle which is very difficult to detect; it is often left out of equations describing beta decay. The following reaction describes what happens when tritium beta decays. We will not pursue the subject of beta decay and weak interactions any further than this:

$$\begin{aligned} {}_0n^1 &\longrightarrow {}_1p^1 + {}_{-1}e^0 + \text{neutrino} \\ {}_1\text{H}^3 &\longrightarrow {}_{-1}e^0 + {}_2\text{He}^3 + \text{neutrino} \end{aligned} \tag{5-9}$$

Suggested reading

1 EISENBUD, MERIL, *Environmental Radioactivity.* New York: McGraw Hill Book Co., 1963.

2 FRIGERIO, NORMAN A., "Your Body and Radiation," U.S.A.E.C., Technical Information Center, Oak Ridge, TN, 37830, 1967. (Books in this series can be obtained free of charge by writing to the above address.)

3 HEWITT, PAUL, *Conceptual Physics.* Boston: Little, Brown and Co., 1971. (See Chapters 34 and 35.)

4 KASTNER, JACOB, "The Natural Radiation Environment," U.S.A.E.C., Understanding the Atom Series, Oak Ridge, TN, 1968.

5 MILLER, FRANKLIN, Jr., *College Physics.* New York: Harcourt, Brace and Jovanovich, 1972. (See Chapter 28.)

6 RICHARDS, J. A., F. W. SEARS, M. R. WEHR, and M. W. ZEMANSKY, Modern University Physics. Reading, MA: Addison-Wesley, Publishing Co. Inc., 1960. (See Chapter 42.)

Questions

1 When X-rays were first discovered in 1895, many people were concerned that fully dressed people might be made to appear unclothed. Was there good reason to fear such an occurrence?

2 What takes place when a hydrogen atom is ionized? Can a 100-eV X-ray photon ionize an atom of hydrogen?

3 Assuming the same starting energy, which will penetrate human tissue further: (*a*) a neutron or a proton? (*b*) an X-ray photon or an electron?

4 Electromagnetic radiation of energy less than that of ultraviolet light strikes your skin. What is the effect?

5 What may happen when a cell nucleus within our body absorbs high-energy radiation? What if the cell is a sperm cell or an ovum?

6 Do you think it is reasonable to assume that most genetic mutations are undesirable?

7 What effect may changes in the earth's magnetic field have had on evolution?

8 What is the composition of cosmic rays? Why is the cosmic-ray dose less at sea level than at the altitude of Denver?

9 Why was it necessary to define the rem when the rad had already been defined for radiation measurements?

10 Would it be more disadvantageous to absorb a dose of 10 rads from fast neutron irradiation or a dose of 50 rads from X-ray irradiation?

11 I^{131} has a half-life of 8 days. If you start off with 16 grams of I^{131}, how long will it take before there are only 4 grams left?

12 When a radioactive atom decays it turns into (1) an isotope of the original atom, (2) an ion, (3) a different element, or (4) an alpha particle.

13 The rate of radioactive decay is affected by (1) temperature, (2) pressure, (3) frequency, or (4) none of these.

14 I^{131} beta decays. What is the atomic number, mass number, and name of the resulting element?

15 When U^{238} decays to Ra^{226}, how many beta decays and how many alpha decays take place?

16 (*a*) What are the main sources of natural radiation exposure?
(*b*) Do they depend on geographical location?

17 (*a*) What force keeps nuclei from coming apart?
(*b*) Why are not all nuclei stable?

Chapter 6

Man-made radiation and man's health

Medical and dental X-rays

The largest population exposure to man-made radiation comes from medical X-rays. The dose delivered to a person is usually not calculated or measured and, in general, there are no systematic records kept on the lifetime accumulated dose to an individual. A significant difference between natural background radiation and both diagnostic and therapeutic medical exposures is that the medical exposures normally involve only some part of the body, while the background radiation results in whole-body exposures. Therefore, in evaluating the overall effect of medical exposures it has been customary to consider the genetic effects of exposures to the gonads as the principal problem. Since X-rays can be scattered (deflected at all angles, just as a moving billiard ball can be scattered by a stationary billiard ball) by body tissue, a chest or head X-ray, for example, will lead to some exposure to the gonads even though the X-ray beam is not aimed in that direction. External shielding with lead cannot stop all of this scattered radiation.

Compared with other technologically advanced countries of the world, the average U.S. X-ray dose from diagnostic procedures is quite large as is shown in Table 6-1 which lists the genetically significant dose for several countries in 1970. The genetically significant dose is defined as the dose which, if received by every member of the population, would be expected to produce the same total genetic injury to the population as do the actual doses received by the various individuals. In essence, this means that the total gonadol doses received by the various members of the population are weighted by the age-considered individual child-producing probability of those actually irradiated. This is then averaged over the entire population. If these dose values in Table 6-1 are to be compared to those in Tables 5-1, 5-3, and 5-4, then they should be about doubled to convert to whole-body doses.

Recently attention has been focused on the effort to lower the large diagnostic X-ray exposure received by the U.S. population. Mobile X-ray units, formerly used to screen the population for tuberculosis, are being retired since the radiation risk from these old machines was often found to be greater than the benefit expected from having the X-ray taken. (The incidence of tuberculosis has dropped.)

Table 6-1 GENETICALLY SIGNIFICANT DOSE (MREM/YR) FROM MEDICAL DIAGNOSIS IN VARIOUS COUNTRIES

United States	95
Japan	39
Sweden	38
Switzerland	22
United Kingdom	14
New Zealand	12
Norway	10

(From Sailor, V., "Population Exposure to Radiation: Natural and Man-Made," U.S. A.E.C)

The dose to the gonads from chest X-rays is usually on the order of 1 to 10 mrem per exposure. The actual total dose received from chest X-rays is about 50 to 500 mrem per film. Fluoroscopic examinations produce the highest X-ray exposures with 50,000 mrem being an average total dose for chest examinations.

A typical dental X-ray results in a 2 mrem exposure to the gonads and a 200-mrem exposure to the skin. Under pressure by state regulatory agencies, dentists are now using better X-ray equipment and safer procedures, thereby lowering the patients' X-ray exposure. There is also a trend toward limiting routine dental X-rays for all patients and eliminating dental X-rays of pregnant women. When proper equipment and procedures are used (shielding and short exposures) the radiation exposure from dental X-rays appears to be insignificant compared to that from diagnostic medical X-rays.

When properly used, diagnostic medical X-rays can save lives and alleviate suffering. In most cases the benefits derived from the use of this diagnostic tool far outweigh the risks incurred. However, the use of diagnostic X-rays in the U.S. has grown to the point where many people now question the necessity of many of the X-rays which are almost routinely taken of them or their children. An informed public can help to reduce the exposure to the general population from diagnostic X-rays by questioning the logic behind all X-rays whose need is not obvious. Elimination of diagnostic X-rays is not recommended here, but thoughtful consideration of the added genetic risks associated with increased radiation is asked for.

Fallout

When a nuclear weapon is detonated above the ground a number of radioactive nuclides are produced and become widely distributed in the atmosphere. Some may slowly come to earth far from the site of the detonation. This radioactive debris is known as fallout. Thus far, two types of nuclear weapons have been exploded.

The fission bomb (developed in 1945, but no longer used by itself) derives its energy solely from the fissioning or splitting into two almost equal parts of either U^{235} or Pu^{239} (plutonium) nuclei. Fissioning can take place when a neutron is captured by either of these nuclei, making them more unstable (Fig. 6-1). One of many possible reactions which can take place with U^{235} is:

$$ {}_{92}U^{235} + {}_{0}n^{1} \longrightarrow {}_{92}U^{236} \longrightarrow {}_{54}Xe^{140} + {}_{38}Sr^{94} + {}_{0}2n^{1} + 200 \text{ MeV} \quad (6\text{-}1) $$

Most of this 200 MeV of energy is carried away by the xenon and strontium nuclei in the form of kinetic energy. Both of these nuclides are radioactive and decay by beta emission. The 200 MeV of energy released per fission may not seem like much energy in macroscopic terms (page 43), but when expressed in terms of the mass of material needed to release that much energy it works out to be a million times greater than the energy emitted per pound

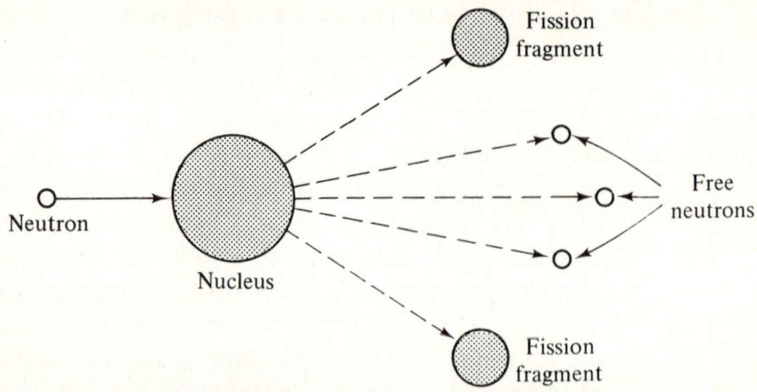

Figure 6-1 A fission reaction

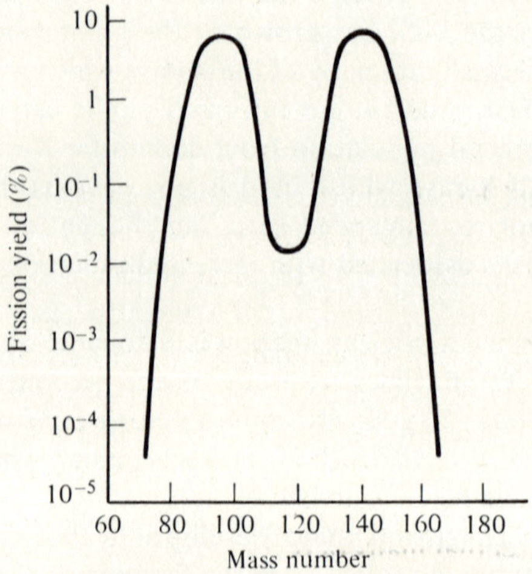

Figure 6-2 Fission yield from U^{235}

from burning coal. Figure 6-2 shows the fission yield from U^{235}; that is, the percentage of the fisssions that lead to the formation of a particular nuclide. We can see that nuclides with mass numbers from about 80 to 152 have fission yields above 0.1%, but that nuclides of mass numbers around 93 and 140 are most probable. Many of these nuclides are radioactive and cause radiation exposures to people and animals living in regions where fallout occurs.

The second type of nuclear weapon is the fusion bomb (also called a hydrogen bomb) which develops its energy from fusion reactions where two light nuclei (such as deuterium and tritium) are joined together to form a heavier nucleus such as helium. Large amounts of energy are released in a fusion reaction, just as in a fission reaction. Very high temperatures are necessary to force a fusion reaction to take place. In a fusion bomb the high temperatures are produced by a fission bomb trigger. Therefore, both types of nuclear weapons involve fission reactions and fallout. Fission and fusion are discussed in more detail in Chapter 11, but as sources of controlled energy.

The radioactive nuclides of interest biologically arise in three ways:

1 Atoms of Pu^{239} or U^{235} from the weapon's core may be scattered (most bombs use Pu^{239} now).

2 Fission products such as Cesium 137 (Cs^{137}), Strontium 90 (Sr^{90}) and Iodine 131 (I^{131}) may be scattered by the blast.

3 Neutrons produced by the weapon may be absorbed by the nitrogen of the air to give Carbon 14 (C^{14}) which is radioactive.

$${}_7N^{14} + {}_0n^1 \longrightarrow {}_6C^{14} + {}_1H^1 \qquad (6\text{-}2)$$

The average dose rate from fallout in the U.S. has been estimated at around 5 mrems per year through the year 2000. This takes into account the decreasing radioactivity of those nuclides of importance. However, the possible increase in fallout from weapon tests by France and China (both nonsigners of the test-ban treaty) is not taken into account. Thus, 5 mrem per year is a minimum dose rate which could increase through weapons testing in the atmosphere or through peaceful applications of nuclear weapons (prospecting for natural gas and harbor building).

The chief danger from fallout lies not so much in its total dose as in the fact that many of the nuclides resemble normal body elements in their chemical behavior. These nuclides are taken up by the body when food and water contaminated by these radioactive nuclides are ingested by humans. Thus as with radium, which was discussed earlier, each nuclide may find its way to a particular organ or tissue and exert an effect out of proportion to its average dose. Sr^{89} and Sr^{90} resemble calcium and accumulate in the bones. I^{131} behaves like stable iodine and concentrates in the thyroid and salivary glands. Pu^{239} finds its way to certain sensitive sites in bone, and C^{14} may lodge within the critical DNA molecules of our cell nuclei. The internal dose from these nuclides depends on such factors as their physical and biological half-lives and on the type of radiation emitted during decay. The physical half-life has been defined and the *biological half-life* is the time required for 50% of a given chemical substance to be excreted from the body. Cesium 137 has a physical half-life of 30 years and a biological half-life of 17 days.

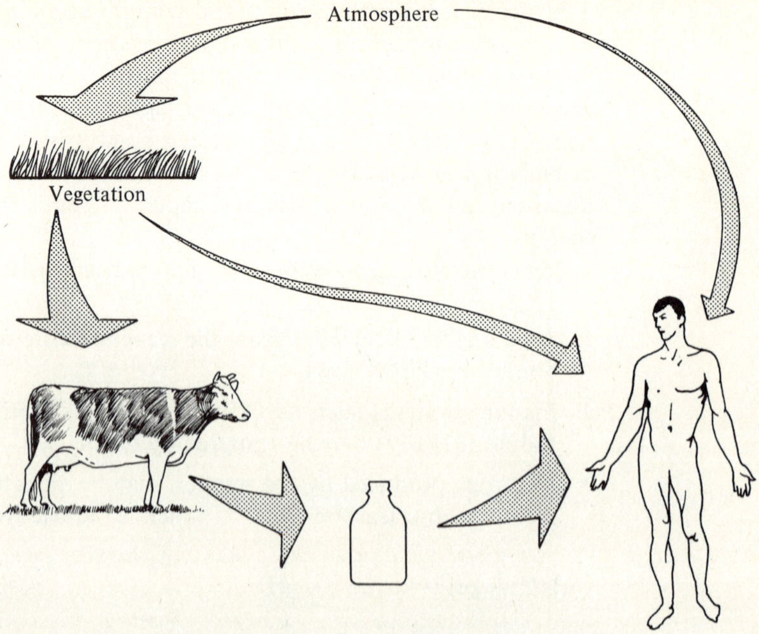

Figure 6-3 Pathways from fallout to our bodies. Most of the I^{131} entering our bodies is found in milk. (From Comar, C. L., "Fallout," Atomic Energy Commission)

Table 6-2 SOURCES OF STRONTIUM 90 IN OUR DIET

Food	***Percent of total*** Sr^{90}	
Milk	47.6	(Milk accounts for about half of the Sr^{90} entering our bodies)
Flour and cereals	22.5	
Root vegetables	5.0	
Leafy and other vegetables	10.1	
Fruits	7.2	
Meat, fish, eggs	3.1	
Water	4.5	
Total	100.0	

From Comar, C. L., "Fallout," U. S. Atomic Energy Commission

Therefore, its residence time in the body will be governed by the smaller half-life which is 17 days. All of the above-mentioned nuclides are beta emitters and would not pose a serious danger if they remained external to the body (beta radiation does not penetrate the skin). However, once inside the body, all of the beta radiation will be absorbed in a small region.

Table 6-3 SOURCES OF CESIUM$_{137}$ IN OUR DIET

Food	*Percent of total Cs^{137}*
Milk	35
Meat	25
Flour and Cereals	20
Vegetables	10
Fruits	10

From Comar, C. L., "Fallout," U. S. Atomic Energy Commission

There have been claims (see Sternglass in Suggested Reading) of increased infant mortality due to radiation received by pregnant women from Sr^{90} contained in fallout. However, these claims have been refuted by such agencies as the U.S. Public Health Service which reports that dispersed fallout has had no detectable adverse effect on human health. On the other hand, localized fallout has definitely had harmful effects on humans in several cases.

In 1954, 64 inhabitants of Rongelap, an island near Bikini in the Pacific, were accidentally exposed to severe fallout from an H-bomb test over Bikini atoll. They received over 2,000 rad of beta radiation to their skins. As Fig. 6-4 shows, they developed dark burn areas on their skin and some loss of

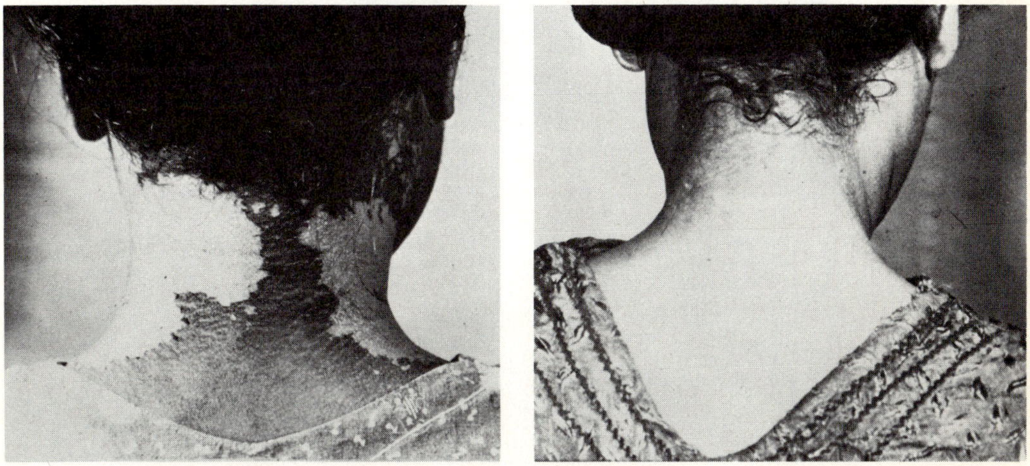

(a) (b)

Figure 6-4 Changes in pigmentation of skin due to beta burns. (a) Burn area on the neck of a Rongelap native one month after accidental exposure to radiation in 1954; (b) The same burn area one year after the accident, showing complete recovery. (Courtesy Atomic Energy Commission) hair. It took several months for healing to be complete. Much more unfortunate were those Rongelap children who were exposed to fallout and who developed thyroid tumors 10 years later. There has even been one death from leukemia which was linked to the fallout.

Hopefully all nations will some day agree to end testing of nuclear weapons in the atmosphere, thereby insuring the human race of no further increase in the dose rate from fallout. In any case, we are stuck with at least an average dose of 5 mrem per year from fallout until the year 2000, and a somewhat smaller dose rate after that.

Nuclear power plants

As of September 1974, there were 52 commercial nuclear power plants (Fig. 6-5) and one fuel reprocessing plant in operation in the U.S. As of the same date 60 new power plants were being built and 105 were planned. Nuclear power

Figure 6-5 Big Rock Point nuclear plant on Lake Michigan near Charlevoix, about 200 miles northwest of Detroit. Boiling-water reactor, 70,400-net electrical kilowatt capacity, designed by General Electric Company. Started commercial operation in 1963. Owned and operated by Consumers Power Company. (Courtesy Atomic Energy Commission)

plants utilize the fission reaction to heat water and produce steam and, eventually, electricity (Fig. 6-6). They release radioactive isotopes into the atmosphere and water. The details on what radioactive isotopes are actually emitted can be found in Chapters 11 and 12 along with a detailed description of how a nuclear power plant works.

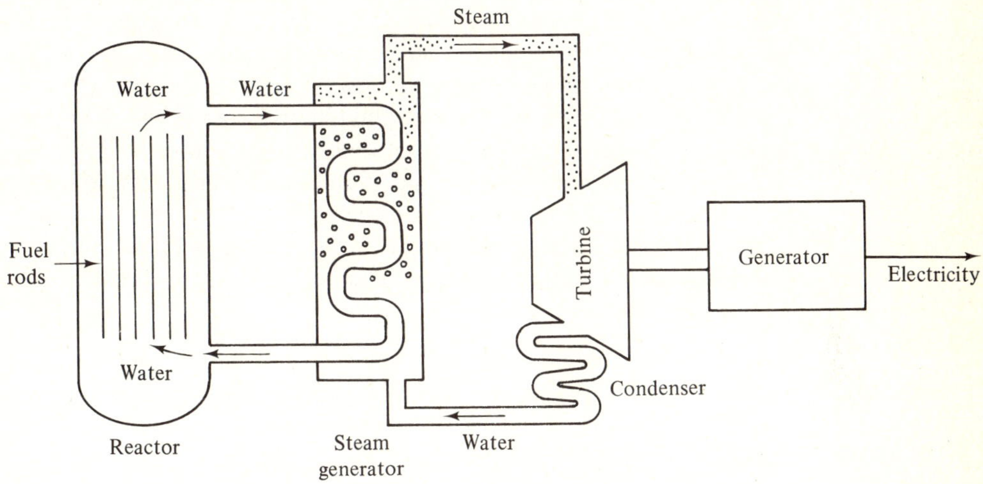

Figure 6-6 In a nuclear power plant the fission of U^{235} releases the energy needed to make steam from water. In a pressurized-water reactor the water that passes through the reactor is under pressure and does not turn into steam, even though it is at a very high temperature. In the steam generator, heat is transferred to a separate water system to make steam. The steam turns a turbine which is connected to an electric generator.

Offsite doses are generally too small to measure in comparison to natural background and, therefore, must be calculated from known emission rates, meteorological conditions, and population distributions. Measurements are made, however, on agricultural products, foliage, wild life, water supplies, and marine biota in the area surrounding nuclear power plants.

Most nuclear power plants produce radiation exposures which are within the proposed new limits (5 mrem/yr) for exposures at the boundaries of new plants. Public hearings were still being held, as of April 1974, on the proposed limits. This dose decreases with distance from the plant and is mostly due to radioactive gases of long half-life such as Krypton 85 (Kr^{85}) and Xenon 133 (Xe^{133}). There are a few nuclear power plants, such as the Humboldt Bay power plant in northern California, which have emissions at the plant boundary above the new limits. For example, in 1970 the emission was 127 mrem at the Humboldt boundary, which was above the new limit. At the present time the estimated whole-body dose to the population from nuclear power plants is less than 0.01 mrem/yr. As is the case with fallout, the main hazard lies with possible accumulation of certain nuclides in body organs or tissue rather than with the external dose to the body. The radiation exposure from nuclear power plants and reprocessing plants can be expected to grow as the number of nuclear power plants increases unless there are further improvements in reducing radioactive emissions from such plants. The reduction in allowable emissions from power plants was a step in the right direction; however, reprocessing plants were not covered by the new limit. Fuel reprocessing plants can emit more radiation than power plants, so it is important to establish radiation limits for them, too.

The estimated radiation exposure from natural and man-made sources is summarized in Table 6-4. These are whole-body doses. The total dose in one year works out to about 230 mrem/yr.

Table 6-4 Estimated Average Annual Whole-body Exposure of the General U.S. Population to Ionizing Radiation for 1970

Source	*Average per capita dose (mrem/yr)*
Natural background	130
Medical diagnostic and therapeutic	95
Weapons' tests fallout	5
Nuclear power plants	0.01

It should be kept in mind that these radiation exposures vary from place to place and person to person and that the above averages are for the U.S. population.

Recognition of danger from radiation

Within several years after their discovery it was recognized that X-rays could produce skin cancers and bone cancers. Many doctors working with X-rays suffered amputations of their fingers and hands in an attempt to prevent the spread of these cancers. By 1914 it was widely agreed that more information was needed on safe exposure limits for diagnostic X-rays.

Becquerel, who discovered radioactivity, received a fairly serious and slowly healing skin ulcer from exposure to the small piece of radioactive material he carried with him. The danger of radioactive materials was made clear by the unfortunate case of the radium dial painters. This incident involved women who were employed in a New Jersey plant during World War I to paint various dials with luminescent material containing radium salts. (These dials were needed for nighttime visibility.) Alpha particles emitted by the radioactive radium excited the luminescent material and light was given off when the luminescent material returned to a stable state again. The women sharpened their brushes with their tongues and thus inadvertently swallowed minute quantities of this radioactive radium. Within a few years some of these dial painters had developed bone cancer at the points of concentration of the radium, and anemia had also been observed (Fig. 6-7). By 1931, 18 deaths had been attributed to internal deposits of radioactive materials. There were other situations which pointed out the danger of radioactive materials (such as the increased incidence of lung cancer among uranium mine workers).

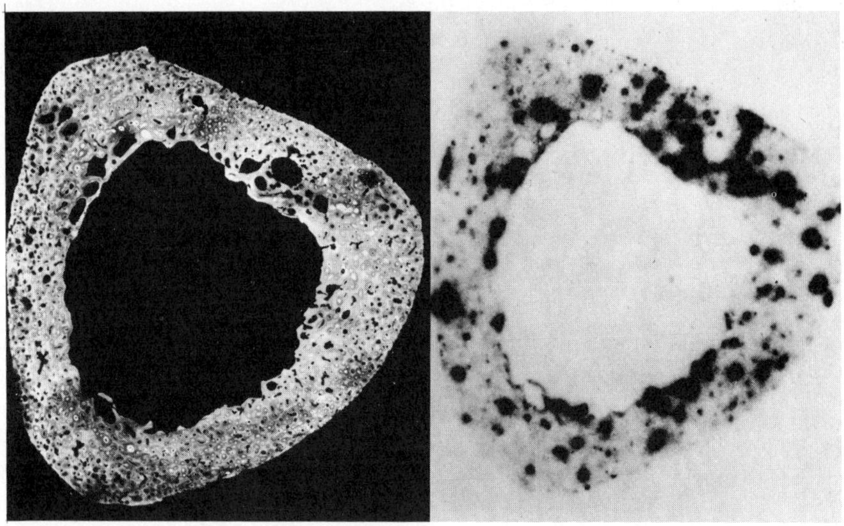

(a) (b)

Figure 6-7 (a) A section of bone from the body of a former radium watch-dial painter who, in order to maintain a fine brush tip, was in the habit of touching the tip with his tongue. This photograph shows darkened areas of damaged bone. (b) An autoradiograph—the bone "took its own picture" by being held against film—showing areas exposed by radium alpha particles. Note that the areas of high alpha activity correspond to the areas of maximum damage in (a). (Courtesy Atomic Energy Commission)

Exposure limits

Because of these dangers from radiation, exposure limits were developed by international and federal agencies to protect workers and the general public from the harmful effects of radiation. These exposure limits have been revised downward several times and there is pressure to lower the permissible exposures once again.

At present there are three distinct but interrelated numerical values which set limits on radiation exposure for the general population:

1 The first applies to individual cases; the limit for any one person's whole-body exposure is put at 500 mrem/yr.

2 The second limiting value is for use when a group of individuals are at risk and not all of the individual exposures can be measured. (This is certainly the case for the U.S. population.) The whole-body radiation exposure for the average of a suitable sample of the population at risk must be less than 170 mrem/yr. It is figured that, if a suitable sample of people (representative of a very large population) receives an average of less than 170 mrem/yr, then only a very small percentage of people within that sample and within the total population will have received more than three times that amount, or about 500 mrem/yr.

3 The third limit is based on considerations of population genetics. The per capita dose limitation for the gonads is 5 rem (5,000 mrem) in 30 years.

The above limits do not include natural background or medical diagnostic radiation. Therefore, they do not effectively limit the total radiation exposure of the population since medical exposure can keep on increasing. In establishing the above exposure limits the assumption is made that the risks (cancer induction and genetic mutations) are outweighed by the advantages of such things as nuclear power plants and other peaceful uses of nuclear energy. Nowhere is it stated that the risk of ill effects is zero if the above limits are adhered to. It is very difficult to evaluate a statement of the type: the risk is worth the benefit. Public reaction to such a statement will depend on individual value systems which may vary from time to time and certainly from person to person.

Short-term effects of acute exposure to radiation

Most of our knowledge of the effects of radiation on man comes from accidents, from earlier medical uses of radiation, and from victims of the nuclear bombs dropped during Warld War II. Since these events all involved large doses of radiation, some extrapolation of this information to lower doses is often necessary when the long-term effects of radiation are sought. Before discussing the long-term effects from radiation exposure we will describe the effects of an acute dose of radiation on an individual.

Table 6-5 gives the currently accepted general effects in man of gamma ray exposures at various levels; by definition these should also be the values in rem for exposures that involve neutrons and other types of radiation. These are whole-body doses and are assumed to occur within a short time, perhaps a few minutes to an hour, but definitely less than a day. An exposure

Table 6-5 SHORT-TERM EFFECTS OF ACUTE RADIATION EXPOSURE

Dose (rems)	*Effect*
25	Perhaps detectable clinically by blood measurements.
100	Nausea in about half of those exposed; fatigue noticed; marked hematologic effects.
200	Nausea in all exposed individuals; fatigue; death possible, especially in the absence of rest and other treatment; increased susceptibility to infection.
400	LD–50 dose in the absence of rest and other treatment; some fatalities even with treatment.
600	LD–100 dose in the absence of rest and other treatment; probably LD–50 to LD–70 with special treatment.
800	Probably LD–100 even with treatment.

NOTE: Although there is some disagreement on the dose values for the various effects noted, the above figures have been rather consistently quoted.

(From *Fundamentals of Radiation Protection*, H. Henry, Wiley-Interscience, New York, 1969)

of 400 rem is estimated to cause death in about half of the population exposed. Thus 400 rem is known as the LD–50 or lethal dose for half of an exposed population.

The main cause of death from exposure in the lethal range is from severe injury to the blood-forming organs in the bone marrow, and death may occur in about 4–6 weeks. The lymphocytes (a type of white blood cell) are the cells in the body most sensitive to radiation. Their life span in the blood varies from a few hours to as much as 200 days. Red blood cells are also very sensitive to radiation; they have life spans of around 100 days. It has been found that cells undergoing rapid change are most susceptible to radiation damage. The early reduction in lymphocytes after acute exposure is one of the major causes of the increased susceptibility to infection of highly irradiated individuals. Lymphocytes are well known as the fighters of infection.

Long-term effects of radiation

Somatic There is very good evidence that leukemia (cancer of the white blood cells) and other cancers can be induced by radiation exposures of 100 rem or greater. The sources of information that shed light on the cause and effect relationship between irradiation and cancer induction in man are:

1 Study of survivors of Hiroshima and Nagasaki by the Atomic Bomb Casualty Commission.

2 Study of patients treated with radiation for nonmalignant diseases who then developed cancer or leukemia later in life.

3 Study of children who received irradiation for treatment of enlarged thymus and then developed thyroid cancer.

4 Study of the occurrence of lung cancer in uranium miners in the U.S.

5 Study of cancer and leukemia in children whose mothers had been given diagnostic X-rays during pregnancy.

Two main problems are associated with the use of these data. First, there are other causes of cancer aside from radiation. For example, leukemia may be more likely in offspring of older women, and it may also be induced by viruses. These other causative agents must be considered when examining statistical data on cancer induction by radiation.

The second problem is that reliable data exist only for doses over 100 rem. For example, Fig. 6-8 charts data concerning the occurrence of leukemia as a function of radiation dose. This is called a dose-effect curve even though it may not be curved.

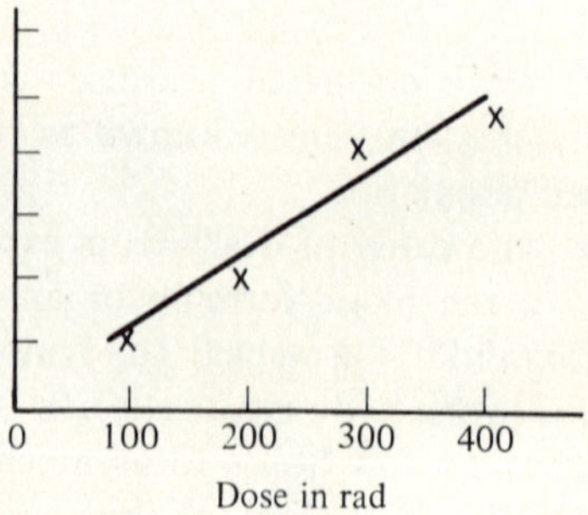

Figure 6-8 Increased incidence of leukemia as a function of radiation dose

If we are interested in the effects of less than 100-rad doses, it becomes necessary to extrapolate the curve which is drawn through the known points to lower exposure values. But no matter how the curve is extrapolated, we are making an inference concerning an unknown relationship based on a known relationship. This inference (or guess) may or may not correctly describe the situation at low doses.

If the relationship between cancer induction and radiation dose is linear at high doses (Fig. 6-8), then a reasonable inference might be to extend the straight line to lower doses. This has been done by John Gofman and Arthur Tamplin of the University of California Berkeley, specialists in medical physics.

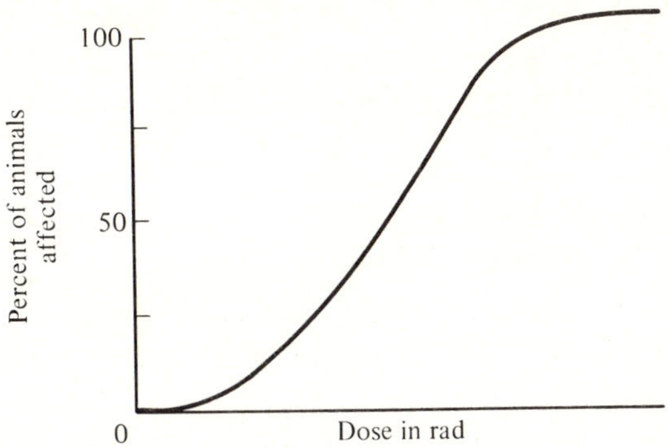

Figure 6-9 Percent of animals infected as a function of radiation dose

However, there are radiation experts who feel that this linear extrapolation leads to incorrect results. They argue that the type of curve in Fig. 6-9 better describes the incidence of leukemia as a function of radiation dose.

Figure 6-9 is a typical dose-effect curve which applies to such different effects as mice dying from radiation exposure or humans cured of infection by penicillin doses. As the dose is increased from zero, very few animals or humans are affected, whichever the case may be. (This means a threshold dose exists.) Then more and more are until 50% of the animals are affected. The middle region of the curve is linear and then greater and greater doses are needed to affect the remaining few animals or humans. Finally, the super-mice die at large radiation exposures or the superbacteria are wiped out by large doses of penicillin.

If the type of dose-effect curve in Fig. 6-9 correctly describes the incidence of cancer due to radiation exposure, then a threshold dose exists. That is, there is no increase in the incidence of cancer above the spontaneous rate at low radiation doses. If a threshold exists for cancer induction by radiation, then linear extrapolation of dose-effect curves from high to low radiation exposures will exaggerate the cancer-inducing ability of radiation. The existence or nonexistence of a threshold for cancer induction is presently a subject of hot debate and of increasing experimental investigation. The effect of dose rate on cancer induction is also not known and is being investigated.

Gofman and Tamplin and radiation standards

Gofman and Tamplin have calculated the increase in the frequency of cancer and leukemia to be expected if the population of this country were to receive the present maximum allowable radiation dosage (170 mrem per year per person exclusive of natural background and medical radiation). The increase would eventually be on the order of 32,000 cases of cancer and leukemia each year after about 30 years.

Gofman and Tamplin reached this result by estimating the risk of various forms of cancer and leukemia per unit of radiation exposure and then calculating the effect if the entire population received the maximum allowable exposure. A controversial assumption which they made is that there is no threshold for cancer induction by radiation. One of their sources of data is shown in Table 6-6.

Table 6-6 OBSERVED AND EXPECTED LEUKEMIA FREQUENCY IN EXPOSED HIROSHIMA POPULATION BY DISTANCE FROM EXPLOSION HYPOCENTER (Persons subsequently resident in Hiroshima City only)

Distance from hypocenter at time of explosion, meters	*Cases with onset in 8-year period 1947–1954*		*Deaths expected among survivors in an 8-year period*	*Ratio of total cases observed to expected*
	Confirmed	*Suspected*		
<1000	15	0	0.15	100.0
1000–1499	28	1	1.32	22.0
1500–1999	6	1	2.33	2.6
2000–2999	6	0	3.96	1.5
≥ 3000	4	2	4.83	1.2
All distances	59	4	12.59	4.7

(From *The Delayed Effects of Whole-body Radiation, A Symposium*, edited by B. Watson, Johns Hopkins Univ. Press, Baltimore, 1961)

These Japanese studies show that there is no question about the relation of dose and leukemia incidence in the higher dose range. It is not at all clear that the persons in the outer zones which received small doses of radiation show an increased incidence of leukemia.

As was indicated before, there are no data for cancer induction at low-radiation doses, and it is certainly possible that there is some threshold dose before cancer will express itself in a human. If this is so, as the A.E.C. believes, then Gofman's and Tamplin's predicted number of additional cancers will be too high. Gofman and Tamplin argue that since we don't know if a threshold exists, we should be conservative in setting radiation standards and assume zero threshold.

It must be remembered that at present the population of the U.S. is receiving far less (~ 5 mrem/yr) than the maximum allowable dose* per person. However, the number of nuclear power plants is expected to grow rapidly, nuclear explosives are being used more often to release natural gas from

*The maximum allowable dose does not include medical or natural background exposure.

beneath the ground, and there is talk of using nuclear explosives to dig harbors and canals. All of these activities can be expected to add to man's radiation exposure. Gofman and Tamplin feel that now is the time to lower the population standards before the radiation exposure increases and while it is still possible to design new nuclear power plants and fuel reprocessing plants to meet lower emission standards. The A.E.C. has decided to lower the standards for radioactive effluents from new nuclear power plants effective 1974. However, there has been no change in the allowable population exposure (170 mrem/yr) and it is still legally possible for that level to be reached. If that happens, there would be some increase in the number of new cancers each year; the only question would be, how many? If there is no threshold, maybe as many as 24,000 each year; if there is a threshold, perhaps less than 10,000 each year.

The A.E.C. has clearly said that they have no intention of letting radiation exposure levels rise to anywhere near the 170 mrem/yr level. Lowering the emission standards for nuclear power plants certainly is a step in the right direction. What remains to be seen is why such high allowable levels for population exposure are needed if there is no intention of ever approaching them.

The last somatic or body effect we will discuss is life shortening. There is evidence that sufficiently high radiation exposures (5 rem/wk) shorten the life span of animals; however, it is not clear whether the reduced life span is due to serious injury or disease from the radiation or if radiation accelerates the aging process. The above dose of 5 mrem/wk is far greater than the background dose (natural plus man-made) of 0.005 rem/wk so it seems that this effect is not important in the general population. In fact, there is some evidence that at radiation exposures less than 5 rem/wk there is some life-lengthening rather than life-shortening in test animals. This observation is probably responsible for the often-heard remark: A little radiation is good for you. Experiments on the genetic effects of radiation have demonstrated the danger of such a statement.

Radiation and genetic mutation

In our modern technological society we are all exposed to physical and chemical agents such as heat, chemicals, and high-energy radiation which can cause mutations in our gonads. As we have already seen in Chapter 4, the body has an effective means for keeping its temperature constant and the gonads are not likely to suffer from exposure to heat. Chemicals are more likely to come in contact with various body cells (skin, lungs, intestinal lining, etc.) than with the cells of the gonads and, therefore, are more of a somatic hazard than a genetic hazard.

Ionizing radiation (such as X-rays and gamma rays) was first shown to be a mutagenic agent in 1927 by Hermann Muller (Fig. 6-10). Using *Drosophila* (a species of fruit flies), he showed that after large doses of X-rays

Figure 6-10 Geneticist Hermann J. Muller studying *Drosophila* in his laboratory. Dr. Muller won a Nobel Prize in 1946 for showing that radiation can cause mutations. (Courtesy Indiana University, Lilly Library)

flies experienced many more lethal mutations per chromosome than did similar flies not exposed to radiation (Fig. 6-11).

Later experiments by Muller and others, indicated that the number of mutations was directly proportional to the quantity of radiation absorbed. That is, doubling the quantity of radiation absorbed doubled the number of mutations, and so on. A straight line is obtained when the number of mutations is plotted against the amount of radiation absorbed (Fig. 6-12). It is generally believed that the straight line continues all the way down to very low radiation absorptions. Therefore, there is no threshold for the mutagenic effect of radiation on flies. This means that no matter how small a dosage of radiation the gonads receive, there is a proportionately increased probability of production of mutated sex cells with effects that may show up in succeeding generations.

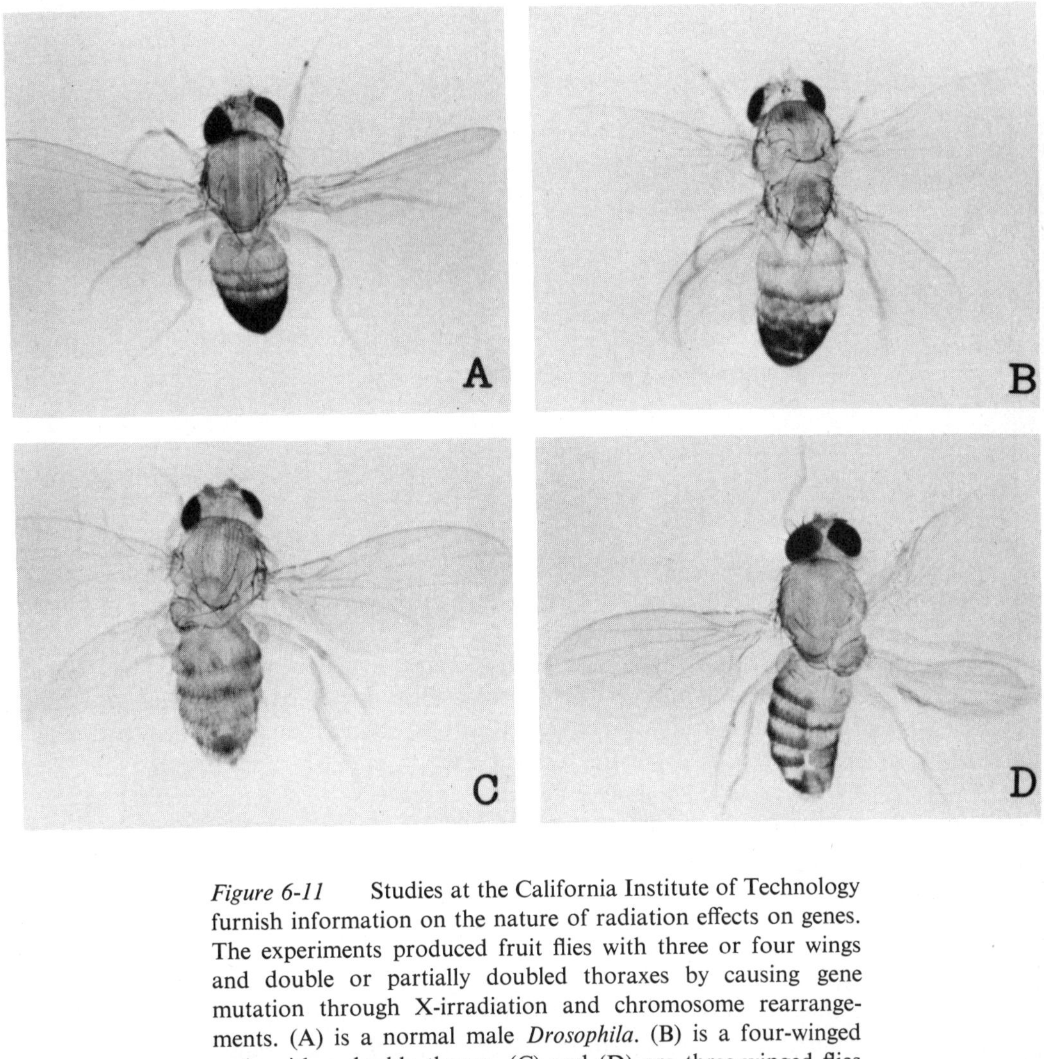

Figure 6-11 Studies at the California Institute of Technology furnish information on the nature of radiation effects on genes. The experiments produced fruit flies with three or four wings and double or partially doubled thoraxes by causing gene mutation through X-irradiation and chromosome rearrangements. (A) is a normal male *Drosophila*. (B) is a four-winged male with a double thorax. (C) and (D) are three-winged flies with partial double thoraxes. (Courtesy Atomic Energy Commission)

It is interesting that the straight line in Fig. 6-12 strikes the vertical axis slightly above the origin. This means that there is a finite mutation rate even when no X-rays are applied. This spontaneous mutation rate is due to the natural background radiation and other effects. It is possible to determine from the above data that background radiation accounts for somewhere between 3% and 10% of the spontaneous mutations that take place (in flies,

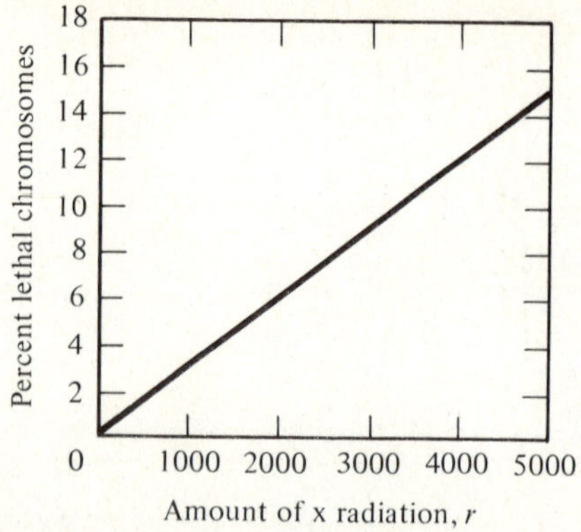

Figure 6-12 Lethal mutations versus radiation dose for fruit flies

that is). The other mutations must arise from random chemical mistakes associated with the statistical nature of all chemical and physical processes taking place within our bodies.

If the earth did not have a magnetic field, the number of mutations caused by radiation would undoubtably increase due to a greater flux of cosmic rays at the earth's surface. The path of a charged particle is bent by a magnetic field and, depending on its energy, it may be trapped in the Van Allen radiation belts or it may pass through the earth's atmosphere altogether. Several times in the past (700,000, 870,000, and 950,000 years ago), the earth's magnetic field decreased to zero and during these times cosmic radiation must have been as effective in changing life forms as X-rays in Muller's experiments with fruit flies.

Effects on Mammals Although genetic findings on such comparitively simple animals as fruit flies and bacteria seem to apply generally to all forms of life, it seemed unsafe to rely on these findings completely in assessing genetic damage to man through radiation. Therefore, in the 1950s and 1960s mice were studied (Fig. 6-13). Since mice, like humans, are mammals, results gained by such experiments are particularly significant. Mice are far closer to man in the scheme of life than any other creature that has been studied genetically on a large scale.

The studies with mice revealed that the induced mutation rate per gene seems to be about 15 times that found in *Drosophila* for comparable X-ray doses. Counteracting this discouraging finding was a later unexpected discovery. In fruit flies and other simple organisms it had been established that the total dose of absorbed radiation was important and that whether this

Figure 6-13 Arrangement for long-term, low dose-rate irradiation of mice used for mutation-rate studies at Oak Ridge National Laboratory. The cages are arranged at equal distances from a cesium-137 gamma-ray source in the lead pot on the floor. The horizontal rod rotates the source. (Courtesy Atomic Energy Commission)

was delivered quickly or slowly did not matter. This was not the case with mice.

In male mice, a radiation dose delivered at the rate of 0.009 rad/min produced only from one-fourth to one-third as many mutations as did the same total dose delivered at 90 rad/min. A possible explanation is that there is present within the cells themselves some process that tends to repair damage to the genes and to counteract mutations. It may be a slow process which could keep up with the damage inflicted at low-dosage rates but not at high ones. High-dosage rates might even damage the repair mechanism.

Assuming that similar processes occur in human beings as occur in mice, it appears that the greater vulnerability of our genes as compared with those of fruit flies is at least partially offset by our greater ability to repair the damage. In the experiments with mice, only seven genes were studied and it is possible that these are not representative of the thousands that exist altogether. One possible conclusion is that more genetic research is necessary and that we should be cautious in establishing radiation protection guidelines which are designed to protect the population from too many genetic mutations.

From fruit fly and other experiments, it appears that a radiation exposure of from 30 to 100 rem will double the spontaneous mutation rate. (Table 6-7). The dose necessary to double the spantaneous mutation rate is known as the doubling dose. The present limit on the total allowable radiation exposure to the gonads for the general population is 5 rem in 30 years. This limit does not include natural background or medical radiation. It does include such sources of radiation as fallout, industrial uses of radiation (luminous watch dials, television tubes, etc.), and radioactive discharges from nuclear power plants.

Table 6-7 SOME EXAMPLES OF DOUBLING DOSES IN EXPERIMENTAL ORGANISMS

Organism	***Mutations***	***Irradiated cells***	***Doubling dose in rad***
Zea mais	4 recessive visibles	Pollen	28
Oenothera	Self-incompatability	Pollen	60
Drosophila	Sex-linked recessive lethals	Post-meiotic germ cells	50 to 100
	Autosomal recessive lethals	Pre-meiotic germ cells	300
	Sex-linked recessive lethals	Oöcytes and oögonia	390
	Autosomal recessive lethals	During spermatogenesis	70
Mouse	7 recessive lethals	Spermatogonia	80
Escherichia coli	3 biochemical mutants		68 to 186

Doubling doses for several organisms under various conditions of irradiation are summarized in Table 6-7. It can be seen that there is considerable diversity but that this is mainly due to the very high doubling doses in immature cells in *Drosophila*. The conditions underlying the doubling dose estimates for these cells, however, are unrepresentative of germ cells as a whole.

(From *Genetic Effects of Radiation*, C. E. Purdom. © C. E. Purdom. Published by George Newnes Ltd., London, 1963)

We can make an estimate of the probable increase in the mutation rate if radiation levels are allowed to reach the maximum permissible levels. The maximum allowable dose to the gonads is 5 rem in 30 years; to this we must add the radiation dose to the gonads from medical uses (presently about 3 rem in 30 years). A total exposure of 8 rem might increase the spontaneous mutation rate by anywhere from:

minimum 8 rem/100 rem = 8%
maximum 8 rem/30 rem = 27%

8% to 27%, depending on whether 100 rad or 30 rad doubles the spontaneous mutation rate in man.

In this calculation we have used the fact that the genetic mutation rate is proportional to the radiation dose. It is difficult to evaluate the cost in lives and suffering of an increase in the spontaneous mutation rate in man of from 8% to 27%. A mutation which takes place in a sex cell can make itself felt only when that cell is one of the pair involved in conception (egg or sperm). It is difficult to relate some number of mutated sex cells in the general population to a probable incidence of mutated offspring of those carrying the mutated genes.

It is not certain that the doubling dose of 30–100 rad applies to man or even that radiation produces genetic mutations in man. Human beings are difficult to study since they have comparatively long lifetimes and few children, and since their mating habits cannot be controlled. There have been attempts to demonstrate that radiation produces genetic mutations in man. The survivors of the Hiroshima and Nagasaki nuclear bomb blasts have been carefully studied. The newborn children of a total of 34,000 men and women who were sufficiently close to the blasts to receive anywhere from 8 to 200 rem of radiation were examined. A somewhat higher frequency of stillborn infants was found among the irradiated parents than among a control group of nonirradiated parents. However, the difference was not statistically significant. The frequency of congenital abnormalities among infants born of irradiated parents was slightly less than for nonirradiated parents. Studies of radiologists (who have been exposed to large amounts of radiation) have also proved inconclusive. Therefore, the most significant evidence that radiation produces mutations in man is that every other organism studied is subject to the mutagenic action of radiation.

The beneficial aspects of nuclear energy and radiation use are numerous; for example, there is the production of electricity by nuclear fission, the diagnostic and therapeutic medical uses of radiation, basic research using radioactive isotopes, and many more. At the same time, the ability of radiation to induce cancers and probably mutations is well known. In order to enjoy the benefits of nuclear energy and radiation we must accept some risks.

Most people who work in the nuclear industries are aware that they are receiving greater radiation exposures than the general population and, therefore, undergoing greater risks. However, they are usually willing to do so for greater monetary compensation than they would otherwise receive in comparable jobs in nonnuclear industries. In general, people are not as aware of these risks when they support the construction of new nuclear power plants or when they agree to have chest X-rays taken to detect possible signs of tuberculosis. It is essential that the risks which would be incurred by allowing man-made radiation levels to reach the maximum permissible doses (established by the National Commission on Radiation Protection and the Environmental Protection Agency) be made known to the general public. Only then can we rationally decide if the risks associated with nuclear power and radiation are worth the benefits.

An incident at Grand Junction

Grand Junction is a town of about 20,000 people in western Colorado (Fig. 6-14). In October 1971 the Joint Committee on Atomic Energy held a hearing on the radiation problem in Grand Junction. Newspapers and magazines published articles about the town that was built on radioactive wastes. It is both interesting and instructive to follow the course of events that led up to this situation.

During the 1950s and 1960s Grand Junction was one location of a uranium mining boom in the western states of America. The uranium was needed for nuclear weapons and as fuel for nuclear power reactors. The Climax Uranium Company began milling operations in Grand Junction in June 1950. Only a very small percentage of the uranium ore (pitchblende) that is mined actually contains uranium. After uranium is separated from the ore in a milling process, a sand-like residue called tailings remains. The tailings that were piled outside of the mills contained small concentrations of decaying radioactive materials, mostly radium (see Fig. 5-11, Chapter 5). Most of the radium could have been removed from the tailings, but it would have been an expensive process.

Radium, with a half-life of 1,620 years emits both alpha and gamma radiation when it decays into radon, which is a radioactive gas (Fig. 5-11). In turn, the radon emits alpha particles as it decays into radioactive isotopes of polonium, bismuth, and lead. These isotopes, which collectively have a half-life of about 30 minutes, are called radon daughters. They attach themselves to the nearest solid, usually an airborne dust particle. Inhaled radon gas diffuses throughout the body where irradiation occurs, although most of it is exhaled before decaying. A large proportion of radon daughters can be inhaled and deposited in the trachea or bronchial tubes. When radon daughters find their way into the human body they become quite dangerous because

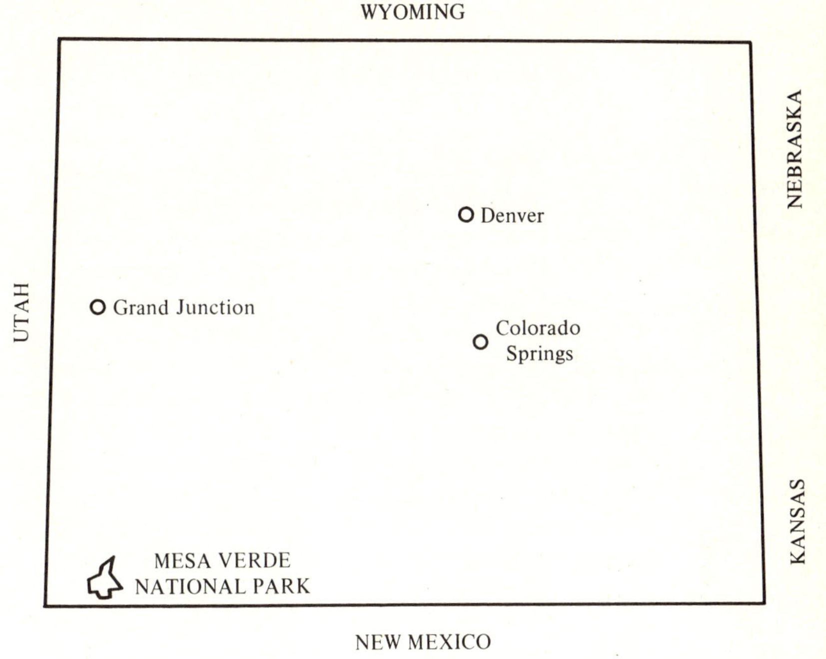

Figure 6-14 Location of Grand Junction in Colorado

their alpha radiation is absorbed in a short distance by a relatively small number of cells. The increased incidence of lung cancer among uranium miners is thought to be due to radiation from radon and its daughters.

In June 1966 the United States Public Health Service (USPHS) and the Colorado Department of Health reported high radon concentrations in a structure in Grand Junction. A couple of years later, an in-depth study was made of radon concentration in buildings that used tailings in their construction. Of the 18,000 habitable structures in the Grand Junction area, about 4,000 have tailings located against or under them. Many more structures have tailings somewhere in the boundaries of their property, such as under backyards or front lawns. The cost of removing the tailings from under and around these buildings has been estimated at $12 to $14 million. The structures would have to be jacked up or have holes cut in their basements. Lawns, gardens, sewers, and utility lines would have to be taken up and replaced.

It was also discovered that many schools in Grand Junction are built on tailings; in fact, some of the highest radon concentrations were found in one elementary school. Principal Frank Folk's reaction was,* "Everyone knows about it. I'd just as soon be here in the clear air with the tailings as in some of those cities with their smog."

Paul Hathaway, editor of the Grand Junction *Daily Sentinel*, said,† "Uranium turned this from a sleepy little cow town to a booming city. They [the people here] accept it as part of their existence. That's why you don't see a lot of immediate concern about the tailings."

Others in Grand Junction were more concerned about the radioactivity from tailings under their city. In most cases they seemed more upset about the expected decrease in real estate values than about the possible effects of the radiation on their health. The Climax Uranium Company closed the mill in 1970. People who were planning to move out of the area were faced with the problem of finding a buyer for a house built on radioactive wastes.

The determination of whether on agency of government or the company is responsible for disposing of the tailings is not an easy task. According to the Atomic Energy Act (AEA) of 1954, the A.E.C. administers a licensing and regulatory program that is designed to prevent or control within established limits the exposure of persons to radiation from a source, a by-product, or from special nuclear materials. Radium and X-rays are specifically under the jurisdiction of State and local authorities. There is a provision in the Atomic Energy Act which allows the states to assume some of the A.E.C.'s regulatory authority. Twenty-three states have entered into an agreement with the A.E.C. whereby they regulate the public health and safety aspects of uranium and other source materials. The agreement between the state of Colorado and the A.E.C. became effective on February 1, 1968.

Prior to February 1968 the A.E.C. regulated the uranium mill operations in Colorado. The discharge of radioactive mill effluents to the environment, and the accumulation of radioactive tailings at mill sites are considered by the A.E.C. to be an integral part of milling operations. However, the transfer of tailings to a new location was not considered by the A.E.C. to be an integral part of milling operations. There is one other way in which the tailings could be considered by the A.E.C. to be their responsibility. If the tailings contain as much as 0.05% by weight (1 pound per ton) of uranium, thorium, or any combination of the two, then they would come under A.E.C. licensing requirements. The uranium content of tailings is far below this level. The component of the tailings that has produced elevated gamma and radon levels in Grand Junction is radium.‡ Under the AEA the A.E.C. has no jurisdiction over radium. Therefore, according to A.E.C. interpretation, the

**The New York Times*, Oct 4, 1971, p. 43.

†*Time the Weekly News Magazine*, Dec. 20, 1971, p. 56.

‡Studies were made by the Colorado Department of Public Health and the Environmental Protection Agency which indicated elevated gamma and radon levels existed in Grand Junction.

A.E.C. is not responsible for radioactivity emitted by tailings removed from the mills.

Doctor Robert Bass of Grand Junction was the first person to claim that evidence had been found to indicate adverse health effects resulted from the high radiation levels found in Grand Junction. In 1970 Dr. Bass reported an increase in the number of cleft palates and other birth defects among his patients in the community. He undertook a more complete study with Dr. Henry Kempe, chairman of the pediatrics department at the University of Colorado Medical Center. Their joint study showed that the incidence of cleft lips and palates was nearly twice as high in Grand Junction as in the rest of Colorado, that the birth rate was significantly lower, and that the death rate from congenital birth defects (those that are not hereditary) was 50% higher than in other areas of the state. There is no way of proving that these conditions were the result of radiation from the tailings. Other environmental, social, or economic factors remain as possible causes of these conditions.

It is expected that the risk of leukemia and lung cancer is higher for residents of Grand Junction than for residents of the rest of Colorado. The increased leukemia risk is due to gamma radiation emitted by radioactive isotopes in the uranium decay series including Ra^{226}, and below. This type of radiation is absorbed uniformly throughout the body. The increased lung cancer risk is due to radon daughters deposited in the trachea or bronchial tubes.

Estimates were made by the U.S. Public Health Service of increased cancer risk as a function of radiation dose. The Surgeon General of the U.S. stated that when the radiation level (gamma rays or radon daughters) was high enough to double the risk of leukemia or lung cancer, remedial action should be taken. Residents of structures with such high radiation levels would have to be moved, or the tailings would have to be dug up and removed. The fate of other structures in Grand Junction was not clearly stated in the Surgeon General's guide lines. It is too early to determine whether there has been an increase in the incidence of leukemia or lung cancer in Grand Junction. It may take as long as 20 years for such cancer to be detectable; the tailings were first used for construction fill in the early fifties.

The natural incidence rate of these cancers is quite low. For example, the natural incidence of leukemia is 0.5 cases per year per 10^4 persons. For a town the size of Grand Junction (20,000), the natural incidence rate is 1 case per year. A doubling of this leukemia incidence rate (which is not actually expected for all residents) would result in an additional 1 case per year. The difficulty of detecting a change of this magnitude when statistical fluctuations in the incidence rate are taken into account is apparent. If data are collected over a period of many years, it may be possible to show that there has been an increase in the incidence of cancer in the Grand Junction area. Of course one can argue that this increase could be due to something other than radiation. It would be far better to protect those who bear a significantly greater risk than does the average population rather than wait for an increase in actual cases of cancer to show up.

In October of 1971 the Joint Committee on Atomic Energy under the chairmanship of Wayne Aspinall, Representative from Colorado, heard testimony from individuals and agencies involved in the tailings problem. Among those individuals and groups were spokesman for the A.E.C., the Environmental Protection Agency, the Colorado Department of Public Health, the Climax Uranium Company, several building contractors, as well as several citizens of Grand Junction, including the mayor. By the end of this hearing it was clear that no individual, business firm, or government agency was willing to take responsibility for what happened to the tailings in Grand Junction or to pay for their removal. During the hearings it was stated that other small towns in Colorado were also built on tailings from uranium mines.

The Outcome The problem of determining legal fault was avoided by the introduction of a bill to Congress to provide Federal financial assistence to limit radiation exposure resulting from the use of tailings in the Grand Junction area. In a speech to the House of Representatives on Feb. 9,1972, Rep. Aspinall stated:

"This bill is neither a reflection of legal liability nor of responsibility for the existing condition on the part of either the AEC or the State of Colorado. It constitutes recognition that at times undesirable situations develop without the legal concept of "fault" being applicable, and seeks to provide a remedy to those affected where no other remedy at law is apparent."

During the summer of 1972 the State of Colorado entered into a joint agreement with the AEC to undertake remedial action to limit further exposure of persons in the Grand Junction area to radiation from uranium tailings used as construction-related material. The AEC was authorized to provide up to 75% of the cost, with the federal share not to exceed $5 million.

Criteria established by the AEC, based upon guidelines prepared by the U.S. Surgeon General, are being used to determine the possible need for remedial action. Generally speaking, dwellings and schools were considered first. During the summer of 1973 work began on the removal of tailings from underneath some homes and schools with high radiation levels.

Suggested reading

1 HENRY, HUGH, *Fundamentals of Radiation Protection*. New York: Wiley-Interscience, 1969.

2 CEMBER, HERMAN, *Introduction to Health Physics*. New York: Pergamon Press, 1960.

3 FRIGERIO, NORMAN, "Your Body and Radiation," U.S.A.E.C., Div. of Tech. Inf., 1967.

4 ASIMOV, ISAAC, and TH. DOBZHANSKY, "The Genetic Effects of Radiation," U.S.A.E.C., Div. of Tech. Inf., 1966.

5 WALLACE, B., and TH. DOBZHANSKY, *Radiation, Genes, and Man.* New York: Henry Holt and Co., 1959.

6 PURDOM, C.E., *Genetic Effects of Radiation.* London: George Newnes Limited, 1963.

7 WATSON, B. ed., *The Delayed Effects of Whole-Body Radiation: A Symposium.* Baltimore, MD: Johns Hopkins Press, 1961.

8 GOFMAN, JOHN, and ARTHUR TAMPLIN, "Radiation: The Invisible Casualties," *Environment*, **12**: 3 (April 1970).

9 LIEBERMAN, JOSEPH, "Ionizing Radiation Standards for Population," *Physics Today* (Nov. 1971).

10 GOFMAN, JOHN W., and ARTHUR R. TAMPLIN, *Poisoned Power.* Emmaus, PA: Rodale Press, 1971.

11 STERNGLASS, ERNEST J., "Infant Mortality and Nuclear Tests," *Bulletin of the Atomic Scientists* (April 1969), 18–20.

12 SAGAN, LEONARD A., "Infant Mortality Controversy: Sternglass and his Critics," *Bulletin of the Atomic Scientists* (Oct. 1969), 26–32.

Questions

1 Why does a chest X-ray result in some radiation exposure to the gonads?

2 (*a*) What is the largest source of man-made radiation exposure to the general population?

(*b*) What might be done to reduce this exposure?

3 (*a*) What is the main danger from fallout?

(*b*) Is the radiation exposure from fallout increasing, decreasing, or staying the same?

4 If all the food in the city in which you live were contaminated by fallout from an atomic explosion, which foods would you avoid eating and which would be safest to ingest?

5 Since nuclear power plants contribute such a small amount of radiation to the total man-made radiation exposure, why are many people so concerned about this additional exposure?

6 What is the main difference between fission and fusion? Why do you think fission bombs were developed before fusion bombs?

7 Do you feel that the radiation exposure limits for the U.S. population provide an effective limit to the total dose one person can receive from natural plus man-made radiation?

8 In your opinion, do you think that the benefits derived from the use of atomic energy and diagnostic X-rays have outweighed the risks? Why?

9 If a person were exposed to 400 rem of radiation over a period of a few hours, what would be his probability of surviving? If he were to die, what would be the probable cause of death?

10 What are some of the main problems in determining the long-term effects of low levels of radiation exposure in humans?

11 Assume that the dose-effect curve for lukemia incidence as a function of radiation dose is linear and that there is zero threshold for leukemia production by radiation.

(*a*) If the increased leukemia incidence is 2 cases/1,000 people at risk at a dose of 100 rad, what is it at a dose of 1 rad?

(*b*) If the type of curve shown in Fig. 6-9 describes the situation more correctly, then will the answer already given to (a) be too big, too small, or just right? Explain your answer.

12 Discuss Gofman's and Tamplin's argument concerning the effect of radiation on cancer induction. Do you think it is reasonable? Why?

13 Derive the ratios in the last column of Table 6-6 from the rest of the data.

14 Why do you think there is a spontaneous genetic mutation rate?

15 What new findings resulted from radiation experiments with mice as far as genetic mutations are concerned? How do you feel about extrapolating to man data obtained from experiments with mice?

16 If the doubling dose in man was determined to be 50 rem, by how much would the spontaneous mutation rate increase if the gonads of each person in the population were exposed to a total of 10 rem of radiation (background radiation is not included in the 10 rem)? Assume that the mutation rate increases linearly with dose and that there is zero threshold for genetic mutations produced by radiation.

Part Four

Noise in the environment

Chapter 7
Noise and man

Introduction

In the past, noise was considered to be a symbol of progress in the same way that chimneys spouting black smoke symbolized a community hard at work. The increasing sound levels which we experience in most of our cities has caused a change in this way of thinking. The frustrating aspect of this situation is that we often cannot escape the noise in our environment even when we enter our homes. The noise produced by both surface and air transportation in particular is audible within our homes and can lead to annoyance and loss of sleep. Faced with no easy retreat from ubiquitous noise, citizens have organized groups to make their feelings known to federal and local governments and, in some cases, have even brought law suits against airports. This pressure has led to the adoption of more stringent anti-noise regulations on federal and local levels. However, it is difficult to enforce regulations concerning noise because of its intangibility and variability. In addition, there is far from universal agreement as to which sounds are unwanted (we call these noise) and which are wanted or necessary.

In this chapter we first consider the properties of sound waves in general and then how we hear sounds. These sections are followed by discussions of the adverse effects of sound on man, sonic booms, and legislation concerning noise in the environment.

Physics of sound waves

In Chapter 2 we discussed wave motion in general and noted that waves carry energy from place to place. Sounds are a form of wave motion produced by the vibrations of material objects. A vibrating string or air column will produce waves which can be heard as sound. When we speak, our vibrating vocal cords produce sound waves which travel up our throats and through our mouths into the air. Sounds are heard when these waves enter ears.

Sound waves are different in one respect from the other types of waves which we studied in Chapter 2. Sound waves are *longitudinal* waves. This means that the particles of the medium through which the wave passes vibrate to and fro in the direction of propagation of the wave (Fig. 7-1). However, as is the case with all wave motion, there is no actual transfer of material in the direction in which the wave travels. When a sound wave passes through air, the air molecules move back and forth (along the direction of propagation) about their equilibrium positions as the wave passes them by.

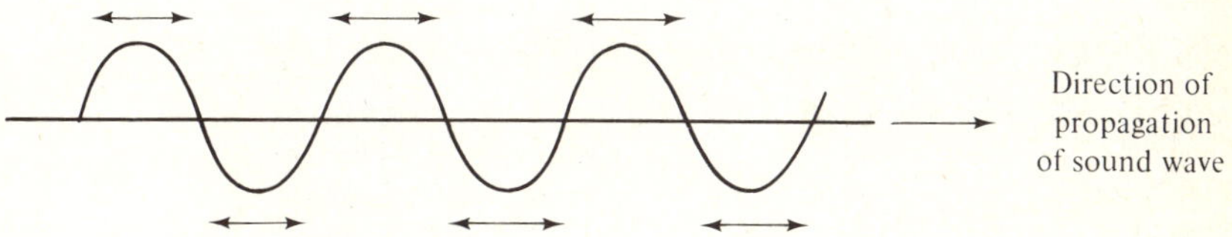

Figure 7-1 Arrows show direction of motion of individual molecules of air.

The other waves (water and electromagnetic) that we studied are both called *transverse* waves. For these waves the individual particles of a medium vibrate back and forth in a direction perpendicular to the direction in which the wave travels (Fig. 7-2). You may remember that no medium is required for electromagnetic waves to travel through and, therefore, it is not particles which are vibrating back and forth but electric and magnetic fields.

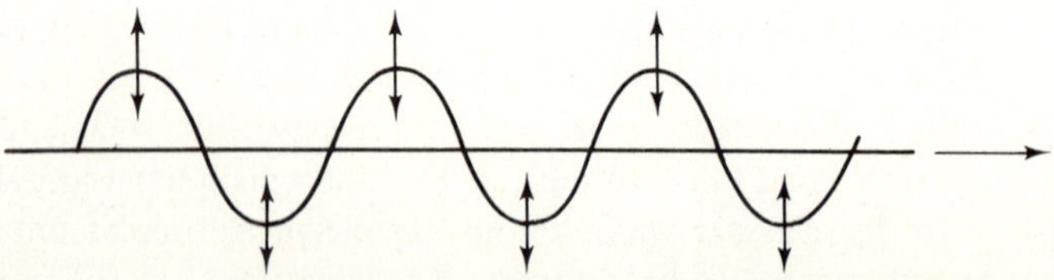

Figure 7-2 Arrows show direction of motion of individual molecules of water.

The speed of sound in air depends on temperature and humidity. For dry air at a temperature of 32°F (0°C) the speed of sound is about 1,100 feet per second or 750 miles per hour. Sound can also travel through solids and liquids, but not through a vacuum. The speed of sound in several materials is listed in Table 7-1.

Example 1: The speed of sound in dry air is around 331 m/sec. Suppose a sound wave of a frequency equal to 1,000 cycles/sec (a frequency humans can hear) travels through air. What will be the wavelength of this sound wave?

The wave equation, $v = \lambda f$ can be used for sound waves as well as for any other kind of wave. So.

$$\lambda = \frac{v}{f} = \frac{331 \text{ m/sec}}{1{,}000 \text{ cycles/sec}} = 0.33 \text{ m} \qquad \text{about 1 foot}$$

Table 7-1

Material	*Speed of sound (meters/sec)*
Aluminum	6,420
Steel	5,960
Glass, Pyrex	5,640
Lucite	2,680
Water, distilled	1,498
Water, sea	1,531
Air	331
Neon	435

This wave has a wavelength which is much greater than that of visible light or X-rays.

We can gain a better feeling for the way in which a sound wave propagates through air by considering how music reaches our ears from a loudspeaker. Electrical signals from an amplifier cause the cone of a loudspeaker to vibrate. Air molecules which come in contact with the vibrating cone are made to accelerate in the direction of motion of the cone. If the cone is moving outwards, then air near the cone is pushed outwards, causing other nearby molecules of air to move out from the speaker, too. This process results in a wave of increased pressure (or compression) which moves out from the speaker. When the speaker cone moves inward, a reduction in pressure (or a rarefaction) takes place near the cone. Air which is near the cone moves into this zone of reduced pressure. Molecules which are further away from the speaker replace those which have moved towards the left. In this way, a reduction in pressure (or rarefaction) travels outwards from the speaker. If the speaker cone moves in and out at a constant frequency (due to a pure musical note, for example), then the resulting pressure wave looks as in Fig. 7-3. The air pressure changes at the same rate at which the cone moves in and out and we hear a sound corresponding to that frequency. In Fig. 7-3, once again it is the disturbance or wave which moves to the right and not the air. The air molecules return to their equilibrium position after the disturbance passes them by.

Sound level measurements The greater the pressure variations within a sound wave, the greater the sound intensity. The basic instrument for the

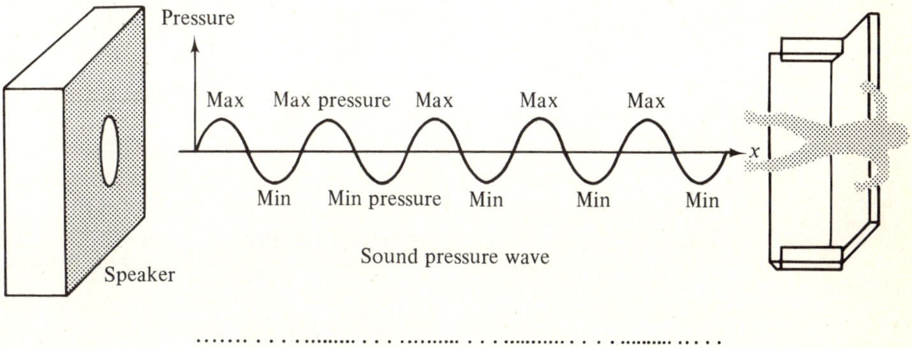

Figure 7-3 Air pressure as a function of distance at some point in time. The dots represent positions of air molecules. There is a greater concentration or density at a maximum in air pressure than at a minimum in air pressure.

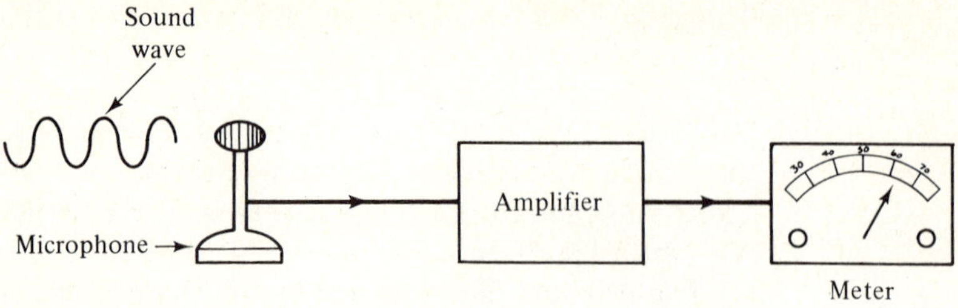

Figure 7-4 A sound wave impinging on a sound-level meter

objective measurement of sound intensity consists of a microphone, an amplifier, and an indicating device which gives visible readings (Fig. 7-4). The microphone transforms sound pressure waves into electrical signals which are then amplified sufficiently to move a pointer on a meter. The meter is usually calibrated so as to read sound intensity in terms of decibels. This basic unit of sound intensity will be defined after the general meaning of an intensity is made clear.

We can use the example of a loudspeaker once again in order to explain the meaning of intensity. Suppose a loudspeaker radiates energy at the rate of 10 joules per second. Since 1 joule/sec is equal to 1 watt, this speaker is radiating 10 watts of sound power. If we were to sit 10 feet from the speaker, how loud would the music sound? In order to answer this question we make some approximations. Therefore, the answer we get will not be exact, but we will save a lot of time in reaching it. We assume that the speaker is a point source of energy in comparison to the room dimensions, that sound energy is emitted equally in all directions, that there is no reflection of sound energy, and that no energy is absorbed by the air.

If we know the rate at which energy is emitted by the speaker (the power, P), then we can calculate the intensity, I, of the sound at a distance r meters from the source. Intensity is defined as the rate of energy flow through a unit area, and in the metric system this is expressed in joules per second per square meter. We can calculate the intensity at any distance from the source by using Fig. 7-5. Since energy is radiated uniformly in all directions, the sound intensity is the same at any point on a sphere of radius r. The intensity r meters from the source is equal to the power radiated by the source divided by the surface area of the sphere of radius r.

Surface area of a sphere of radius r is $4\pi r^2$:

$$1 \text{ ft} = 0.30 \text{ meter}$$

$$r = 10 \text{ ft} = 3.0 \text{ meters}$$

$$I = \frac{P}{4\pi r^2}$$

$$I = \frac{10\ \text{W}}{4\pi(3\ \text{m})^2} \cong \frac{10\ \text{W}}{12(9)\ \text{m}^2} = \frac{10\ \text{W}}{108\ \text{m}^2} \cong 0.1\ \frac{\text{W}}{\text{m}^2} = 0.1\ \text{joules/sec-m}^2$$

Now that we have made an approximate calculation of the intensity 10 ft from the speaker, we still do not have any idea about the perceived loudness. That is, does it sound like a steam locomotive is about to come through your living room or more like a rustle of leaves in the wind?

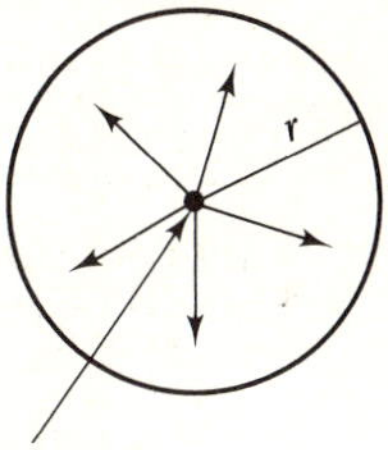

Figure 7-5 Sound energy spreading out from a point source

The lowest intensity which some people are able to hear is around 10^{-12} watt/m^2, which is an extremely small amount of power per unit area. The intensity which we just calculated is 0.1 watt/m^2. Because of this very large range of intensity values for sound waves it has become customary to express sound intensity in terms of decibels (dB). The decibel scale depends on the use of ratios. The intensity of a sound can be expressed in decibels as follows:

$$\text{sound intensity (dB)} = 10 \log_{10}\left(\frac{I}{I_0}\right)$$

where I = intensity of the sound in watts/m^2
I_0 = reference intensity = 10^{-12} watt/m^2
$\log_{10}$ = logarithm to the base 10

Logarithms are discussed in Appendix A, but we can briefly show how they are manipulated by calculating the intensity in decibels of a sound of intensity 0.1 watt/m^2. When we express a sound intensity in decibels we are comparing it to the reference intensity which represents the threshold of hearing. For the example which we just calculated, $I = 0.1$ W/m^2. So

$$\frac{I}{I_0} = \frac{0.1\ \text{W/m}^2}{10^{-12}\ \text{W/m}^2} = 10^{11}$$

Therefore, we need to determine the $\log_{10}(10^{11})$. In order to do this we must ask to what power do we have to raise 10 in order to get 10^{11}, or $10^x = 10^{11}$. What is x? In this example the answer is obviously 11.

$$\text{sound intensity (dB)} = 10 \log_{10}(10^{11}) = 10(11) = 110 \text{ dB}$$

We can get a feel for what 110 dB of sound intensity is equivalent to by looking at Table 7-2. It is quite loud!

Table 7-2

Source of sound	*Sound level (dB)*
Threshold of hearing	0
Rustle of leaves in a gentle breeze	10
Whisper at a distance of 4 feet	20
Quiet automobile tens of feet away	50
Conversation at 3 feet	60–70
Noisiest spot at Niagara Falls	80–90
Riveter 35 feet away	95–100
Jet airplane, 1,000 feet overhead	103
Jet airplane, 100 feet away	140

It is important to point out that if two sounds, each of 110 dB intensity, are added together the resulting intensity in decibels is not the sum of the two individual intensities (that is, not 220 dB). Intensities expressed in watts per square meter can be added together in the normal way, but not intensities expressed in decibels. Therefore, we must convert from decibels to watts per square meter in order to add intensities. Let us say that we are sitting 10 feet away from two identical speakers which radiate 10 watts of power each, and that the intensity from each is 110 dB at the location where we are sitting. We can work backwards to find the intensity of each speaker in watts:

$$110 \text{ dB} = 10 \log_{10}\left(\frac{I}{I_0}\right)$$

$$\log_{10}\left(\frac{I}{I_0}\right) = \frac{110}{10} = 11$$

Therefore,

$$\frac{I}{I_0} = 10^{11}$$

since

$$I_0 = 10^{-12}\ \text{watt/m}^2, \qquad I = 0.1\ \text{watt/m}^2$$

As we already knew:

$$I(\text{total}) = I_1 + I_2 = 0.1\ \text{W/m}^2 + 0.1\ \text{W/m}^2 = 0.2\ \text{W/m}^2$$

Now that we have calculated the intensity in terms of watts per square meter, we can convert the total back to decibels:

$$I\,(\text{dB}) = 10\log_{10}\frac{(0.2\ \text{W/m}^2)}{(10^{-12}\ \text{W/m}^2)} = 10\log_{10}(2 \times 10^{11})$$

$$I\,(\text{dB}) = 10(\log_{10} 2 + \log_{10} 10^{11}) \quad \left(\text{since} \quad \log_{10}(ab) = \log_{10} a + \log_{10} b\right)$$

$$I\,(\text{dB}) = 10(0.301 + 11) = 10(11.301) = 113\ \text{dB}$$

$\log_{10} 2$ is found in the table in Appendix A

We find that the intensity in watts per square meter has been doubled, but not the intensity as expressed in decibels. In fact, the intensity in decibels has increased only by 3 dB, from 110 dB to 113 dB. Any time two equal intensities are added together the result in decibels is 3 dB greater than each individual intensity expressed in dB. For example, 60 dB + 60 dB = 63 dB.

Loudness and intensity Sound intensity can be measured by various instruments. Loudness, on the other hand, is a subjective sensation which depends on intensity in a complicated way. In order to make a judgment that a sound has doubled in loudness, the intensity of the sound would have to increase by a factor of about ten (or 10 dB). The loudest sounds we can tolerate (about 120 dB) have intensities 10^{12} (a million million) times greater than the faintest sounds we can detect (0 dB). However, our subjective reporting of loudness and softness of sounds is not directly proportional to intensity. For example, if we double the intensity of a sound, people report that it sounds only a little bit louder. If we go from the faintest to the loudest sounds, people report that the loudest seems *not* 10^{12} times as loud as the softest, but only 1,000 times.

This may be a good place to summarize some of the concepts concerning sound waves. When a sound wave propagates through some region of space it causes pressure fluctuations in that region. The greater the pressure fluctuations (which are generally very much smaller than the normal atmospheric pressure), the greater the intensity of the sound wave. The intensity of a wave, which is equal to the rate of flow of energy perpendicular to a unit area, decreases with distance from the source of energy. In the metric system we can express intensity in joules per second per square meter or in watts per square meter. Loudness, which is a subjective attribute of a sound, is not directly proportional to intensity. The intensity of sounds which we can hear without pain range from about 10^{-12} W/m^2 to 1 W/m^2, a factor of 10^{12}. However, the loudness of sounds varies by only a factor of about 1,000. When sound intensity is expressed in decibels, the range of values decreases significantly (0–120 dB) and the sound intensity in decibels more closely approximates the loudness of a sound. Sound intensity can be readily measured with a sound-level meter which is calibrated in decibels.

Pitch and frequency Thus far we have talked about sound waves in terms of intensity or loudness. Another important attribute of a sound wave is its frequency. The number of cycles per second, or frequency, is closely related to what we subjectively call pitch. The greater the frequency, the higher the pitch and vice versa. However, things are not quite that simple since pitch is also affected by the intensity of a sound wave. If the frequency of a pure tone is kept constant and the intensity is varied, the pitch will also vary. For frequencies below 1,000 cycles/sec, pitch decreases with intensity, while for frequencies above 3,000 cycles/sec, pitch increases with intensity. In between 1,000 and 3,000 cycles/sec there is very little variation of pitch with intensity.

Frequency response of the ear The ears of a normal young adult are sensitive to frequencies between about 20 and 20,000 cycles/sec. Children can sometimes hear sounds of 30,000 cycles/sec frequency, but this ability to hear high-frequency sounds diminishes with age. The sensitivity of the ear as a function of the frequency of the sound reaching it has been determined by the U.S. Public Health Service for a typical group of Americans. The results are shown in Fig. 7-6. Since there is more than one curve in this figure (and three variables—intensity, frequency, and percent of the population), the situation appears more complicated than it really is. If we look at one curve at a time in Fig. 7-6, then we have to deal with only the usual type of graph with two variables (intensity and frequency). The various numbers labelling the curves indicate what percentage of the population could hear a sound as weak or weaker than the level indicated by the curve.

For example, if we look at the curve labeled 75%, we find that it crosses the 1,000 cycles/sec line at an intensity of about 25 dB. Therefore, 75% of the population can hear a sound with a frequency of 1,000 cycles/sec if the intensity level is around 25 dB. Still looking at the same curve, we find that it takes an intensity level of about 50 dB before this same 75% of the population can hear a 100-cycles/sec sound. From this same curve (which dips down around 3,000 cycles/sec) we can see that the ear is most sensitive to

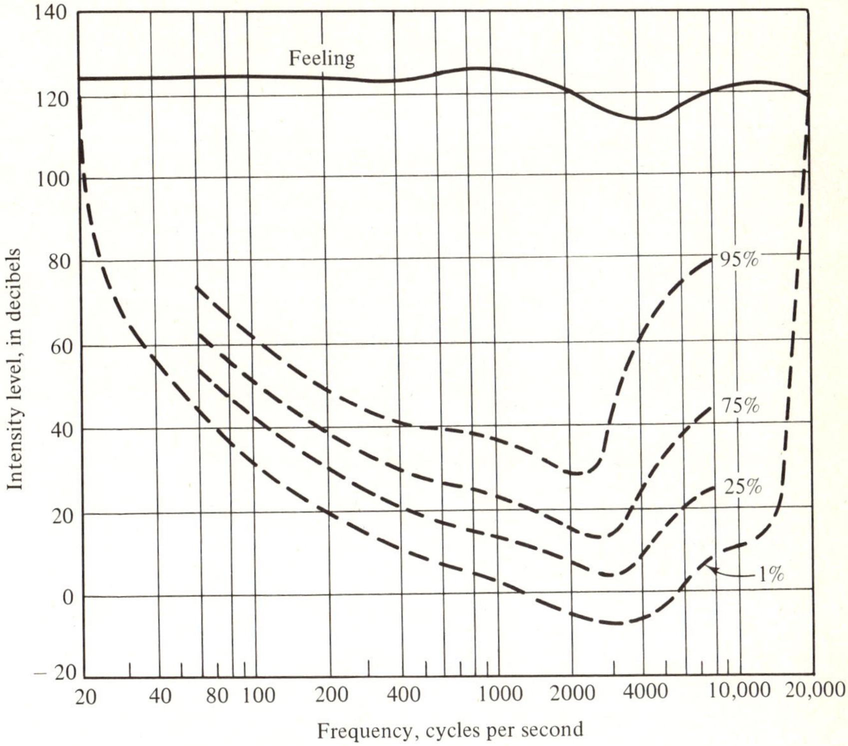

Figure 7-6 Minimum sound intensity needed to hear a sound as a function of frequency

sounds with a frequency near 3,000 cycles/sec. Later we discuss the physical reasons for this maximum in sensitivity around 3,000 cycles/sec.

In order for frequencies above 3,000 cycles/sec to become audible, the sound level must be increased above its minimum value at 3,000 cycles/sec. For example, an intensity level of around 40 dB is needed before 75% of the population will hear a frequency of 6,000 cycles/sec, whereas an intensity level of only 15 dB is necessary for 75% of the population to hear frequencies around 3,000 cycles/sec. As we might expect, if the sound intensity level is increased at any particular frequency, the percentage of the population that can hear that sound increases. This holds true until an intensity of around 120 dB is reached. At this intensity, essentially the whole population will begin to feel the sound no matter what the frequency may be (if the frequency is still in the audible range) and pain may even result.

Since the human ear responds unequally to sounds of different frequencies, it is important to know the frequency distribution of sounds whose intensity we are measuring. This can be done with a sound-level meter, which measures sound level in decibels, by inserting electrical filters before the meter. These filters pass electrical signals of only certain frequencies. In this way, the frequency range from 45 cycles/sec to 11,000 cycles/sec is divided into eight frequency bands, according to international convention. Sources vary in their frequency characteristics. Most of the fundamental sounds of speech are between 300 and 3,000 cycles/sec in frequency. Propeller aircraft emit most of their sound energy in the frequencies below 300 cycles/sec, and an ambulance siren emits most of its sound energy between 600 and 5,000 cycles/sec. The two curves in Fig. 7-7 show the frequency distribution of two different sound sources when the frequency range from 37.5 Hz to 9600 Hz is divided into eight frequency bands. As can be seen from the curves, in the power house low frequency sounds predominate whereas the sound energy from the circular saw is more evenly distributed throughout all the frequencies at which energy is emitted by the saw.

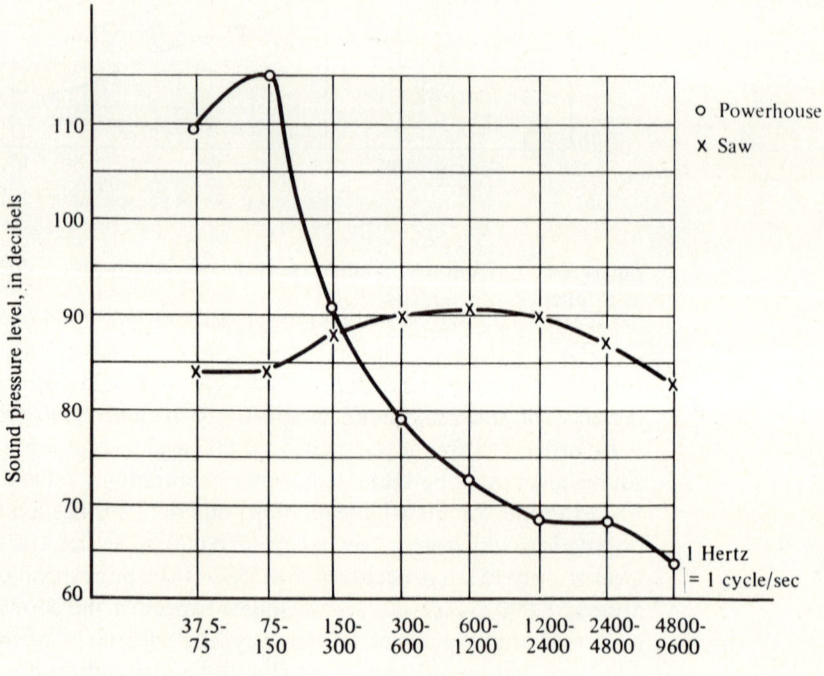

Figure 7-7 Sound level versus frequency for a powerhouse and a circular cut-off wood saw.

Often such detailed information is not necessary and a different measurement scheme is used. Most sound-level meters have three scales, called the A, B, and C scales. The C scale is essentially a linear scale, that is, the response of the meter is almost independent of the frequency. The response of the A scale is similar to that of the human ear; it is greatest around a frequency of 3,000 cycles/sec and falls off to either side of this frequency. The B scale is between these two scales. If we want to know the total sound level without regard to how the ear perceives it, then we should use the C scale and express the decibel reading as dBC. On the other hand, if we want to take the frequency response of the human ear into account, we should use the A scale and express the sound levels as dBA. When community noise or occupational noise is being measured, the A scale is usually used since we are interested in the effects of noise on man, and the A scale gives us a rough idea of the impact of this noise on the human hearing mechanism.

The ear and hearing

The ear is an extremely sensitive sound detector. At some frequencies, amplitudes of vibration of the eardrum as small as 10^{-11} meter (one-tenth the diameter of a hydrogen atom) can be detected. Fortunately, the ear is not *so* sensitive that it can detect the random pressure fluctuations of the air around it. If it were, we would be continuously bombarded by random background noise.

It is interesting that the ear has evolved in such a way that we normally are unaware of our own vibrations. Contraction of the muscles of our bodies and bodily movements produce low-frequency vibrations. The human ear has evolved in such a way as to be less sensitive to these low-frequency sounds than to higher-frequency speech sounds. It is possible for us to listen to the vibrations produced by our muscles when they are contracted and to imagine what it would be like if we could always hear these sounds. We can do this by putting a finger in each of our ears, thereby shutting out external sounds which travel through the air, and at the same time allowing body vibrations to be conducted through our finger bones to the ears. Our bones are pretty good sound conductors.

The ear can be divided into three principal regions: the outer ear (or auditory canal), the middle ear which consists of three small bones called the ossicles, and the inner ear which is composed of a complex system of fluid-filled canals. The sensation of hearing is produced by a complicated series of events after the arrival of sound waves at the auditory canal (Fig. 7-8). When these sound waves reach the ear drum (or tympanic membrane) they cause it to vibrate at the frequency of the sound wave.

The vibrations of the eardrum are transmitted through three contiguous bones (called the ossicles), one of which is in contact with the eardrum and one of which is in contact with the inner ear. These three bones, the maleus,

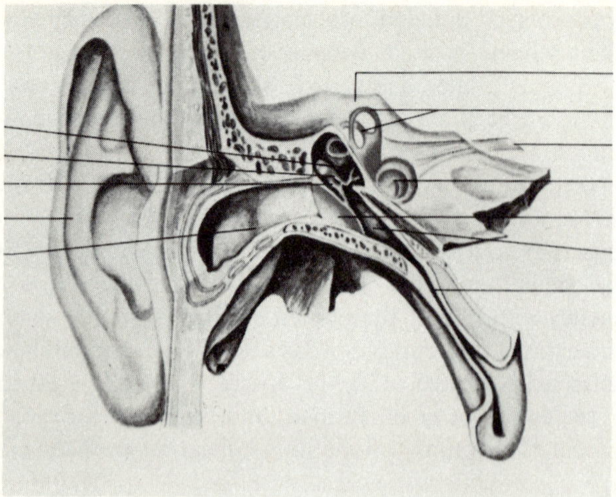

Figure 7-8 Outer, middle, and inner ears (From *Human Anatomy and Physiology*, by James E. Crouch, Jr. and J. Robert McClintic, © 1971 by John Wiley & Sons, Inc., New York).

incus, and stapes, are found in the middle ear. The stapes (also called the stirrup because of its shape) acts on the fluid of the inner ear like a piston, moving it back and forth at the frequency of the sound wave. In addition to the fluid, structures in the inner ear such as the cochlea (a spiral tube containing nerve endings essential for hearing) are also sent into vibration by the motion of the stapes. Finally, the vibrations reach the organ of Corti (situated in the cochlea), which is the end organ of hearing, where mechanical vibrations set up electrical impulses in the fibers of the auditory nerve. These nerve impulses travel to the brain where they are perceived as a sound.

Conduction of sound in the ear and resonance The ear canal has a length of about 2.7 cm in an adult. It is open to the atmosphere at one end and closed at the other end by the eardrum. Therefore, the structure of the outer ear is similar to that of a pipe which is closed at one end. It has been known for a long time that such pipes have certain natural frequencies of vibration which depend on their length. Instruments such as recorders, flutes, and organs depend on the principle that air columns of finite length have specific natural vibration frequencies.

Many objects have natural frequencies of vibration. By natural, we mean that these frequencies depend on the object's shape and composition, rather than on any external factors. If we strike a steel drum with a rubber mallet and a wooden drum with the same mallet, the sounds we hear from each drum are quite distinct from each other. This is because each object, when struck, vibrates at its own natural frequency and sets the air into vibration at this special frequency. As a result, we hear two very distinct sounds with different frequencies or pitch.

If we blow air into a tube of length L (Fig. 7-9) which is closed at one end, we find that the natural frequencies of vibration for the air in this tube are odd multiples of $v_s/4L$, where v_s is the speed of sound in air. Therefore, the sound frequencies which we would hear, if we blew loud enough, are

$$f_1 = v_s/4L \qquad f_2 = 3v_s/4L \qquad f_3 = 5v_s/4L \qquad \text{etc.}$$

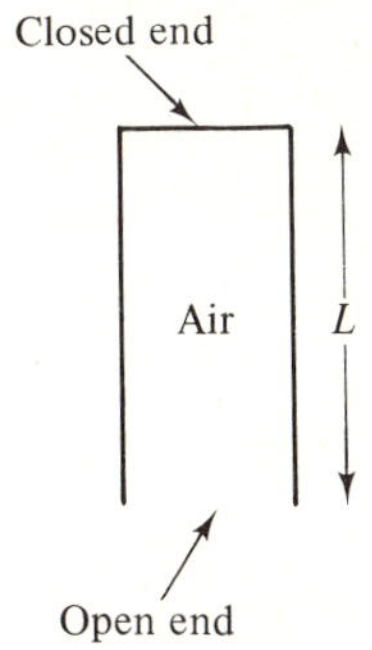

Figure 7-9 Air-filled tubes have more than one natural frequency of vibration. For a tube open at one end, $f_n = nv_s/4L$ (n is an odd integer).

The lowest of the natural frequencies is called the fundamental, and the higher ones are called the overtones. Most of the energy output of musical instruments, such as clarinets and French horns, is found within the first few possible natural frequencies. The quality or timbre of a musical instrument is determined by the relative intensities of the fundamental and overtone frequencies. The overtone structure accounts for the difference between a piano and clarinet that are producing sounds of the same fundamental frequency.

Now that we know how to obtain the natural vibration frequencies of a tube that is open at one end, we can apply this knowledge to the outer ear. This calculation will be only approximate since the ear canal is curved and the tympanic membrane is not rigid. The fundamental natural frequency of the ear canal is:

$$f = v_s/4L$$

where $L = 2.7$ cm and $v_s = 331$ m/sec. So

$$f = \frac{331 \text{ m/sec}}{0.108 \text{ m}} = 3{,}060 \text{ cycles/sec or } 3{,}060 \text{ Hz}$$

This frequency is in the region of highest sensitivity for human hearing (Fig. 7-7). When we say that the ear is more sensitive to sounds which have frequencies near 3,060 Hz we are invoking the principle of *resonance*. Resonance is the enhancement of the response of a body to a periodic driving force when the driving frequency is equal to the natural frequency of the body.

When a sound wave enters the ear canal, the air within the canal vibrates. The wave travels through the canal, finally reaching the eardrum which then vibrates at the frequency of the incoming sound wave. If the frequency of the incoming sound wave is close to one of the natural vibration frequencies of the air column within the canal, then the small amplitude vibrations of air molecules at the outer end of the canal are amplified to larger amplitude vibrations by the time the disturbance reaches the inner end of the ear canal. As a result of this amplification or resonance, the fluctuations in air pressure due to the incoming sound wave are greater at the ear drum than at the opening of the ear canal. This increase in pressure results in a greater displacement of the ear drum for frequencies which are near the resonant frequency of 3,060 Hz.

All of the natural frequencies of an object are possible resonant frequencies. It is easier to increase the amplitude of vibration of an object when it is excited at one of its natural frequencies than when it is at some other frequency. For example, a simple pendulum (Fig. 7-10) has a single natural

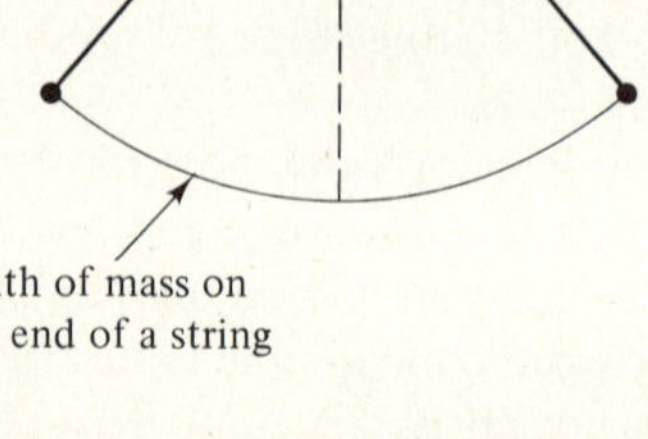

Figure 7-10 A pendulum is a mass suspended so that it is free to swing in a vertical plane about some fixed axis. The fixed axis is perpendicular to the page, through the point of suspension.

frequency which depends on its length. If we want to increase the amplitude of vibration (the angle θ in this case) of this pendulum, it would be advantageous to push the pendulum at just the right time during each cycle (that is, at a frequency equal to its natural frequency).

The above is true of the eardrum also. If we wish to obtain large amplitudes of vibration of the eardrum, we should excite it with pressure waves which have a frequency at or near the natural frequency of the air column in the ear canal. Then there is resonance between the incoming sound pressure wave and the vibrating air column in the ear canal, with a corresponding increase in the air pressure at the eardrum.

The middle ear The eardrum separates the ear canal from the middle ear. There is a connection between the upper part of the throat and the middle ear via the Eustachian tube (Fig. 7-8). When we swallow, the Eustachian tube opens to the middle ear, allowing atmospheric air to come in, thus equalizing the pressure on both sides of the eardrum. The malleus (the largest of the three tiny bones called the ossicles), which is connected to the eardrum, transmits the vibrations to the incus which transmits them to the stapes (Fig. 7-11). The footplate of the stapes is connected to the oval window

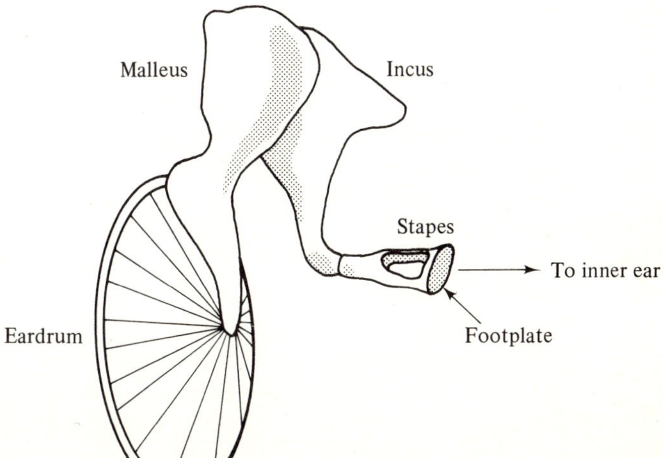

Figure 7-11 The vibrations of the eardrum are transmitted by the ossicles to the fluid of the inner ear. The eardrum (or tympanic membrane) has a much larger area than the oval window (not shown) which is a flexible membrane in contact with the footplate of the stapes. (Adapted from *Human Anatomy and Physiology*, by James E. Crouch, Jr. and J. Robert McClintic, ©1971 by John Wiley & Sons, Inc., New York).

(a flexible membrane) of the cochlea. When the stirrup vibrates it sets up waves in the liquid (lymph*) of the inner ear.

The ossicles increase the pressure which is exerted at the eardrum by the incoming sound wave. The area of the eardrum is about twenty times greater than the area of the footplate of the stapes (which is in contact with the oval window; Fig. 7-11). If the same force were applied at the eardrum as at the oval window, then the pressure would be twenty times greater at the oval window than at the eardrum. This is due to the dependence of pressure on area through the equation: $P = F/A$; that is, pressure is equal to force divided by the area the force acts on. A decrease in area, A, by a factor of twenty while the force, F, remains constant results in an increase in pressure, P, of twenty times its original value.

One of the main functions of the middle ear is to transfer the sound energy which enters the outer ear into the cochlea of the inner ear. If the middle ear were not present, then much of the energy of the incoming sound wave would be reflected off the ear drum, and would never enter the inner ear. The ossicles of the middle ear, however, make the energy transfer from the outer ear to the inner ear more efficient.

The inner ear The semicircular canals and the cochlea are located in the inner ear. The semicircular canals are associated with the maintenance of body equilibrium and with the detection of acceleration. The cochlea, which is a snail-shaped cavity, is divided along its length by the basilar membrane. When a pressure wave passes through the lymph of the cochlea the basilar membrane begins to vibrate. Sound energy is converted to electrical energy in the organ of Corti, which is supported by the basilar membrane. Sensory hair cells are located in the organ of Corti. Movement of the basilar membrane eventually results in the bending of these hairs and finally in excitation of the auditory nerve. Nerve impulses travel along the auditory nerve to the brain.

In the organ of Corti, frequency discrimination takes place. The complete details are not known, but a lot of work has been done, especially by Georg von Bekesey, who spent over thirty years investigating the nature of the ear and hearing. He received the Nobel Prize for medicine in 1961 for his work on the ear.

In this section we have indicated how the ear functions as a detector of sound waves. The details of the motions of various structures of the ear have been avoided, but these are not necessary for a general understanding of hearing. We can now discuss the effects of noise on man's health, both physical and psychological.

**Lymph:* a liquid similar to water in density and viscosity; contains white blood cells.

The effects of noise on man

The hazards of noise were first recognized in industrial situations. There are many work environments that are particularly noisy, such as building construction sites and steel mills. Experience has shown that workers will suffer hearing losses if noise levels reach certain values for specified time periods (we will discuss details later). This realization has led to the institution of strict limits on noise exposure for all workers.

Noise-induced hearing loss usually occurs in the organ of Corti. Exposure to excessive noise at first produces deterioration and then destruction of hair cells.

There are many other effects of noise which are felt long before hearing loss occurs. We now discuss these effects and point out the uncertainties.

Physiological effects of noise on man Noise has its greatest effect upon the circulatory system of our bodies (excluding the ear). Experiments with human volunteers have shown that noise causes constriction of the peripheral blood vessels. This results in either a decrease in blood circulation to the extremities or a comparable increase in arterial blood pressure. If this constriction lasts long enough, the decreased blood volume reduces the oxygen available for human activity. Noise-induced fatigue may result from this situation.

However, if a particular type of noise is repeated very often, then some adaptation occurs. Just how much man can adapt to noise is still a subject of great debate among workers in this field.

Research in the USSR and England indicates that noise interferes with proper digestive functioning, particularly when the noise is sudden and unexpected. Noise of this type produces a decrease in bowel activity, saliva flow, and digestive juices. There is also some adaptation to this type of stress.

Other physiological responses have been detected. Heartbeat is often slowed and respiratory functions such as breathing may quicken. Unexpected noise tends to cause muscle contraction and the release of adrenalin, which stimulates other physiological changes.

Sound intensities as low as 70 dB may cause noticeable responses such as pupil dilatation and blood vessel constriction. The least sound intensity which brings about hazardous side effects is not definitely known, but the general feeling is that intensities greater than 90 dB for an eight-hour day are dangerous.

Sleep interference It seems safe to say that interruption of sleep is annoying to almost everyone. In the United States, each year we purchase billions of barbiturate capsules to help us get enough sleep. Many studies have been made to determine the effects of prolonged loss of sleep. The reported results range from mere irritation to visual hallucinations and disorientation for time and place.

There are several stages of sleep, all of which can be interrupted by noise. One of these stages is characterized by rapid eye movement, and is called REM sleep. This dream stage occurs five to six times each night with approximately a 90-minute cycle, and accounts for about 20% of adult sleep time. People are most likely to be awakened by noise when they are in this REM stage of sleep. Initial studies of REM deprivation* indicated an increase in anxiety, irritability, and inability to concentrate. However, considerable doubt has been cast on these studies by other experimenters. Some sleep researchers feel that sleep deprivation in general caused the above effects (not REM sleep in particular). Others believe that these effects did not even occur.† Therefore, it is not at all clear if behavioral effects are caused by deprivation of REM sleep.

Noise can also disrupt other phases of sleep. It is even possible for noise to cause changes in the brain wave patterns of sleeping subjects without waking them. A sleeping person can have his level of sleep shifted from one sleep stage to the next lighter one without being aware of it. Experiments have shown that vehicle noise levels of 60 dBA will cause either awakening or a change in sleep level for 50% of the people exposed. Aircraft noise of about the same level awakens more than 50% of those sleeping subjects who are exposed to it.

The time it takes to fall asleep is also affected by environmental noise. Experiments indicate that the threshold for optimum sleeping conditions is around 35 dBA. Therefore, many of the people living near airports and busy thoroughfares are having the quality of their sleep diminished. Even if the loss of sleep leads to no effects other than annoyance and irritation, these are reason enough to reduce environmental noise levels at night time.

Interference with communication Noise-induced hearing loss initially occurs in the frequencies above 3,000 Hz. In the past, the loss of hearing in the higher frequencies was not thought to be detrimental to hearing speech. Most speech sounds are below 3,000 Hz in frequency, but a number of sounds involving consonants are above this frequency; therefore, it is essential to hear frequencies greater than 3,000 Hz. Listening to speech becomes fatiguing when less than 90% of the words spoken are correctly heard. Since it is necessary to hear such a high percentage of the total number of words spoken, the loss of hearing even above 3,000 Hz can be very important.

The community noise level is an important factor in speech discrimination. When it reaches 80 dBA, hearing accuracy drops even for distances of less

*Dement, W., "Recent Studies on the Biological Role of Rapid Eye Movement Sleep," *The American Journal of Psychiatry*, **22**: 10 (Oct. 1965), 404–408.

†Vogel, G. W., "REM Deprivation," *Archives of General Psychiatry*, **18**: 3 (March 1968), 287–329.

than one foot from the speaker. It has been found that a background noise level of 70 dBA is a maximum acceptable level at which good voice communication can be maintained. At sound intensities above this level, people complain or feel that they are less able to perform their jobs.

Work efficiency The degree to which noise interferes with task performance seems to depend on the type of work considered and on the particular survey looked at. It is not at all clear if mechanical type tasks, such as assembly line work, are measurably affected by noise until sound levels of around 120 dB are reached. At this sound level there is pain and damage to the ear, so protective hearing devices must be worn. Mental tasks, such as problem solving and creative thinking, are much more easily influenced by noise. Mental performance is usually affected in some way at noise levels between 90 and 110 dB.

Although there may be no serious effects of noise on performance of mechanical tasks (for sound levels less than 120 dB), there is some evidence that workers compensate for this stress in other ways. A German study* indicates that workers in noisy jobs might be more quarrelsome, both at work and away from it, than those performing equivalent jobs that are not subject to similar noise. A study of this type is difficult to carry out since there are many other factors affecting the behavior of the workers. It appears that simple studies of job performance as a function of noise levels is not enough. Psychologists must consider the possibility of other effects from stress due to noise.

There is fairly good agreement as to the levels of noise which cause damage to the hearing mechanism, but much less agreement as to what noise levels precipitate undesirable psychological and physiological side effects. Most people agree that excessive noise can cause loss of sleep, can interfere with communication, and can be annoying. What is not clear is the extent to which humans can adapt to increasing noise levels and at what price.

Sonic booms Most of us have probably heard the sharp crack of a sonic boom at least once. Sonic booms are capable of shaking houses, breaking windows, and badly startling those who hear them. In two national parks (Mesa Verde in Colorado and Bryce Canyon in Utah) prehistoric cliff dwellings and natural sandstone formations were destroyed by sonic booms produced by supersonic military aircraft. Aircraft that travel faster than the speed of sound (760 miles/hr at sea level) are called supersonic aircraft. Sonic

*Lehman, D. et al., "An Investigation of the Effects of Various Noise Levels as Measured by Psychological Performance and Energy Expenditure," *Journal of School Health*, **35**: 3 (March 1965).

Figure 7-12 This is a full-size mockup of the SST which was to be built by Boeing Aircraft Company. It would carry 300 passengers and fly at 1,800 miles per hour, more than twice the speed of sound in air. (Courtesy Boeing Commercial Airplane Company)

booms have even been implicated in the collapse of an old farm building in France, which resulted in the death of three occupants.

While the United States decided in 1972 not to build a supersonic transport (SST), England, France, and Russia have gone ahead and built their own supersonic airliners. There are also many military aircraft that travel at supersonic speeds, so the problems created by the sonic boom are likely to remain with us. At this time, it is not clear if overland flights of the SST will be allowed in the United States. If they are, we will be subject to a significant increase in noise pollution.

Generation of sonic booms Whenever an object (a bullet or an airplane, for example) moves through the air, it displaces air molecules and a sound wave is generated. Bodies that travel through the air at less than the speed of sound produce spherical waves which spread equally in all directions. In these cases, the sound wave is always ahead of the moving object and a continuous sound is heard. One could look at this situation and say that the sound waves which are generated by the subsonic body move ahead of it at all times, thereby setting the air ahead in motion before the moving object reaches that volume of space. On the other hand, when the speed of an object is greater than the speed of sound, it gets ahead of its own pressure disturbances. Therefore, the air must move out of the way abruptly, without any previous movement in contrast to the subsonic case. A wave of suddenly compressed air, called a shock wave, builds up and is transmitted in a sideways direction to the motion of the supersonic object (Fig. 7-13). The top picture (a) shows shock waves produced by aircraft moving at supersonic speeds. Bottom picture (b) shows the change in air pressure (due to shock waves) measured at ground level. The two sharp increases in pressure each result in an audible boom. As long as the aircraft moves at a supersonic speed, sonic booms are heard continuously beneath its flight path. It is estimated that SSTs in level flight will produce sonic booms along a 50-mile wide path on the ground. This shock wave is analogous to the surface wave (called a bow wave) produced by a speedboat as it moves across the water. When a sonic bow wave passes us we hear a sharp crack, a sonic boom.

What happens on the ground An aircraft traveling at a speed greater than the speed of sound sets up multiple shock waves which are related to the shape of the aircraft. The nose, wings, tail, and other surfaces produce

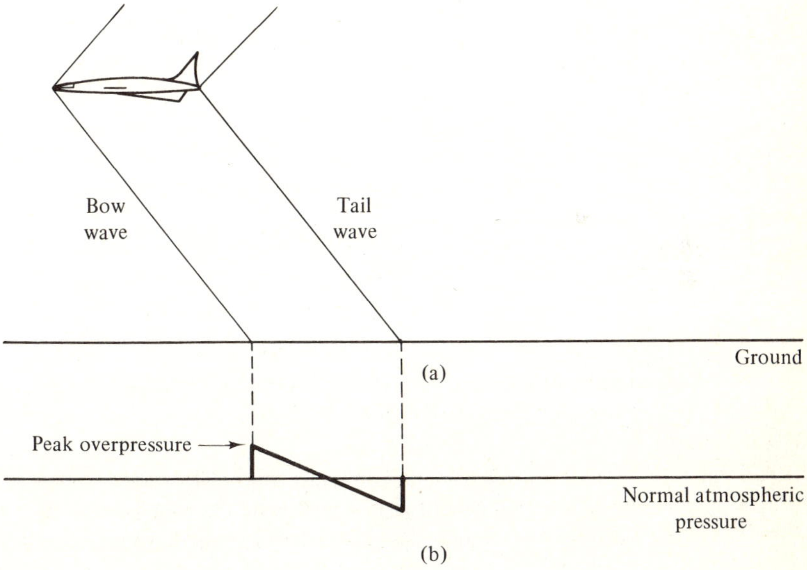

Figure 7-13 (a) Shock waves produced by aircraft moving at supersonic speeds. (b) The change in air pressure (due to shock waves) measured at ground level. Each of the two sharp increases in pressure results in an audible boom.

shock waves. Therefore, within a few hundred feet of the airplane a succesion of booms are heard. At greater distances, these multiple booms cannot be distinguished and only two booms (the bow and tail booms) are heard, as shown in Fig. 7-13. These shock waves travel with the aircraft, the bow shock wave leading the pattern and the tail shock wave ending the pattern.

Figure 7-13 also shows the change in air pressure at the ground due to the arrival of these shock waves. At first there is a steep increase in pressure due to the arrival of the bow wave, then a slow decrease in pressure, and finally another sharp increase in pressure back to normal atmospheric pressure when the tail wave arrives. At the ground we hear two distinct booms in succession.

A measure of the sonic boom's intensity is the peak overpressure, which is the maximum increase in air pressure over normal atmospheric pressure. This pressure change depends on the speed, size, and altitude of the aircraft producing it. The peak overpressure is measured in pounds per square foot (psf). A fighter plane at a few thousand feet produces a boom which is usually around 1 psf. At lower altitudes the boom intensity will be greater. One psf is still fairly small when compared to normal atmospheric pressure (14.7 pounds/inch^2 at sea level), but quite large when compared to average sounds in the environment. Normal atmospheric pressure $= 14.7$ lb/in.^{2} (144 in.^{2}/ft^2) $= 2{,}120$ psf. The conversion factor is (12 in./ft)^2 $= 144$ in.^{2}/ft^2. Ordinary conversation at a distance of 3 feet generates a noise level between 60 and 70 dB. The peak noise level in a 1-psf sonic boom is around 130 dB. This intensity is a million times greater than that of ordinary conversation at 70 dB. We must remember that the decibel scale is not linear and that each 10-dB increase is equivalent to a 10 times increase in intensity. Six 10-dB increases result in a 10^6 increase in intensity.

Previously we indicated that a sound level greater than 120 dB can be painful. This is true only for sounds lasting on the order of minutes. In the case of sonic booms, we are talking about sounds of very short duration (less than a second) and thus far there is no evidence of damage to the ear from such sounds. In addition to being of short duration, sonic booms are of low frequency. Most of the energy in a sonic boom is concentrated at low frequencies (centered around 10 cycles/sec), to which the ear is very insensitive. This is another reason why no physiological effects from sonic booms have been recorded.

This energy which is concentrated at low frequencies is responsible for the window- and door-rattling and wall-shaking associated with the sonic boom. Although our ears are insensitive to very low frequency sound waves, structures such as our homes are not. They absorb a great amount of this energy and vibrate at or near their natural frequencies of vibration.

People and booms In the 1960s supersonic military aircraft were flown over a number of United States cities as part of a military training program.

Surveys were conducted in St. Louis and at Oklahoma City. Over a ten-month period residents of St. Louis were exposed to an average of one boom every two days. Peak overpressures varied from 0.4 to 3 psf.

Thorough interviews as part of a general community survey were conducted with about 1,000 families by the National Opinion Research Center. It was found that the noise inside homes was considered to be more objectionable than that outside homes. This was attributed to the longer duration of the sonic boom-induced noise inside a house and to the rattling and shaking of the house.

The interviewees were asked if the sonic booms were interfering or annoying, where "annoying" was defined as a stronger reaction than "interfering." In about 40% of the interviews, residents reported annoyance at the booms, and in 90% of the interviews they experienced interference with normal activities. In families where the booms were considered interfering, 93% reported the house shook, 74% were startled, 42% were wakened abruptly, and about 20% had their rest and relaxation, conversation, or radio and television reception interrupted. Structural damage occurred to some extent, plaster and glass cracks being the most numerous. No adverse physiological effects were reported. There is not enough information to judge whether or not adaptation to sonic booms is possible. It is even possible that the opposite may occur; that is, we may become sensitized to sonic booms and our reactions to them may worsen with time.

One thing about sonic booms is clear: they cannot be eliminated because they are an integral part of supersonic flight. However, sonic booms can probably be reduced somewhat by proper design of aircraft and by careful planning of flight paths. It is also quite clear that if SSTs were to fly overland, millions of people would be affected by each flight. It appears that the public has become educated enough about sonic booms not to allow overland flights unless some technological breakthrough occurs which reduces the intensity of sonic booms.

What is being done about noise

People vary in their sensitivity to noise. Many famous intellectuals were so disturbed by noise that they took extreme measures to assure their concentrating in silence.* French novelist Marcel Proust (1871–1922) hermetically sealed his studio windows against noise (and germs). American poetress Amy Lowell (1874–1925) so disliked noise that she slept by day and worked by night. When she went to the theatre she bought tickets to the seats on each side of her and left them empty; when staying in a hotel she took the room above hers to assure its being empty to prevent the noise of conversation and footsteps above her head.

*From the book, *The Fight for Quiet*, by Theodore Berland, © 1970 by Theodore Berland, Published by Prentice-Hall, Inc., Englewood Cliffs, N. J.

Noah Webster (1758–1843) compiled his first dictionary at home with nine children running about. He did it by working in his second floor study, which he had specially designed with foot-thick walls lined with cork.

In 1897, Giuseppe Verdi (1813–1901), at 84, moved from his estate to spend his final years in a second-floor apartment of the Grand Hotel in Milan, Italy, not far from his beloved La Scala. He worked a little every morning on sacred music. Each afternoon he walked and spoke to the people he met on the street. When he was ill and didn't appear for his walks, the city placed straw mats on the street below his room to muffle the sound of horse's hooves on the pavement. This is somewhat reminiscent of Jerusalem's erecting a road sign near the home of the late Israeli Nobel Prize writer S.Y.Agnon stating: QUIET, AGNON IS WRITING.

Concern about noise is not a new phenomenon. As far back as the 13th century many towns in England had enacted laws prohibiting blacksmiths from working in early morning hours because of bothersome noise. In Beverly, England, a fine was imposed on persons who drove iron-wheeled carts over stone pavements. Before our present emphasis on visual communication (TV, magazines, books, newspapers), audible communication, some examples of which are town criers and church bells, was most important for community survival. Therefore, not too surprisingly, prohibitions on noise, such as those mentioned above, were taken to reduce community noise levels.

Along with the industrial revolution there was an increase in the noise levels throughout most cities in the industrialized nations. In the 20th century the greatest contributor to the rising noise level has been the transportation industry, although the construction industry is a significant contributor too. Noise pollution has not received as much attention from the public or from law makers as more visible environmental problems such as air and water pollution. For example, air pollution from aircraft has received more attention than noise pollution from aircraft. The black smoke from aircraft taking off and landing seems to annoy a greater number of people than the noise produced by these same aircraft (Fig. 7-14).

Increasingly, however, there have been new attempts at reducing rising noise levels through legislation at state and local levels. When legislation fails to reduce noise to an acceptable level, people and communities often resort to bringing suits against the offenders.

Sources of noise We are exposed to noise in our homes, at work, and in our communities. Some of the noisier objects which can be found around most homes are vacuum cleaners, blenders, lawnmowers, dishwashers, and, of course, radios and phonographs. Unfortunately, there has always been a strong connection between power (or efficiency) and noise in consumers' minds as far as the above products (and many others, such as automobiles)

Figure 7-14 The smoky, air-polluting trail from a jet plane taking off from Love Field, Dallas, blackens the sky for miles around, as well as adding to the noise pollution of the area. (Courtesy EPA—Documerica)

are concerned. For example, many people feel that a vacuum cleaner that is not producing a loud roar is not functioning well. Manufacturers are now designing less noisy products (at slightly higher cost) in addition to their conventional products. These less noisy products can be sold if the public is shown that they really can do the same job as the noisier versions, but with less of a hazard to the users' health.

For more than a hundred years it has been known that there are certain occupations which are dangerously noisy. Men in those industries (boilermakers and miners, for example) came to accept noise deafness as part of their jobs. Boilermakers and excavation machine operators are often exposed to noise of over 100 dBA, truck drivers to 95 dBA, and bus drivers to 90 dBA. White-collar job environments are substantially quieter, but these noise levels are not always in the safe or sanity-promoting ranges. A secretary sitting near a copying machine may hear 75 dBA all day. An accounting office din is about 79 dBA, a keypunch room 84 dBA, a computer operations room 85 dBA. Surprisingly, a U.S. Public Health Survey found that "hearing impairment was found most frequently among farmers." With the mechanization of the farm our romantic vision of the peaceful farm seems to be unfounded.

Even music can be dangerous to our hearing. Electronic rock bands often produce noise levels that exceed those considered safe for prolonged exposure. Sound levels from 100 dBA to 115 dBA have been measured in rock and roll auditoriums. Temporary hearing loss and ringing in the ears often result after exposure to such high sound levels for a couple of hours. There is even evidence of permanent hearing loss among rock musicians who use electronic amplifiers. Professionals in the field (doctors and researchers) are concerned that young people frequenting these concerts will suffer permanent hearing loss. Many teenagers say they like the noise level in spite of the hearing-loss dangers, and manufacturers of amplifying equipment say they are only giving the kids what they want.

Occupational health In 1970, Congress passed the Williams-Steiger Occupational Safety and Health Act. The declared congressional purpose and policy of the act is "to assure so far as possible every working man and woman in the Nation safe and healthful conditions and to preserve our human resources." The provisions of the law apply to every owner of a business in which there are employees.

Even before this law was passed, the American Conference of Industrial Hygienists (ACGIH) published each year a list of threshold limit values of airborne contaminants and physical agents for the place of work. These threshold limit values were for airborne concentration of substances such as asbestos and mercury, and for physical agents such as sound and radiation. The words *threshold limits* refer to conditions under which it is believed that nearly all workers may be repeatedly exposed day after day without adverse effect. Because of wide variability in individual susceptibility, however, a small percentage of workers may develop an occupational illness from exposure to contaminants or physical agents at the threshold limit values.

Although these standards produced by the ACGIH were not binding in a court of law, they were generally followed by industry.

By publication in the Federal Register in 1971, most of the threshold limit values promulgated by the ACGIH were made official federal standards. The permissible noise exposures are in Table 7-3.

Table 7-3 PERMISSIBLE NOISE EXPOSURES

Duration per day in hours	*Sound level dBA*
8	90
6	92
4	95
3	97
2	100
1.5	102
1	105
0.75	107
0.50	110
0.25	115

The sound level in decibels is measured on a standard sound-level meter operating on the A-weighting network with slow meter response. The above values apply to total time of exposure per working day, regardless of whether there is one continuous exposure or a number of short-term exposures, but does not apply to impact or impulsive type of noises. It is recommended that exposure to impulsive or impact noise should not exceed the 140-dB peak sound pressure level.

It is important to remember that not all workers will be protected by these noise standards. Some small percentage (those most sensitive to noise) will suffer permanent hearing loss if, for example, workers are exposed to 90 dBA for an 8-hour day, even though this exposure is within the threshold-limit values for noise.

Community noise and legislation Noise in our communities is much more difficult to control than in workplaces. In the community many sources of noise contribute to the overall noise level, and many of these are mobile, such as automobiles and aircraft. At present there are no national noise standards for our communities in general, although several cities have established noise ordinances of their own.

There are three places at which noise can be limited: at the source, along the path from source to receiver, and at the receiver. It is best to treat noise at the source, but communities do not always have the jurisdiction, as in the case of aircraft. Least desirable is the use of such devices as ear plugs and ear mufflers to limit the noise reaching the receiver.

Some ways to reduce noise levels in our communities are:

1 Educate the public as to the hazards of noise.

2 Build quieter machines.

3 Through land-use planning, separate industrial, commercial, and residential zones; separate airports from residential areas.

4 Muffle noise through better building construction and placement; use trees and shrubs as sound absorbers.

5 Make and enforce laws against excessive man-made noise, for example, aircraft and automobiles.

Aircraft A lot of attention has been focused on method (5) for noise reduction. The Federal Aviation Administration (FAA) has several purposes, two of which sometimes conflict with each other. These are: encouraging the use and development of the aircraft industry and providing safety and welfare to the public. The nation's first federal regulation limiting the noise of new commercial and civil aircraft was established by the FAA in 1969. All aircraft built before 1970 will not have to meet the standards set by the FAA. Although noise certification of jet aircraft constructed since 1970 is a positive step, the noise levels airline manufacturers are required to meet do not remove the community annoyance problem. Moreover, the FAA regulations also do not cover helicopters and vertical takeoff and landing (VTOL) aircraft.

Because of weather or for other reasons pilots cannot always keep their airplanes on scheduled courses or at planned altitudes. This can result in an excessive noise exposure for those people living or working near airports. The effectiveness of the FAA regulations is limited by the fact that they do not monitor the noise made by individual aircraft.

Residents near the Los Angeles International Airport have been awarded several million dollars in property devaluation damages by California courts. Noise from aircraft landing and taking off at the airport represented a taking or pre-empting of property without just compensation. The compensation was sought by the plaintiffs to retrieve their property loss due to noise. Hundreds of similar suits have been filed in other parts of the country. Since aircraft are responsible for the most annoying source of community noise, it is not surprising that there is community resistance to construction of new airports or enlargement of old ones.

Building construction The U.S. Department of Housing and Urban Development has published some interior noise standards for new and rehabilitated construction. In particular, noise levels for sleeping quarters were established. The difficulty with these standards is that they are not legally enforceable and thus very little attention is given them.

In 1968, New York City became the first major city in the U.S. to approve a building code requiring acoustical materials that limit noise transmission within a building. Abroad, particularly in Scandinavia and elsewhere in Europe, there are stringent building codes that minimize the noise transmission problem. Unfortunately, the N.Y.C. code applies only to multiple-family dwellings and does not limit noise penetrating from outside the building. A project of high priority for the federal government would seem to be the establishment of noise-reduction standards for building materials on a national basis.

Motor vehicles Several states have established regulations to limit noise production by motor vehicles. For example, the California legislature has adopted the following standards:

"No person shall sell or offer for sale a new motor vehicle which produces a maximum noise exceeding the following noise limit at a distance of 50 feet from the centerline of travel under test procedures established by the department," as shown in Table 7-4.

As can be seen from Table 7-4, before 1975 motorcycles were allowed to make more noise than automobiles (vehicles less than 6,000 pounds). This was changed, and autos and motorcycles are now treated alike in this respect. The general trend in these regulations is for vehicle noise to decrease with later models, and all vehicle types to reach the same level of 70 dBA by 1987. The exception to this trend are snowmobiles, which can produce the

Table 7-4

Motorcycles		
manufactured before 1970:	92 dBA	
manufactured 1970–1972:	88 dBA	
manufactured 1973–1974:	86 dBA	
manufactured 1975–1977:	80 dBA	
manufactured 1978–1987:	75 dBA	
manufactured after 1987:	70 dBA	
Snowmobiles		
manufactured after 1972:	82 dBA	
Motor vehicles of 6,000 pounds or more		***Less than 6,000 pounds***
manufactured 1968–1972:	88 dBA	86 dBA
manufactured 1973–1974:	86 dBA	84 dBA
manufactured 1975–1977:	83 dBA	80 dBA
manufactured 1978–1987:	80 dBA	75 dBA
manufactured after 1987:	70 dBA	70 dBA

Figure 7-15 Motor vehicle noise affects large numbers of people, as can be seen in this photograph of traffic on a freeway that passes through a heavily populated area of west Los Angeles. (Courtesy EPA—Documerica, Gene Daniels)

same sound level now as after 1972. It takes a large number of trained personnel with sound-level meters to enforce these regulations. However, since motor-vehicle noise is a close second to aircraft noise in community annoyance, the money and personnel needed to reduce this annoyance would be well used.

City programs In addition to state and federal regulations, cities are now establishing programs to curtail noise. Chicago has the most comprehensive program to limit noise of any American city. Its ordinance sets noise limits for sources from bulldozers to garden tools. Particularly useful is the 55-dB limit which was established as the maximum noise that may emanate from a residence. In most communities, the only law which relates to excessive noise from a residence is the subjective one prohibiting the disturbance of the peace. Breaking the law (in Chicago) can result in a $500 fine and a six-month jail term.

In Chicago, three-man teams of two noise inspectors and a police officer cruise the city. When violations are spotted, citations can be issued on the spot. In the first year of operation (1971–1972), convictions were obtained in about 80% of those cases brought to court and completed (1,000 cases), and compliance was obtained in most of the others. Many other cities are adopting similar legislation. There is no question that a lot can be done to make our cities quieter. One way of achieving this is through adequate enforcement of comprehensive anti-noise legislation.

Method (2) for reducing noise has already been discussed briefly. Both consumers and governments can put pressure on manufacturers to build quieter products. This is already happening with autos for example. Advertisements now often speak of the quietness of a particular model automobile. Unfortunately the main effort is to reduce the noise reaching the inside of the auto, not necessarily noise emitted by the car itself.

Local governments in Chicago and New York are now involved in reducing the noise that construction machines make. In addition, the Environmental Protection Agency (EPA) has been directed by Congress to establish national noise emission standards for four kinds of machinery. These are: transporta-

Figure 7-16 This jackhammer, boring into a concrete sidewalk, is an example of the noisy environment which has been reported to Congress by EPA, together with studies of the effects on the health of persons living in a noisy environment. (Courtesy EPA—Documerica)

tion equipment, construction equipment, motors and engines, and electrical equipment. The EPA is authorized to assess civil fines of up to $25,000 for violation of these standards by manufacturers and distributors.

Methods (3) and (4) are not being used as much as they might. Land-use planning, which in theory could separate noisy and quiet areas of a city, is difficult to apply effectively in cities where growth took place rather haphazardly and there is little open space left to be used in planning. However, in younger cities, which are still growing or in the still growing suburbs surrounding many cities, land-use planning could be very effective in separating industrial from residential areas. Land-use planning is, of course, used for many other purposes than limiting noise pollution.

The main idea of land-use planning is to consider all the possible uses of each parcel of land in some geographical area, the needs of the particular community in that area, and then to arrive at the best possible use for each parcel of land. Under community needs, the following factors might be considered: population growth rate, recreation facilities, open space, schools, tax base, and environmental pollution. Assigning values to these various needs is undoubtedly the most difficult part of land-use planning. So noise reduction is only one of many factors that are considered in land-use planning, and its importance to the community must be weighed against all the other needs.

As we already mentioned, the muffling of noise through better building construction is an area which needs lots of attention. There is a lot of room for improvement in the use of sound-absorbing materials in building construction. Careful landscaping with trees and shrubs can also reduce the noise entering our homes (not large apartment buildings) from the surrounding environment.

Of the five methods for noise abatement mentioned earlier, education of the public may well be the most important way of achieving noise reduction. Public outcry against the sonic booms from SSTs succeeded in halting construction of these aircraft in the United States. Therefore, a project that may have resulted in a significant addition to noise pollution in the U.S. was never realized. Now that environmental impact statements (discussed in Chapter 13) must be filed for almost any construction project on a state or federal level, the public has an important opportunity and responsibility to give input to the decision-making processes for these projects. This represents another opportunity to limit the noise levels in one's environment, since noise level is one of the many factors which must be considered in impact statements.

Aside from fighting against projects that would tend to increase noise levels excessively in our communities, we can also make an impact on the noise problem by reducing the amount of noise we are personally accountable for each day.

Suggested reading

1 Bekesey, Georg von, "The Ear," *Scientific American* (Aug. 1957).

2 Berland, Theodore, *The Fight for Quiet*. Englewood Cliffs. NJ: Prentice Hall, Inc., 1970.

3 Bragdon, Clifford, *Noise Pollution, The Unquiet Crisis*. Philadelphia: University of Pennsylvania Press, 1970.

4 Burns, William, *Noise and Man*. Philadelphia: J.B. Lippincott Company, 1969.

5 Green, David M., "Sonic Booms," *Psychology Today* (Nov. 1968).

6 Kryter, Karl D., *The Effects of Noise on Man*. New York: Academic Press, 1970.

7 Littler, T.S., *The Physics of the Ear*. New York: Macmillan Company, 1960.

8 Saunders, Frederick A., "Physics and Music," *Scientific American* (July 1948).

9 Welch, Bruce L., and Annemarie S. Welch, *Physiological Effects of Noise*. New York: Plenum Press, 1970.

10 Perera, Thomas, and Gretchen Perera, *Louder and Louder: The Dangers of Noise Pollution*. New York: Franklin Watts, 1973.

Questions

1 Sound waves, like all other waves, do not transport material from place to place, but they do transfer energy from one place to another. Describe how a sound wave (or pressure disturbance) propagates through the air, transferring energy as it does so.

2 On a very cold night an apartment dweller bangs on his radiator (both the radiator and pipes are made of steel) to inform the manager that he needs more heat in his apartment. If the apartment dweller lives two floors above the manager, how long will it take for the banging sound to reach the manager's apartment? (Assume the height of a room is 3 meters.)

3 A source of sound emitting energy at the rate of 1 joule/sec (1 watt) if placed about a foot away from a human ear is energetic enough to cause pain. What is the intensity in watts per square meter and in decibels at a distance of 1 foot from this 1-watt sound source?

4 Are these statements true or false? Explain your answers.

(*a*) A 2-watt sound source is twice as loud as a 1-watt sound source when standing the same distance away from each.

(b) A 60-dB sound is twice as loud as a 30-dB sound.

5 At a frequency of 1,000 cycles/sec a 4-dB sound can be heard by about 1% of the population (Fig. 7-7). If we wanted 25% of the population to hear a sound of 1,000 cycles/sec frequency, what intensity should the sound have?

6 What percent of the population can hear a 40-dB, 1,000-cycles/sec sound?

7 In Fig. 7-7, would you expect the saw or the powerhouse to give a higher reading on the A scale of a sound-level meter? Why?

8 What would the fundamental natural frequency of a 2.7-cm ear canal be if it were filled with water (while swimming, for example) rather than air. (See Fig. 7-9.) What effect would this have on hearing?

9 How could application of the principle of resonance aid us in pushing a swing efficiently?

10 From your own personal experience, which aspect of noise pollution is most bothersome? Have you found any way of dealing with this problem?

11 (*a*) Sonic booms cause a temporary increase in atmospheric pressure at the ground. A boom with a peak overpressure of 2.12 psf increases the normal atmospheric pressure (at sea level) by what percent?

(*b*) A supersonic aircraft can induce a peak noise level on the ground of about 130 dB. Why doesn't damage to the ear result from such a loud noise?

12 Do you feel, from your own experience, that adaptation to random sonic booms and other unexpected noises in our environment is likely? Why or why not?

13 (*a*) The present occupational noise standards do not protect all workers. If you were given the responsibility of drawing up the occupational noise exposure standards, would your aim be to protect every worker? Why?

(*b*) What difficulties might arise in attempting to protect *all* workers from hearing loss?

14 Does it seem reasonable to you to have national community noise standards, or should the amount of noise a community wishes to tolerate vary from place to place?

15 (*a*) List all the sources of automobile noise that you can think of. Which of these is responsible for the greatest noise input to the environment?

(*b*) Can you think of any ways of reducing auto noise?

16 (*a*) In your opinion, would your own actions to reduce the amount of noise in your environment have much impact?

(*b*) What actions would you take?

Part Five

The atmospheric environment

Chapter 8

Pollution of the air—history and sources

Introduction

We tend to consider air pollution as a recent phenomenon. But since the beginning of the industrial revolution, people in many cities have endured levels of smoke pollution that would be considered intolerable today. In the late 1800s, a number of aroused citizen groups protested the smoke-filled air of London. Not much attention was given to the protests, however, since they came at a time when the majority of people favored industrial development at any price.

As far back as the 17th century, John Evelyn,* one of the founders of the Royal Society in England, was moved to write the following about London:

> "It is this horrid Smoake which obscures our Churches, and makes our Palaces look old, which fouls our Clothes, and corrupts the Waters, so as the very Rain, and refreshing Dews which fall in the several Seasons, precipitate this impure vapour, which with its black and tenacious quality, spots and contaminates whatever is exposed to it."

Figure 8-1 Smoke from a fossil-fuel power plant. Smoke from power plants and industrial sources is not as visible as it used to be. However, air pollution continues to be a problem. (Courtesy EPA—Documerica)

*Evelyn, John, *The Smoake of London*. Elmsford, N.Y: Maxwell Reprint Company, 1969.

In the United States air pollution became a significant problem in the late 19th century. Some smoke control laws were passed, but there was little enforcement of them. From the 1930s through the 1950s, smoke pollution aroused a public outcry in many of the industrial cities of the United States. Some anti-smoke legislation was enacted and partially enforced during this time, and there was visible improvement in the air of some industrial cities. Pittsburgh, PA, which used to be known as the Smoky City, was a notable example of a city where the air quality improved markedly. In Pittsburgh it was only after World War II, when some of the industrialists supported anti-smoke legislation, that real progress was made in reducing the smoke in that city. Before that time, the opposition from industrial and railroad companies was too strong for government leaders and environmentalists to overcome. The increased use of natural gas instead of coal as a fuel for space heating (residential and industrial) and the use of control devices helped to reduce air pollution in Pittsburgh and some other cities during the 1950s.

Air pollution incidents

Several disastrous incidents in the 20th century focused attention on air pollution as a public health problem. These incidents made it obvious that the air quality of a city may deteriorate enough to damage the health of its inhabitants. All of these episodes took place about the same time of year (Fall) under very unusual meteorological conditions which caused stagnation of the air in those locations where the disasters occurred.

The first of these disasters took place in Belgium in 1930. Beginning December 1, there was severe air pollution which lasted for about five days in a highly industrialized area of Belgium called the Meuse Valley. A large number of people became ill with respiratory tract complaints and, before the week ended, sixty deaths were attributed to the smoke and fog (called smog). Older persons, especially those with previous histories of heart and respiratory diseases, experienced the greatest mortality.

An episode that had a strong impact on Americans took place in Donora, Pennsylvania, (28 miles south of Pittsburgh) in 1948. Berton Roueché,* in *Eleven Blue Men*, described Donora as follows:

> "Its mills are no bigger or smokier than many, but their smoke, and the smoke from the passing boats and trains, tends to linger there. Because of the crowding bluffs and sheltering hills, there is seldom a wind, and only occasionally a breeze to dispel it."

The mills Roueché was referring to were the steel mills, a wire plant, and a zinc and sulfuric acid plant.

*Roueché, Berton, *Eleven Blue Men*. Boston: Little, Brown & Company, 1953.

On October 26, 1948, an extremely dense fog descended on Donora. The fog and smoke were so thick that it was just possible to see across the street and the air had a sickening smell. Just like in the Meuse Valley, there was fog, smoke, and a condition known as a temperature inversion* which lasted four or five days.

During this air pollution incident about 43% of the 14,000 people living in the valley in and around Donora became ill. The symptoms were chest constriction, headache, vomiting, nausea, and coughing. Twenty deaths occurred during these few days, whereas only one or two deaths would occur normally. Most susceptible to the effects of the air pollution were those with previous respiratory or heart disease.

The worst air pollution disaster took place in London, England, in 1952. From December 5 through December 9, most of England was covered by a fog and a temperature inversion. Because of the cold weather, more coal was burned for heat than usual, and large quantities of smoke and sulfur-containing gases were released into the atmosphere. In London the fog was so thick that airports closed and even automobiles could not be used. Hospital admissions for treatment of respiratory and heart diseases climbed markedly. Mortality increased among all ages, but most significantly among the aged and infirm. The total number of excess deaths in the greater London area, for several weeks (during and following the temperature inversion), was between 3,500 and 4,000. A search of past mortality and weather records for London indicated that periods with excess mortality had occurred several times before.

Since the London disaster of 1952 there have been other periods of severe air pollution around the world. Among other cities which have been severely afflicted are New York, Tokyo, and Los Angeles. In the 1950s it was discovered that the "smog" in Los Angeles was quite different from the smog of London or Donora. Los Angeles did not use very much coal and was not very foggy; nevertheless, the air-pollution problem appeared there during World War II and worsened afterwards.

Dr. A. J. Haagen Smit, of the California Insitute of Technology, found that the principal sources of "smog" in Los Angeles are hydrocarbons and nitrogen oxides from automobile exhausts. This type of air pollution is called photochemical smog, since sunlight causes certain pollutants to undergo chemical reactions in the atmosphere. Classical smog (as found in London) requires smoke and fog, whereas photochemical smog requires automobiles and sunlight. Photochemical smog was at first thought to be limited to Los Angeles. However, we now know that most large cities are affected to some degree by photochemical smog as well as by other forms of air pollution.

*A temperature inversion is a condition of the atmosphere that leads to the trapping of pollutants. It will be explained shortly.

Temperature inversions

We mentioned that a temperature inversion existed during each of the air pollution disasters we have discussed. The connection between a temperature inversion and the trapping of air pollutants is complex.

The temperature of the atmosphere usually decreases with height in the troposphere (Fig. 8-2). The shape of the curve in Fig. 8-2 varies with location and time of day or year. During a thermal inversion there is a region (Fig. 8-3) of the troposphere where the temperature increases with height. In Fig. 8-3 this region extends from altitude *A* to altitude *B*, but it may begin at the surface, too.

A thermal inversion prevents mixing of the air in the vertical direction, thus concentrating pollutants in a smaller region of the atmosphere than

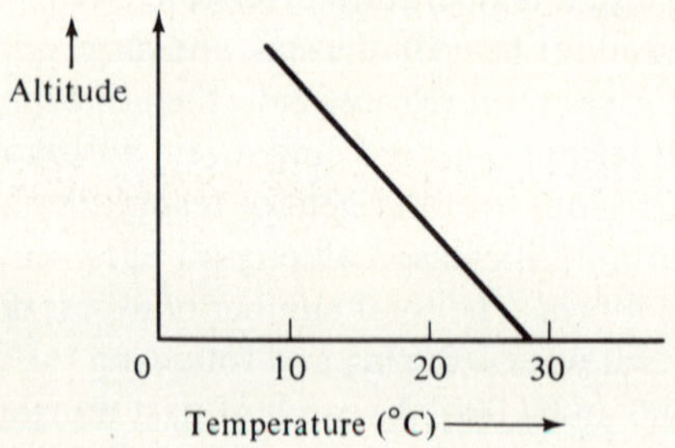

Figure 8-2 The rate at which the temperature of the atmosphere changes with altitude is called the environmental lapse rate.

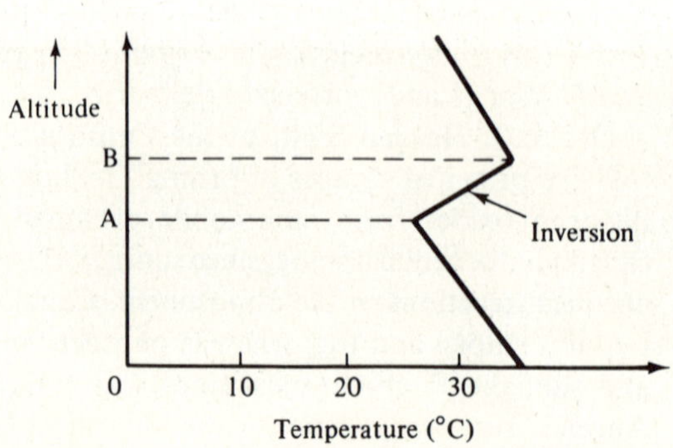

Figure 8-3 The region where the slope of the temperature-versus-altitude curve is positive in sign is known as an inversion layer.

would occur if the inversion did not exist. In the example in Fig. 8-3, air pollutants generated near the surface will be trapped between the surface and the inversion layer. The lower the altitude at which the inversion begins (*A* in this situation), the smaller the volume of clean air available for pullutants to mix with. In Fig. 8-3 pollutants generated near the surface can rise until they reach some altitude between *A* and *B*. Since they cannot penetrate the inversion layer they must spread out laterally within the region between the surface and the inversion layer. If there are no horizontal winds to move air pollutants out (as was the case in London and Donora), then most of the air pollutants generated by a city will be trapped overhead (Fig. 8-4).

Figure 8-4 Air pollutants are trapped over Denver by a temperature inversion. Above the inversion and the smog, the sky is clear and the Rocky Mountains can be seen in the distance. (Courtesy EPA—Documerica, Charles E. Glover)

The atmosphere may be called stable, unstable, or neutral. A thermal inversion causes a condition known as a stable atmosphere. In a stable atmosphere a mass of air given a small vertical displacement tends to return to its original altitude. This is analagous to a marble in an upright bowl (Fig. 8-5).

In an unstable atmosphere a mass of air given a small displacement does not return to its original altitude. Instead, it keeps moving in the direction of the original displacement. This situation is similar to that of a marble on top of an inverted bowl (Fig. 8-6).

In a neutral atmosphere a displaced mass of air remains at the altitude to which it is moved. It neither accelerates in the original direction of motion nor returns to the original altitude. A marble on a flat table illustrates neutral equilibrium (Fig. 8-7).

The stability of the atmosphere can be determined by comparing the temperature variation of the atmosphere to the temperature variation of a displaced volume of air. A volume (or parcel) of air can become displaced by colliding with rocks, trees, buildings, etc. If a displaced volume of air rose to

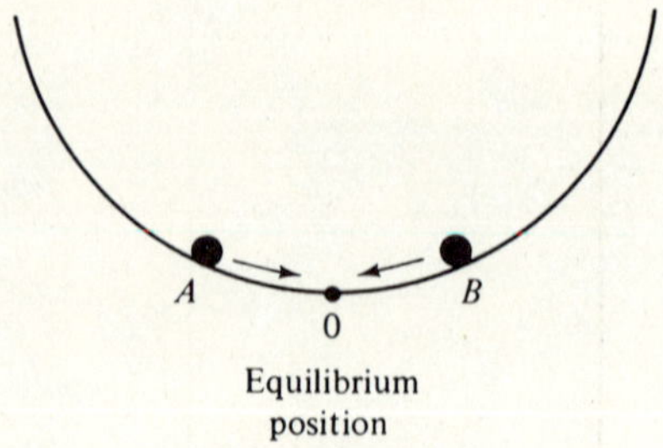

Figure 8-5 Stable equilibrium. If a marble is displaced from its initial equilibrium position at *O* to either *A* or *B*, it tends to return to position *O*.

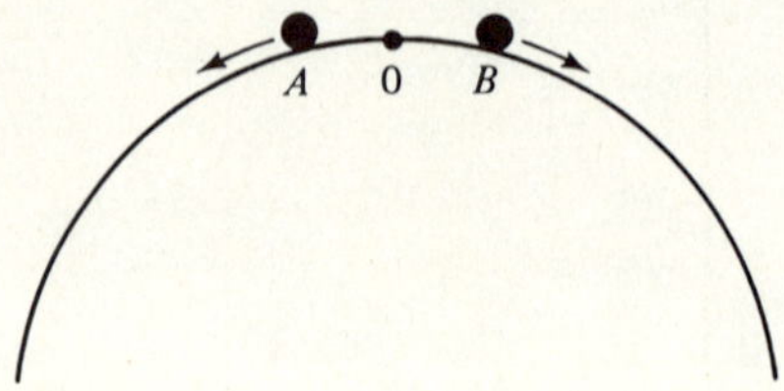

Figure 8-6 Unstable equilibrium. If a marble is displaced from its initial position at *O* to either point *A* or *B*, it tends to keep moving in the original direction of the motion.

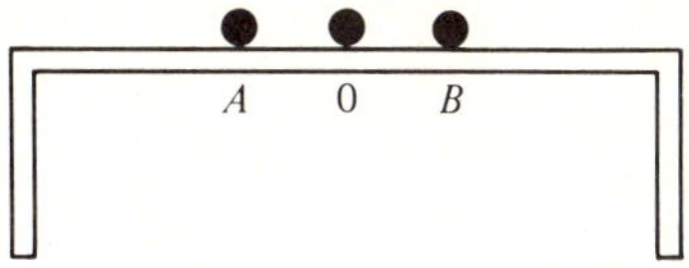

Figure 8-7 The marble can be moved to *A* or *B* and remains there.

an altitude where it was warmer than the surrounding air, then it would be less dense, and because it is lighter than the surrounding air it would continue to accelerate upward. Warmed air expands and becomes less dense since the same mass is spread over a larger volume. This situation produces an unstable atmosphere.

If, however, the displaced parcel of air rose to an altitude where it was cooler than the surrounding air, then it would be denser than its surroundings and would sink back towards the ground. This is called a stable atmosphere. The atmosphere is termed neutral when the rising air parcel has the same temperature and density as its environment. In this case the air parcel would remain at its new position.

As a volume of air rises it expands due to the fact that the surrounding air is now at a lower pressure. The same thing happens to a balloon; as it rises, the air within it expands, forcing the rubber to expand also. Small cohesive forces between molecules must be overcome when air expands. The work required to separate the molecules comes from the kinetic energy of the gas molecules and, therefore, the gas cools during expansion.

The rate of temperature change when a volume of dry air is displaced vertically is constant and equal to 1°C/100 meters. This is called the dry adiabatic lapse rate and is the theoretical rate of fall of temperature with altitude. *Adiabatic* means no heat is added to or lost by the air parcel. If we sketch both the dry adiabatic lapse rate and the actual temperature distribution (called the environmental lapse rate) for the particular time and location of interest, then we can determine the stability of the atmosphere.

Figure 8-8 is representative of an unstable atmosphere. A rising parcel of air (see adiabatic lapse rate) becomes progressively warmer than its environment (see environmental lapse rate) and, thus, less dense than its surroundings. Therefore, it tends to continue rising. In this situation air pollutants will tend to disperse in the vertical direction.

By analyzing Fig. 8-9 we can see why a thermal inversion leads to a stable atmosphere and little vertical mixing. Let us now consider a parcel of air which strikes some obstacle near the ground and begins to rise. We can determine what happens to it by following the dry adiabatic lapse rate (the dashed curve in Fig. 8-9).

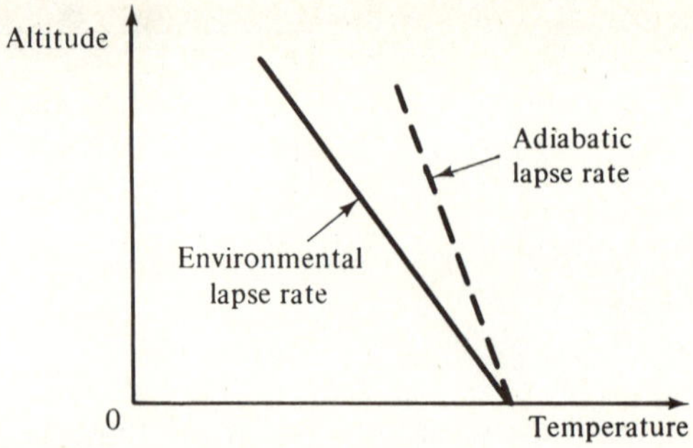

Figure 8-8 An unstable atmosphere

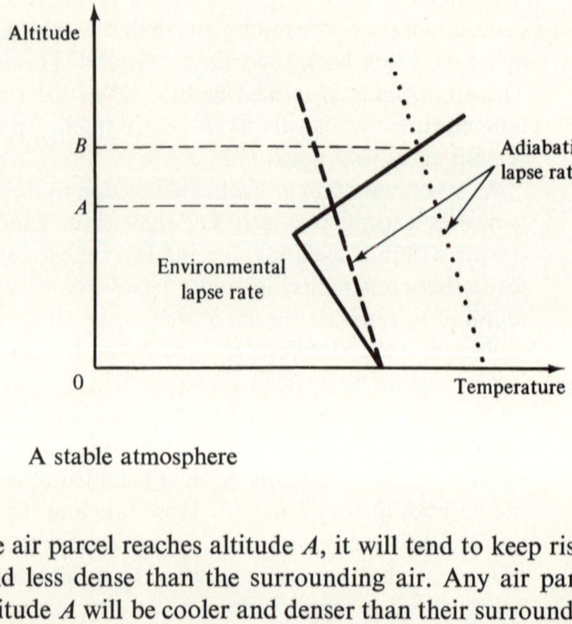

Figure 8-9 A stable atmosphere

Until the air parcel reaches altitude *A*, it will tend to keep rising since it is warmer and less dense than the surrounding air. Any air parcels that rise beyond altitude *A* will be cooler and denser than their surroundings and tend to sink back to altitude *A*. If an air parcel were displaced below altitude *A*, then it would be warmer and less dense than its environment and tend to rise back towards altitude *A*. Therefore, altitude *A* is a position of stable equilibrium, and during this thermal inversion there will be little mixing of the air above altitude *A* with the air below altitude *A*.

Air pollutants which are emitted by automobiles or smoke stacks will have a higher temperature than their surroundings. Therefore, given some initial velocity in the upwards direction, they will start off with a higher temperature (dotted curve) but will follow the same adiabatic lapse rate as cooler air parcels which are thrust in the vertical direction. Consequently, air pollutants will be able to reach altitude *B* which is higher than altitude *A*. The ultimate effect of the inversion is to trap air pollutants between the surface and the inversion layer, as if there were a lid placed over the city where the inversion was located. Above altitude *B* the sky might be clear, while below *B* it would be darkened with pollutants. The interaction between pollutants and the atmosphere is a complicated one. We have simplified our discussion in order to get a rough idea of how inversions trap pollutants.

Finally, now that we have discussed the mechanics of an inversion, we are ready to examine some of the meteorological conditions that give rise to it. High-pressure systems are generally regions where the air is sinking or subsiding. As the air moves to lower altitudes and higher pressure, it is compressed and its temperature rises. This causes a temperature inversion. The subsidence usually does not go all the way to the ground since there is more resistance from denser air below. This warming of air above the surface can result in an inversion of the type illustrated in Fig. 8-10. (Also refer again to Fig. 8-9.)

High-pressure systems also often bring clear skies which allow inversions to form close to the earth's surface. When the sky is clear and dry, heat can be

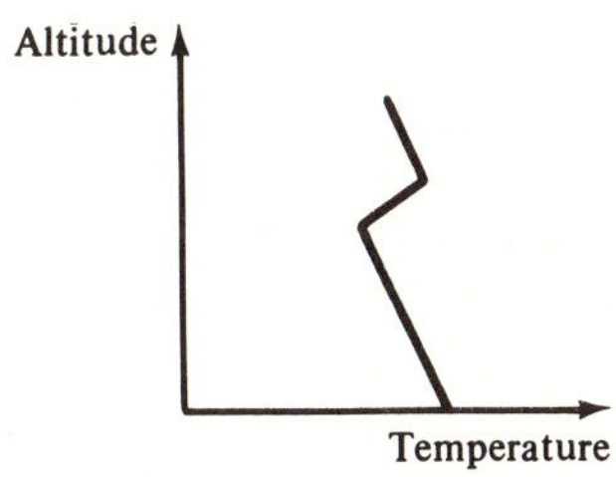

Figure 8-10 A subsidence inversion

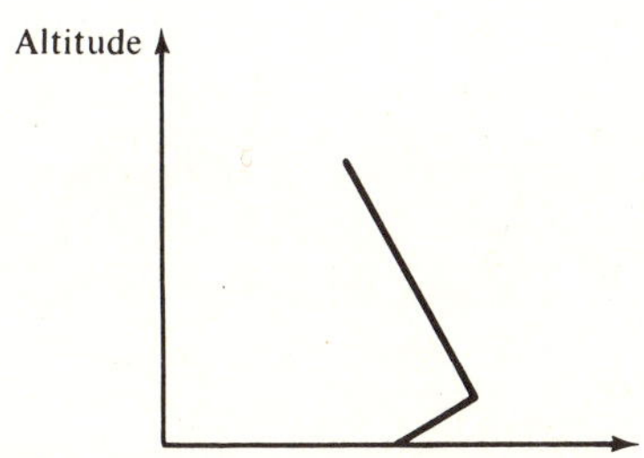

Figure 8-11 A temperature inversion due to radiative cooling

radiated from the ground without much being absorbed above (by clouds or water vapor) and radiated back down. Under these conditions, the air close to the ground (the first few hundred feet) cools off quickly after sunset and an inversion will form close to the ground (Fig. 8-11). This radiative cooling accounts for clear nights being colder than cloudy nights in general. Luckily, the emission of pollutants usually decreases after sunset so this type of inversion, which is quite common, is not as dangerous as the daytime inversions which are rarer. Inversions coupled with weak winds have been responsible for all of the air pollution disasters thus far.

Gaseous air pollutants and their sources

Each year over 200 million tons of pollutants are discharged into the air over the United States. The five main classes of pollutants are: carbon monoxide, particulates, sulfur oxides, hydrocarbons, and nitrogen oxides. Table 8-1 lists the sources of these pollutants and their estimated contributions for the year 1971. Carbon dioxide (CO_2), which is not found in this table, is not considered to be a pollutant since it does not appear to have any direct effect on human health, safety, or comfort. However, we will discuss CO_2 emission in Chapter 9 since the accumulation of carbon dioxide in the atmosphere may affect the world's climate.

Table 8-1 AIR POLLUTANT EMISSIONS IN MILLIONS OF TONS PER YEAR—1971

Source	*CO*	*Particulates*	*SO_2*	*HC*	*NO_x*
Transportation	77.5	1.0	1.0	14.7	11.2
Fuel combustion in stationary sources	1.0	6.5	26.3	.3	10.2
Industrial processes	11.4	13.6	5.1	5.6	.2
Solid waste desposal	3.8	.7	.1	1.0	.2
Miscellaneous	6.5	5.2	.1	5.0	.2
Total	100.2	27.0	32.6	26.6	22.0
Percent change 1970 to 1971	−.5	+5.9	−2.4	−2.6	0

Source: U.S. Environmental Protection Agency

Carbon monoxide Carbon monoxide (CO) is a colorless, odorless, toxic gas that is produced by the incomplete burning of carbon in fuels. Carbon monoxide emissions can be prevented by supplying enough air to insure complete combustion of the fuel. If this can be done, then CO_2 is produced instead of CO. A molecule of carbon monoxide (CO) contains 1 oxygen atom and 1 carbon atom, whereas a molecule of carbon dioxide (CO_2) contains 2 oxygen atoms and 1 carbon atom. When a fossil fuel or wood is burned in air, both of the following reactions are possible:

$$C + O_2 \longrightarrow CO_2 \quad \text{complete combustion} \tag{8-1}$$

$$2C + O_2 \longrightarrow 2CO \quad \text{incomplete combustion} \tag{8-2}$$

Reaction (8-1) can be read as: 1 atom of carbon combines with 1 molecule of oxygen to form 1 molecule of carbon dioxide. From reaction (8-2) we see that 2 atoms of carbon combine with 1 molecule of oxygen to form 2 molecules of carbon monoxide. The ratio of oxygen to carbon during incomplete combustion is half of its value during complete combustion. So we can see why increasing the oxygen supply when fuel containing carbon is burned results in a greater production of carbon dioxide. In the above reactions, oxygen appears in molecular form since that is its natural state in the troposphere where these combustion reactions take place.

Carbon monoxide is the most abundant pollutant found in the atmosphere; it generally exceeds the emission of all other pollutants combined (excepting CO_2). As given in Table 8-1, transportation activities (mostly the automobile) account for about 77% of the carbon monoxide released to the atmosphere. No large natural sources of carbon monoxide have been definitely identified. The total amount of carbon monoxide released by all activities in a year (in the U.S., 1971) is about 25% of the total amount of carbon monoxide found in the atmosphere. However, even with this relatively large yearly input of carbon monoxide into the atmosphere, the background levels of CO in clean air (Greenland, for example) do not appear to be increasing. Therefore, there must be one or more processes by which carbon monoxide is removed from the atmosphere.

The concentration of CO in metropolitan areas varies widely with time and place, and depends on human activity and meteorological factors. The CO levels in most communities correlate well with traffic volume; the highest levels are found most often in places where motor vehicle traffic is greatest (Fig. 8-12). Figure 8-13 is a typical curve of CO concentration in parts per million (ppm) versus the hour of the day. Carbon monoxide concentration reaches its peaks during the morning and afternoon rush hours.

By parts per million (ppm) we mean the number of molecules of a particular type found in 1 million (10^6) molecules of air.* For example, a CO concentration of 10 ppm means that if we were to take a sample of air containing 1 million molecules (mostly N_2 and O_2), then we would find that, on the average, 10 of these molecules are CO molecules. Another way of looking at

*There is no such thing as an air molecule. We really mean the various molecules comprising air, such as nitrogen and oxygen.

Figure 8-12 Most of the CO produced by man's activities is emitted into the atmosphere by automobiles. CO levels can often climb to 50 ppm or more in heavy traffic such as pictured in this photograph of the Hollywood Freeway near Ventura, CA. (Courtesy EPA—Documerica)

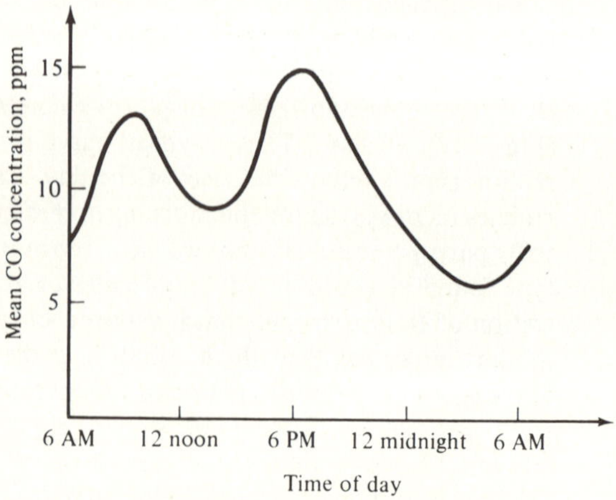

Figure 8-13 Variation of CO levels with time of day, during the week in a typical city.

parts per million is in terms of volume. That is, 10 ppm of CO also means that CO occupies 10 cm^3 out of each 10^6 cm^3 of air. The two approaches are equivalent since equal volumes of gases (at the same temperature and pressure) contain the same number of molecules.

Unpolluted air contains from 0.01 ppm to about 1 ppm of carbon monoxide. That is, in the cleanest air (0.01 ppm) 1 molecule in 10^8 is a CO molecule. Carbon monoxide measurements in a number of cities in the United States indicate that CO levels range from 1 ppm to greater than 140 ppm, the latter being present only at brief peaks in heavy traffic. Commuters in heavy traffic are often exposed to CO levels of 50 ppm or more.

The health hazards associated with the various pollutants are discussed in Chapter 9, so for the present we will only briefly state the effect of CO on the body. Hemoglobin, which is found in red blood cells, carries oxygen to the cells of the body. Carbon monoxide has a much stronger affinity for hemoglobin than oxygen; therefore, when the CO concentration rises in the air being inhaled, the number of red blood cells carrying oxygen will decrease and the cells of the body will receive less oxygen. The symptoms of CO poisoning are dizziness, blurred vision, and rapid breathing.

Sulfur dioxide Fossil fuels (coal, oil, natural gas) often contain large amounts of sulfur that is burned and oxidized to sulfur oxides, mainly sulfur dioxide (SO_2):

$$S + O_2 \longrightarrow SO_2 \tag{8-3}$$

Sulfur dioxide is a colorless gas which produces a pungent irritating odor (like rotten eggs) for concentrations greater than about 3 ppm in air. The sulfur content, by weight, of commercial coal varies from about 0.3% to 5%. Most of the coal mined in the eastern states contains more than 1% sulfur. Most of the coal mined in such western states as Montana and New Mexico is now in great demand due to its low sulfur (less than 1%) content.

From Table 8-1 we can see that most of the SO_2 emission is due to the burning of fuel. Electric utilities (Fig. 8-14) and industrial plants are the principal producers of sulfur dioxide since their most abundant fuels are coal and oil which contain sulfur as an impurity. Natural gas contains much less sulfur than coal or oil, but there is not enough to meet the demand. The burning of coal is responsible for most of the sulfur oxides emission, but oil burning and industrial processes that use sulfur (such as petroleum refining and copper smelting) also contribute significant amounts to the atmosphere. Seven northeastern industrial states account for almost half of the national total of sulfur oxides. This situation is now being changed by the use of low-sulfur fuels (less than 1% sulfur). However, the supply of low-sulfur coal and oil is not great enough to meet the demand of all eastern states.

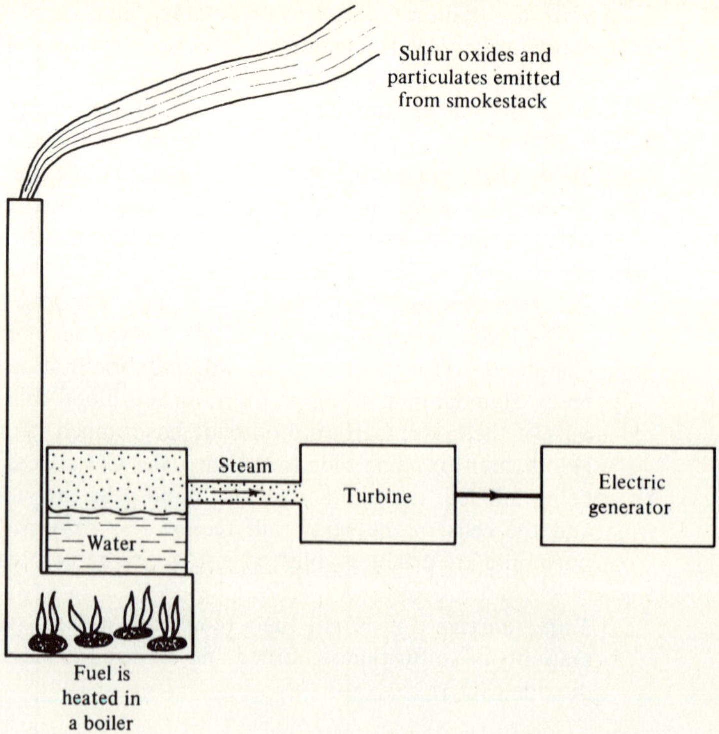

Figure 8-14 In a fossil-fuel power plant, coal, oil, or gas is burned to provide heat for turning water into steam in a boiler. Pollutants resulting from the combustion of fuel are released into the atmosphere via the smokestack. The steam which is produced in the boiler turns a turbine which runs an electric generator.

The sulfur compounds in the atmosphere come from both the natural environment and from air pollutant emissions. About one-third of the sulfur reaching the atmosphere comes from pollutant sources, mainly as sulfur dioxide. Natural sulfur compound emissions come mostly from sulfate (SO_4) salts [such as ammonium sulfate, $(NH_4)_2SO_4$] which are sent into the atmosphere by sea spray, and the hydrogen sulfide (H_2S) from decomposition of organic matter in swamp areas, bogs, and tidal flats. There seems to be no buildup of SO_2 in the atmosphere beyond emission areas. The concentration of sulfur dioxide in remote areas of the world is about 0.001

ppm or 1 ppb (one part per billion). In large cities the sulfur dioxide concentration may average, for a 24-hour day, anywhere from 0.01 ppm to 0.70 ppm.

The fate of sulfur dioxide (SO_2) in the atmosphere is of great importance insofar as its ultimate effect on people and materials is concerned. Sulfur dioxide is transformed to sulfuric acid (H_2SO_4) in a matter of days by several different reactions in the atmosphere. One of these reactions uses nitrogen dioxide (NO_2) as a catalyst, and another proceeds as follows:

$$2SO_2 + O_2 \longrightarrow 2SO_3 \tag{8-4}$$

$$SO_3 + H_2O \longrightarrow H_2SO_4 \tag{8-5}$$

In these reactions, and in the one that uses NO_2 as a catalyst, water vapor from the atmosphere is necessary to convert sulfur trioxide (SO_3) into sulfuric acid. Sulfuric acid is highly corrosive and can erode marble and corrode metal. Sulfur dioxide and sulfuric acid are both capable of irritating the respiratory system of humans and animals. Sulfur dioxide in combination with other pollutants has been implicated in the air pollution disasters which occurred in the Meuse Valley, Donora, and London. We discuss the evidence for SO_2 damage to people, vegation, and building materials in Chapter 9, along with the air quality standards for sulfur dioxide.

Nitrogen oxides Nitrogen oxides (NO_x) are produced when fuel is burned at very high temperatures and cooling occurs fast enough to prevent their decomposition. Nitrogen oxides are written as NO_x where x is an integer that indicates the number of oxygen atoms. The nitrogen and oxygen that combine to form nitrogen oxides both come from the atmosphere and not the fuel. Only NO (nitric oxide) and NO_2 (nitrogen dioxide) are formed in the atmosphere in significant quantities. They are both gases:

$$N_2 + O_2 \rightleftharpoons 2NO \quad \text{nitric oxide formation} \tag{8-6}$$

$$N_2 + 2O_2 \rightleftharpoons 2NO_2 \quad \text{nitrogen dioxide formation} \tag{8-7}$$

The arrows pointing to the left in reactions (8-6) and (8-7) indicate that they are reversible reactions. If the exhaust gases produced when fuel is burned do not cool off very quickly, then the NO or NO_2 that has been formed will decompose back to nitrogen (N_2) and oxygen (O_2). However, if cooling does occur quickly, then the reverse reaction does not take place.

In Table 8-1 we see that transportation vehicles and stationary combustion sources account for almost all of the man-made NO_x production. Internal combustion engines operate at very high temperatures, and so do large electric power and industrial boilers. Nitrogen that is inert at lower temperatures combines with oxygen in high temperature flames and tends to stay combined if the exhaust gases are cooled too quickly. The control of NO_x from stationary sources requires careful adjustment of flame and stack gas temperatures. The control of nitrogen oxides from automobiles is even more difficult since a reduction in combustion temperature to decrease NO_x production tends to increase the production of other pollutants such as carbon monoxide.

Most of the NO_x produced during fuel combustion is NO. NO is not an irritant and appears to have no adverse health effects at concentrations that occur in urban areas. However, NO reacts with O_3 (ozone) in the atmosphere to form nitrogen dioxide (NO_2).

$$NO + O_3 \longrightarrow NO_2 + O_2 \qquad (8\text{-}8)$$

NO_2 (a yellowish-brown gas) is responsible for the brownish color of smog. It is a toxic gas which can be lethal to humans and animals at concentrations of 100 ppm or more for a few minutes. At much lower concentrations (5 ppm), NO_2 can have a corrosive effect on lung tissues and may lead to an increase in respiratory disease in humans. The evidence for the adverse effects of nitrogen dioxide on humans (at levels already existing in urban air) is discussed in Chapter 9.

Both NO and NO_2 are involved in the formation of what we call photochemical smog. NO, NO_2, and hydrocarbons (mostly from auto exhausts) interact in the presence of sunlight to form eye-irritating chemicals. The production of photochemical smog is discussed later in this chapter.

Hydrocarbons Hydrocarbons are compounds containing only hydrogen and carbon. The light hydrocarbons are gases at room temperature. Some examples are methane and ethylene:

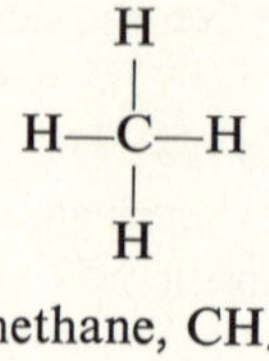

methane, CH_4

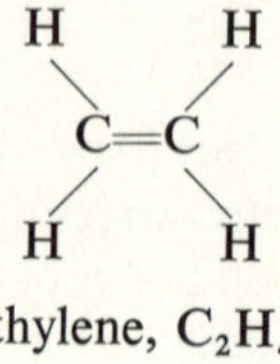

ethylene, C_2H_4

Methane, a colorless and odorless gas, is the principal constituent of natural gas. It was formerly known as marsh gas and was first observed as bubbles rising from the decomposition of vegetable matter in swamps.

There also exists a large family of hydrocarbons that are derivatives of benzene. These may be liquids or solids and are referred to as aromatic hydrocarbons:

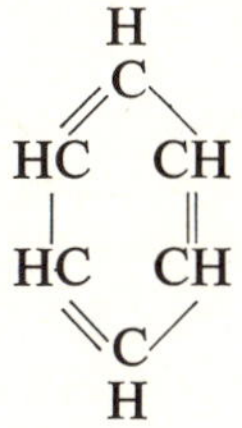

Benzene, C_6H_6

Hydrocarbons are of importance as a pollutant mainly because of their involvement in photochemical reactions leading to photochemical smog. There is also concern over some of the aromatic hydrocarbons which are thought to be carcinogenic, that is, causes of cancer. Transportation is the main source (man-made) of hydrocarbons in the United States (Table 8-1). Gasoline (which is a hydrocarbon) is derived from petroleum, a complex mixture of hydrocarbons together with varying percentages of other compounds which contain nitrogen, oxygen, and sulfur. Crude petroleum is refined by a process known as fractional distillation. This process is based on the fact that the different components of petroleum have different boiling points and can, therefore, be separated from one another, within certain limits. This is done by heating the crude oil to successively higher temperatures and collecting the portions that boil off within certain temperature ranges. Gasoline, kerosene, and lubricating oil are some important compounds that are collected in this manner. The incomplete combustion of gasoline in automobile engines results in the release of a wide array of hydrocarbons to the atmosphere.

*Photochemical smog**

Photochemical smog is usually associated with California, especially Los Angeles. Los Angeles was the first city in the United States to experience this type of smog and also the first to initiate research aimed at understanding its causes and effects. Photochemical smog, which is initiated by the absorption of sunlight by nitrogen dioxide, is the result of a complex series of reactions between the oxides of nitrogen, ozone, and hydrocarbons. These reactions lead to the formation of substances called oxidants. Oxidants are chemical compounds detrimental to biological systems and destructive to certain materials.

The term *oxidation* originally was used to describe chemical reactions in which certain atoms combine with oxygen (O_2) to form compounds known

*The document, "Air Quality Criteria For Photochemical Oxidants," NAPCA Publication No. AP-63, was used as a source of material for this section.

as oxides. The complete combustion of carbon in air to form carbon dioxide (CO_2) is a typical oxidation reaction: $C + O_2 \longrightarrow CO_2$. Now the term oxidation is more generally used to describe the loss of one or more electrons by an atom, ion, or molecule. In the above reaction the carbon atom loses two electrons to each oxygen atom, so it is oxidized. Each oxygen atom gains two electrons; we say it is reduced. *Reduction* means the gain of electrons by an atom, ion, or molecule. In any chemical reaction one of the reactants is oxidized while the other is reduced.

The combustion of fossil fuels is an oxidation-reduction reaction which is responsible for most of the air pollution in urban atmospheres. If a fossil fuel could be completely burned, it would produce only water and carbon dioxide. Since combustion processes are not 100% efficient, the exhaust gases also contain small amounts of the original fuel plus some partially oxidized fuel, such as carbon monoxide (CO) and certain hydrocarbons. At the high temperatures characteristic of combustion processes, small amounts of the nitrogen (N_2) in the atmosphere are oxidized to NO and NO_2 (mostly NO). Most of this NO, however, is converted to NO_2 by reactions with ozone (O_3).

When all of these contaminants are released into the atmosphere, they may react chemically to produce other contaminants, quite different from those originally released. Nitrogen dioxide (NO_2) is the most efficient absorber of the sun's ultraviolet light which reaches the earth's surface. Nitrogen dioxide is broken down by ultraviolet light into NO and atomic oxygen:

$$NO_2 \xrightarrow{\text{ultraviolet light}} NO + O \tag{8-8a}$$

Atomic oxygen (O) combines with molecular oxygen (O_2) to form ozone (O_3). Ozone is the oxidant found in greatest concentration in photochemical smog:

$$O + O_2 \longrightarrow O_3 \tag{8-8b}$$

Finally, NO and O_3 react to generate new NO_2 and O_2:

$$O_3 + NO \longrightarrow NO_2 + O_2 \tag{8-8c}$$

and the cycle continues.

This series of chemical reactions explains the initial formation of ozone in polluted air. However, if there were no other reactions taking place, most of the ozone created by reaction (8-8b) would be destroyed by reaction (8-8c), and the ozone levels would never reach the observed values. The emission of hydrocarbons into the atmosphere (mostly from gasoline combustion in autos) changes this situation. Some O and NO produced in reaction (8-8a)

Figure 8-15 Fossil-fuel power plants are major sources of air pollution. They are responsible for more than 50% of all sulfur oxide emissions, and about 30% of the particulate and nitrogen oxide emissions. (Courtesy EPA—Documerica)

reacts with these hydrocarbons in a series of complex chemical reactions to form NO_2 and O_3. Thus, the production of ozone by reaction (8-8b) begins to outweigh its destruction by process (8-8c) as the concentration of NO

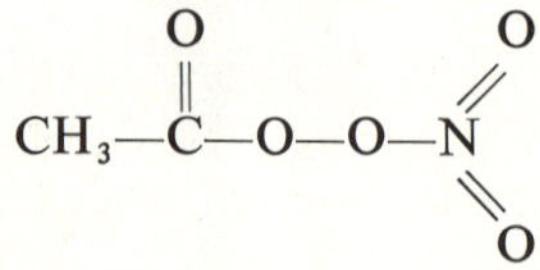

peroxyacyl nitrate (PAN)

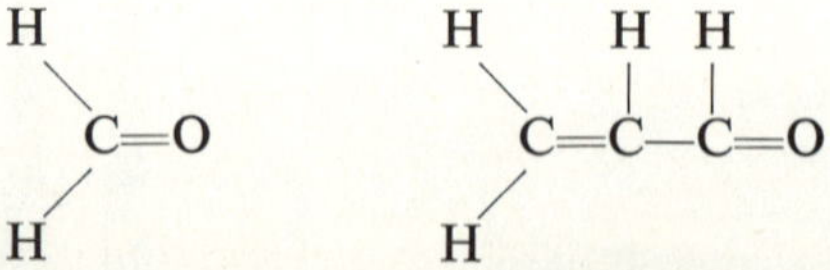

Formaldehyde Acrolein

(a)

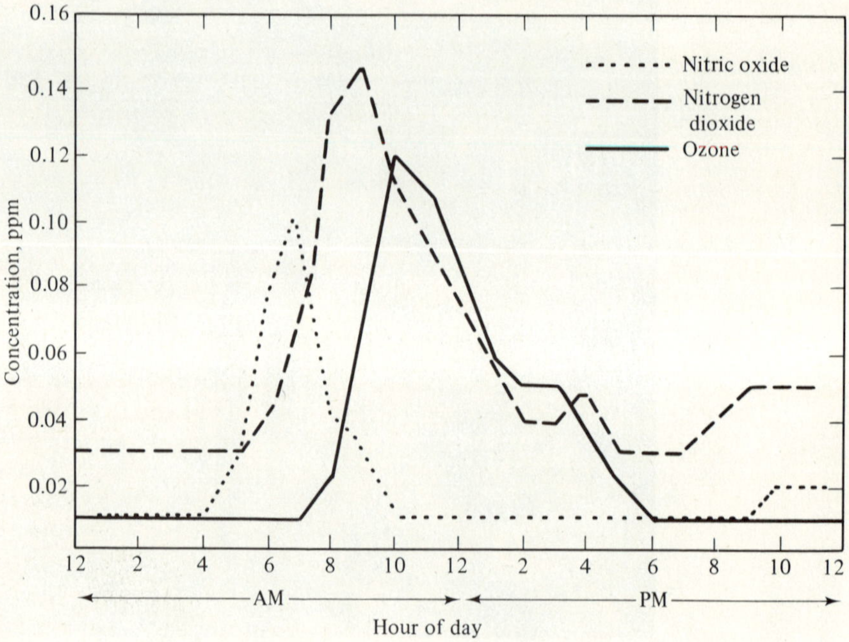

(b)

Figure 8-16 (a) Some of the eye irritants found in photochemical smog. (b) The variation of NO, NO_2, and O_3 concentrations in Los Angeles, July 19, 1965. (From "Air Quality Criteria for Photochemical Oxidants," NAPCA Publication No. AP-63) drops and that of NO_2 rises. Some other undesirable chemicals (in addition to ozone) which are produced in these reactions with hydrocarbons are formaldehydes, peroxyacyl nitrates (PAN), and acrolein (Fig. 8-16). These are all eye irritants; ozone is not.

Because of the importance of solar radiation on photochemical reactions, the level of oxidants rises during the summer months (June to October in Los Angeles) and is a strong function of the time of day on any particular day. Figure 8-16(b) shows the variation of NO, NO_2, and O_3 concentrations in Los Angeles on a typical summer day. The concentration of ozone, which is the main oxidant, usually peaks around noon and then slowly decreases. The same pattern is repeated in many other U.S. cities and in other parts of the world.

Particulates

The term particulates includes a variety of pollutant types, such as smoke, dusts, fumes, and mists.

Smoke, which is composed of carbon particles (about 10^{-5} cm in diameter), results from the incomplete combustion of carbonaceous materials such as coal, oil, tar, and tobacco.

Dusts are solid particles generated by crushing, grinding, detonation, etc., of materials such as rock, ore, coal, wood, grain, etc.

Mists are liquid droplets suspended in the air.

Fumes are solid particles (usually metals) generated by condensation from the gaseous state.

When any of these particles (liquid or solid) are suspended in the air, they are called aerosols. Therefore, the terms particulate and aerosol are usually used interchangeably.

The major man-made sources of particulates in the United States are electric power generation plants, the crushed stone industry, the forest products industry, agriculture and related operations, the cement industry, and the iron and steel industry. Transportation accounts for only a very small percentage of particulate production.

The concentration of particulate matter in the atmosphere is usually stated in terms of a mass density. The density of an object is equal to its mass divided by the volume it occupies. When we speak of the mass density of particulates in the atmosphere, we mean the mass of particulates found in some volume of air, divided by that volume of air:

$$\text{Mass density} = \text{mass/volume} \qquad (8\text{-}9)$$

The most often used unit for mass density of particulates is millionths of a gram per cubic meter or $\mu g/m^3$.

In most American cities the average annual particulate concentration is between about 30 $\mu g/m^3$ and 200 $\mu g/m^3$. Even at a concentration of 200

$\mu g/m^3$ we are not talking about very much total mass:

$$\text{mass density} = 200\ \mu g/m^3 = 200 \times 10^{-6}\ g/m^3 = 2 \times 10^{-4}\ g/m^3$$

It would take an air sample of 5,000 m³ to collect 1 gram of particulates even in one of the more smoky or dusty of cities:

$$2 \times 10^{-4}\ g/m^3 \times 5{,}000\ m^3 = 1\ g$$

In our present system of weights 1 gram is equal to about $\frac{1}{30}$ ounce. This volume (5,000 m³) is equivalent to that of a large lecture hall or auditorium about 3 m × 40 m × 40 m or 10 ft × 130 ft × 130 ft. So if we were to leave the windows open in such a room, in a city where the particulate concentration was 200 $\mu g/m^3$, we would have a total of 1 gram of particulates in the air. This may not seem like very much to be concerned about, but the number of particles involved is tremendous, as we shall see shortly, and many of them can be inhaled and deposited in the respiratory system where they cause irritation.

Example 8-1: (*a*) How many particles are there in this very large room of 5,000 m³ volume? We make a few assumptions to simplify the calculation. First, we assume that the particles are cubes since the volume of a cube is easy to calculate. Because most of the particles in the atmosphere are about 10^{-5} cm in size (see page 207), we take this as equal to the length of each side of the cube. Finally, since water has a density of 1 gm/cm³, and smoke and dust particles will probably just barely float on water, we take the density of the particles as 1 g/cm³.

The volume, V, of each particle is $(10^{-5}\ cm)(10^{-5}\ cm)(10^{-5}\ cm) = 10^{-15}\ cm^3$. The mass, m, of each particle is given by $m = \rho V$ where ρ is the density of the particle (assumed to be 1 g/cm³):

$$m = (1\ g/cm^3) \times (10^{-15}\ cm^3) = 10^{-15}\ g$$

So the total mass, M, of all the particles is given by $M = mN$, where N is the number of particles with mass m in the room. We know that the total mass, M, is equal to 1 gram; so:

$$1\ g = 10^{-15}\ g \times N$$

Therefore, $N = 10^{15}$ particles, which is 1 followed by 15 zeros. There are a quadrillion particles in this room;

(*b*) What is the number density of these particles? There are 10^{15} particles in a volume of 5×10^3 m^3

$$\text{Number density equals } N/V = \frac{10^{15} \text{ particles}}{5 \times 10^3 \text{ m}^3}$$
$$= 2 \times 10^{11} \text{ particles/m}^3$$

We can convert this to particles per cubic centimeter as follows:

$$1 \text{ meter} = 100 \text{ cm} = 10^2 \text{ cm}, \quad \text{so} \quad 1 \text{ m}^3 = 10^6 \text{ cm}^3 \quad \text{and}$$

$$2 \times 10^{11} \text{ particles/m}^3 = \frac{2 \times 10^{11} \text{ particles/m}^3}{10^6 \text{ cm}^3/\text{m}^3} = 2 \times 10^5 \text{ particles/cm}^3$$

Therefore, 1 cubic centimeter of air will contain about 200,000 dust and smoke particles.

Settling velocities of particulates In these calculations we have assumed (incorrectly) that each particle is a cube of the same size. Actually, particles suspended in the air will have different sizes and shapes. The size distribution of suspended particles is pretty much the same from city to city, as shown in Fig. 8-17. There are many more small particles than big particles and most of the particles have diameters less than 2×10^{-5} cm. We are assuming now that the particles are spherical in shape. We can see (from Fig. 8-17) that choosing an average size of 10^{-5} cm for our particles was a reasonable assumption.

So far we have said nothing about the fate of these suspended particles. That is, do they remain suspended indefinitely or do they settle down rather quickly? Obviously, some of the particles do settle down. We know this is so

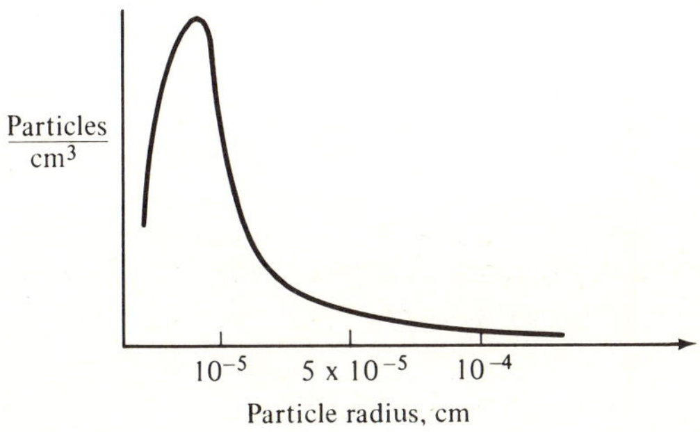

Figure 8-17 The typical size distribution of particles suspended in the atmosphere. The particles are assumed to be spheres.

because of the soot and dust we find on our windowsills, furniture, and just about everywhere else we look.

If gravity were the only force acting on a suspended particle, then it would fall in the same amount of time no matter what its size or density. This conclusion follows from Newton's famous second law of motion:

$$F = ma \tag{8-10}$$

There are three laws of motion all together. Equation (8-10) can be read as follows: A body of mass m acted upon by a force F will experience an acceleration a which can be computed from Eq. (8-10). If force is expressed in newtons and mass in kilograms, then acceleration is measured in meters/sec/sec (read meters per second per second). Acceleration is the time rate of change of velocity,* so the units of acceleration are the velocity divided by time or meters per second divided by seconds.

Example 8-2: Suppose that a block of wood (mass 5 kg) is resting on a highly polished wooden floor. If a force of 10 newtons is applied to the block, assuming that friction between the block and floor is negligible, how fast will it accelerate?

$$F = ma$$

Rearranging, we get

$$a = F/m$$

$$a = 10 \text{ newtons}/5 \text{ kg} = 2 \text{ m/sec/sec}$$

in the same direction as the force.

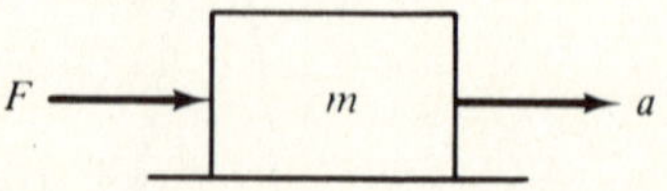

Figure 8-18 A constant force F is applied to a block of mass m resting on a frictionless surface.

*The velocity of an object is not specified until both its speed and direction of motion are known. The speed at which an object moves may, for example, be 50 miles/hr, whereas its velocity is 50 miles/hr in a northerly direction. In this section we are using speed and velocity interchangeably. However, we should keep in mind that they are not equivalent. We don't have to keep mentioning the direction of motion since we are always talking about downward motion.

An acceleration of 2 m/sec^2 means that the block's speed will increase by 2 m/sec each second. At the end of 1 second its speed will be 2 m/sec; at the end of 2 seconds it will be 4 m/sec, etc., until the force is removed or changed.

Returning to the falling particle, if the only force acting on a particle of mass m is that of gravity, then the force, F, is equal to the particle's weight, W:

$F = W$ the weight of the particle

The weight of an object is proportional to its mass, but they are not equal to each other. Weight and mass are related by this equation:

$$W = mg \tag{8-11}$$

where W is the weight of an object of mass m, and g is the acceleration due to gravity. The acceleration due to gravity, g, is essentially a constant near the earth's surface and can be taken to be equal to 9.8 m/sec/sec or 32 ft/sec/sec. Weight has the same units as any other force, so it is expressed in newtons in the mks system.

Since $W = mg$, $F = W = mg$. According to Newton's second law, $F = ma = mg$. Therefore, $a = g$ for a freely falling body, independent of its mass. We can conclude that, independent of air resistance, all particles which start at the same height will fall in the same amount of time if they have the same initial velocity.

If we take air resistance into account, the problem becomes more complicated. The falling particle will experience a retarding force, F_r, which depends on its velocity. As its velocity increases, the retarding force will increase until, finally, the retarding force equals the weight, W, of the particle. At this time the particle stops accelerating and reaches a constant velocity called the terminal velocity. The resultant force acting on the particle of mass m is its weight, W, less the retarding force, F_r. So $F = W - F_r = ma$. When the retarding force equals the particles weight, $F_r = W$. At this time, $F = 0 = ma$. So $a = 0$.

When the acceleration is zero, the velocity, v, is not changing with time, so $v =$ constant. This constant velocity is called the terminal velocity, v_t. If we plot the particle's velocity as a function of time, we get the curve in Fig. 8-19.

We can determine the relationship between the particle's radius, r, and the terminal velocity, v_t, by considering the following relationship between the particle velocity, v, and the retarding force, F_r. In this experimentally determined relationship k is a constant:

$$F_r = krv \tag{8-12}$$

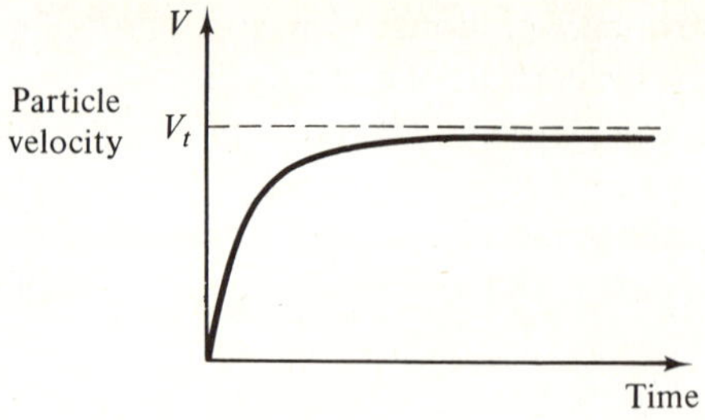

Figure 8-19 The speed of a falling particle as a function of time.

When the particle velocity reaches the terminal velocity ($v = v_t$), the retarding force is equal to the particle's weight ($F_r = W = mg$) and we can substitute mg for F_r.

Making the proper substitutions into Eq. (8-12), we get, $mg = krv_t$. Rearranging,

$$v_t = mg/kr \tag{8-13}$$

The mass of a particle can be written as its density, ρ, times its volume, V: $m = \rho V$. If we take the particles to be spheres of radius r, then the volume is equal to $4\pi r^3/3$.

Substituting for m in Eq. (8-13), we get

$$v_t = \frac{4\pi r^3 \rho}{3kr} = \frac{4\pi r^2 \rho}{3k}$$

Finally, we have the terminal velocity proportional to the square of the radius:

$$v_t \propto r^2 \tag{8-14}$$

From Eq. (8-14) we can see that big particles fall faster than small ones. Table 8-2 shows actual settling velocities of particles of different size in still air. Air currents will change these values. We can see that the settling velocity increases with particle size, but not as fast as Eq. (8-14), $v_t \propto r^2$, would predict. When r increases by a factor of 10, v_t should increase by a factor of 100, but it doesn't increase quite that fast (Table 8-2). It should not be very surprising that the experimental values in Table 8-2 are not exactly equal to the values predicted by Eq. (8-14). Equation (8-12), which was used to derive

Table 8-2 APPROXIMATE SETTLING VELOCITIES IN STILL AIR OF PARTICLES HAVING A DENSITY OF 1 GRAM/CM^3

Radius (10^{-4} *cm*)	*Settling velocity* (*cm/sec*)
0.05	8×10^{-5}
0.50	4×10^{-3}
5.0	3×10^{-1}
50.0	25
500.0	390

(From "Air Quality Criteria for Particulate Matter," National Air Pollution Control Administration Publication No. AP-49)

Eq. (8-14), is not exact, and we have made other assumptions such as sphericity of the particles.

Example 8-3: A particle with a radius of 0.5×10^{-5} cm exits from a chimney that is 50 meters above the ground. Assume that it has no initial vertical velocity. How long will it take the particle to reach the ground if there are no air currents operating in the vertical direction? We can assume that the particle reaches its terminal velocity in a negligibly short time compared to the total fall time.

The distance, s, traveled by an object moving at a constant speed, v, during a time interval, t, is given by $s = vt$. Rearranging,

$$t = s/v = 50 \text{ meters}/8 \times 10^{-5} \text{ cm/sec}$$

The terminal velocity of a 0.5×10^{-5} cm particle is 8×10^{-5} cm/sec.

In order to compute t we must convert either s or v so that all quantities are in the same system of units. So we will change 50 meters to centimeters:

$$50 \text{ m} = 50 \text{ m} \times 100 \text{ cm/m} = 5{,}000 \text{ cm}$$

$$t = 5{,}000 \text{ cm}/8 \times 10^{-5} \text{ cm/sec} = \tfrac{5}{8} \times 10^{8} \text{ sec}$$

$$t = 6.25 \times 10^{7} \text{ sec}$$

$$t \cong 2 \text{ years!}$$

Of course, in a real situation this result is inconsequential since there will be winds and precipitation to help remove the particles from the air. However, the point is that particles can remain suspended in the air for long periods of time. They can be spread by atmospheric circulations over very wide geographical areas and thus influence the climate on a global scale. More about this in the next chapter.

Other pollutants

Many other pollutants in addition to those already mentioned are found in urban air. Some examples are rubber particulates from motor vehicle tires, insecticides from agricultural spraying, asbestos from building construction and brake linings, and lead from gasoline. The health hazards produced by such air pollutants were mostly discovered through industrial experience, where workers were exposed to much higher concentrations of these chemicals than found in urban air. We discuss two of these air pollutants in detail.

Lead In the early 1970s production of gasoline antiknock additives represented about 20% of the total lead used in the United States. In contrast, the exhaust from gasoline-fueled vehicles accounted for more than 90% of the airborne lead. Gasoline is composed of hundreds of different compounds which will ignite at different temperatures. In high-compression engines the probability of pre-ignition* of one of these compounds is high. One way of preventing pre-ignition is by adding lead compounds to the gasoline. Tetraethyl lead, $(C_2H_5)_4Pb$, has usually been added in amounts of about 3 grams per gallon.

When leaded gasoline burns, the lead is converted to inorganic compounds such as lead oxide. Some lead oxide remains in the engine, fouling the valves and spark plugs. Chlorides and bromides are added to gasoline to help remove lead oxide from the engine by changing it to a more volatile† compound such as lead bromide (PbBr). Some of this gas (lead bromide) condenses in the exhaust system (where the temperature is cooler than in the engine) and comes out as very fine particles which may remain dispersed in the air or may stick together and settle out. Studies have shown that most of the lead particles are less than 10^{-4} cm in diameter. Some of these small particles will be inhaled and deposited in the lungs. A small percentage of the lead deposited in the lungs accumulates in the body where its toxicity can lead to adverse health effects.

Lead is one of the oldest metals used by man. It was used by the Romans for pipes and also in an alloy (lead and tin) for soldering. There is speculation that the Romans may have been harmed on a large scale by the use of lead containers for storage and drinking and lead pipes for conveyance of water.

Lead is a dense metallic element which has many uses. Some of the main uses of lead are in storage batteries, in antiknock additives in gasoline, in pigments for paints, in ammunition, and in solder. The concentration of lead in the ocean and atmosphere has risen sharply since about 1940 when there was a large increase in the use of tetraethyl lead in gasoline as an antiknock

*Pre-ignition results in an explosion rather than smooth burning within the engine. An audible noise, called engine knock, is heard.

†A volatile compound is one that vaporizes easily.

additive. Even such remote areas as Greenland and the Antarctic continent have recorded increasing levels of lead in the ice.

Lead poisoning is mostly known as an industrial and childhood disease. There have been many cases of children ingesting paint chips which contain lead. In many of the buildings built before 1939, paints containing lead pigments were used on interior walls and never completely removed. In the United States, during the late 1950s and early 1960s an average of about 10 children died each year from lead poisoning.

The symptoms of lead poisoning are gradual weakness, constipation, abdominal colic, anemia, and often a typical paralysis of the wrists and ankles. Children under six years of age are the main victims of lead poisoning, and those between 1 and 3 years comprise 85 to 90% of the childhood cases. A high percentage of those children who are treated for lead poisoning are left with some kind of neurological handicap (for example, seizures, mental retardation, or cerebral palsy). So even though the fatality rate from lead poisoning has been dropping (now less than 5%), many of the survivors are left severely handicapped. Lead pigments are no longer used in interior paints, but the walls of many substandard older buildings are still covered with pealing lead paint, thus presenting a hazard to the children living there.

Elements that comprise less than 0.01% of an organism are called trace elements. Some of these elements, such as copper, iron, and zinc, are known to be essential for optimal biological functioning. Others, such as cadmium, lead, and mercury, are not presently known to be necessary for normal growth and metabolism. Large amounts of either essential or nonessential trace elements may be toxic.

There are trace amounts of lead in the food we eat, the water we drink, and the air we breathe. Most of the lead which enters our bodies does so through food and air. The total amount of lead in our bodies (about 200 mg) is divided into two parts. Most of it is relatively tightly bound in our bones with a biological half-life of about 10 years, and a small percentage ($\sim 5\%$) is found in various soft tissues, including blood. The concentration of lead in human bone apparently increases during life until at least 40 years of age; however, the concentrations of lead in the soft tissues ordinarily remains low and relatively stable throughout life. The known acute toxic effects of lead are associated with increases in the small but mobile fraction of the total body lead burden.

An estimate of the mobile fraction of the body lead burden can be obtained from blood lead concentrations. The normal blood level of lead has a wide range, between 0.05 and 0.40 ppm, with 0.25 ppm usually taken as an average. As blood lead concentration rises above 0.40 ppm, increasing numbers of individuals may be expected to show metabolic evidence of interference with synthesis of hemoglobin, although anemia may not be evident until blood lead concentrations reach 0.70–0.80 ppm. It is not known if there are undetectable effects on the kidneys, nervous system, or other cellular functions in the 0.40- to 0.80-ppm range.

The health effects of the existing blood levels in the general population are not known, and there has been a lot of controversy pertaining to the setting of maximum acceptable concentrations of lead in the air and in the blood. A study* of the amount of lead in the air of three cities and of the amount of lead in the blood of about 2,000 residents of these cities revealed that the annual average concentration of lead in the air was 1.4 $\mu g/m^3$ in Cincinnati, 1.6 $\mu g/m^3$ in Philadelphia, and 2.5 $\mu g/m^3$ in Los Angeles. This urban concentration is about 50 times the concentration of lead in rural air, and about 5,000 times the natural concentration. The blood tests of residents of these cities showed acceptable levels of lead, but not a wide margin of safety. It is likely that increasing concentrations of lead in the atmosphere will result in higher blood levels in the population.†

The Environmental Protection Agency has decided that it would be advantageous for public health to attain an average lead concentration (yearly) of 2 $\mu g/m^3$ or less in urban areas (2–5 $\mu g/m^3$ is now a common average lead concentration in the air of urban areas). In order to achieve this, the amount of lead in gasoline will have to decrease from about 3 g per gallon to 1.25 g per gallon by 1978. In addition, the Environmental Protection Agency has required certain stations to offer for sale one grade of gasoline that is almost lead-free. This type of gas must be used by 1975 and later model cars in order to prevent damage to the catalysts used in pollution-control devices. These two actions taken by the Environmental Protection Agency should significantly reduce the amount of lead found in urban air.

Asbestos Asbestos is a trade name used for any magnesium silicate mineral which is fibrous in structure and more or less resistant to acid and fire. Among the chief products made from asbestos are pipe covering, brake linings, plaster, and plasterboard. A common practice in constructing high-rise office buildings has been to spray fireproofing material containing from 10 to 30% asbestos onto girders and other structural components. Not only are the construction workers exposed to asbestos fibers in this way, but there is also an extensive fallout of fibers in the vicinity of the construction site. In New York City, this use of asbestos is now more carefully controlled.

In the late 1960s a study was made of the Insulation Workers Union of New York. The cause of death of union members who had started work in 1942

*"Survey of Lead in the Atmosphere of Three Urban Communities," U.S. Dept. of HEW PHS Publication No. 999-AP-12, 1965.

†Goldsmith, J. R., and A. C. Hexter, "Respiratory exposure to lead: epidemiological and experimental dose-response relationships," *Science*, **158** (1967), 132–134.

Figure 8-20 This picture of asbestos fibers was taken with an electron microscope. The size of the fibers has been magnified 60,000 times. They are cylindrical in shape and have diameters of 0.03 to 0.05 $\times 10^{-4}$ cm. (Courtesy Jack Murchio, School of Public Health, University of California, Berkeley)

was compared to a sample of men in comparable working circumstances except for nonexposure to asbestos. Since disease associated with asbestos may take as long as 20–30 years to develop, it was necessary to wait this long before beginning the study. It was found that 255 union men had died compared with 203 men in the control group; 45 workers had died of lung cancer or lung disease, compared with 6 or 7 in the control group; 29 had died of stomach cancer, compared with 9 or 10 in the other group; and 12 had died of asbestos disease, while there were no deaths from this cause in the control group.

The use of asbestos is widespread. More than 100,000 people in the United States work directly with asbestos every day, and up to 3.5 million work with asbestos at some time in some capacity in the construction and shipbuilding industries.

In the 1970s there has been an increase in the number of mesotheliomas, a previously extremely rare cancer that develops on the mesothelium or lining of the chest cavity. A connection between high concentrations of asbestos fibers in the air and this type of cancer has been made, and unions have demanded, and industry has generally provided, proper ventilation, respirators, and shielding for asbestos workers. However, people living near asbestos mines and asbestos plants (many of whom had not worked directly with asbestos) have also contracted mesothelioma. Even more frightening are the results of autopsies of people supposedly remote from sources of asbestos. In many cities throughout the world asbestos fibers and asbestos bodies have regularly been found in people's lungs. Asbestos bodies are fragments of asbestos coated with a material containing iron. The asbestos becomes coated when the body reacts to the invasion of this foreign body.

It appears that all of us who breathe urban air are exposed to this very toxic substance. Studies have shown that the air of Philadelphia and New York City contains from 0.01 to 0.1 g/m^3 of asbestos. As little as 0.001 g/m^3 may represent as many as a million asbestos fibers.

There is no linear dose-response curve for asbestos as there is for radiation or lead. People have as many different reactions to asbestos as they do to poison oak, for example. Two people receiving the same dose of asbestos (or poison oak) may have very differing reactions; one may develop a disease while the other one will be unaffected. This is true to some extent of all environmental contaminants, but appears to occur more often with asbestos. The threshold for cancer induction by asbestos fibers is unknown. Perhaps one fiber deposited in the lungs will cause cancer, or maybe two, or a hundred. Nobody knows for sure. Because of this uncertainty, caution is certainly appropriate.

Studies have recently been made to determine if asbestos levels might be higher on the Los Angeles freeways than in other locations in that city. Auto brakes are made of asbestos, and some fibers can be released into the atmosphere during deceleration. However, one study reported that asbestos levels were not significantly higher on the freeways than in other places, suggesting that emission from brake linings is not significant.

The Environmental Protection Agency has recently moved to establish emission limits for asbestos mines and for all manufacturing plants that use asbestos. Thus it appears that the hazard has been recognized and action is being taken. It is also crucial to inform the occasional user of asbestos products (such as plaster board) of the dangers of breathing asbestos fibers. Because of the many products that contain asbestos, consumer protection is an area that still needs lots of attention.

Suggested reading

1 ELWYN, D., "Childhood Lead Poisoning," *Scientist and Citizen*, **10**: 3 (1968), 53–57.

2 GILFILLIAN, S.C., "Lead Poisoning and the Fall of Rome," *Journal of Occupational Medicine*, **7** (1965), 53–60.

3 PATTISON, J.N., "Pros and Cons for Lead in Gasoline," *Journal of the Air Pollution Control Association*, **21**: 6 (June 1971).

4 BRODINE, VIRGINIA, "A Special Burden," *Environment*, **13**: 2 (March 1971), 22–33.

5 SELIKOFF, I.J., "Asbestos," *Environment*, **11**: 2 (March 1969).

6 ANDREWS, WILLIAM, ed., *Environmental Pollution*. Englewood Cliffs, NJ: Prentice-Hall, Inc., 1972.

7 BATTAN, LOUIS, *Unclean Sky: A Meteorologist Looks at Air Pollution*. New York: Doubleday & Company Inc., 1966.

8 ESPOSITO, JOHN C., *Vanishing Air*, The Ralph Nader Study Group Report on Air Pollution. New York: Grossman Publ., 1970.

9 FENSTERSTOCK, JACK C., and ROBERT K. FANKHAUSER, "Thanksgiving 1966 Air Pollution Episode in the Eastern United States," NAPCA Publication No. AP-45.

10 HAAGEN-SMIT, A.J., "The Control of Air Pollution," *Scientific American*, **210** (Jan. 1964), 24–31.

11 LOGAN, W.P.D., "Mortality in the London Fog Incident, 1952," *Lancet*, **264** (1953), 336–338.

12 STERN, ARTHUR, ed., "Air Pollution," Vol. 1, *Air Pollution and Its Effects*. New York: Academic Press, 1968.

13 FAITH, W.C., and ARTHUR A. HEISSON, Jr., *Air Pollution*. New York: Wiley-Interscience, 1972.

Questions

1 The following table gives the temperature distribution of the troposphere (environmental lapse rate) at some particular time and location, and the adiabatic lapse rate. Plot the two curves from these data and determine if the atmosphere is stable or unstable under these conditions.

Altitude	*Environmental temperature*	*Adiabatic temperature*
0 m	70°F	70°F
100 m	60°F	68°F
300 m	50°F	64°F
500 m	40°F	60°F
800 m	30°F	54°F
1,000 m	20°F	50°F

2 (*a*) Is the likelihood of temperature inversions greater in regions which are surrounded by high mountains? Explain.

(*b*) Is your geographical location a likely one for an air pollution disaster? Consider its topography, frequency of temperature inversions, and magnitude of pollutants emitted to the atmosphere.

3 (*a*) From Table 8-1 compute the percent of CO emission that transportation was responsible for in 1971.

(*b*) Explain why a CO concentration of 1 ppm in air can be looked at as either 1 molecule per million molecules of air or as 1 cm^3 of CO per million cm^3 of air. Remember that molecules are units of individual particles of matter, whereas cm^3 are units of volume.

4 (*a*) How much sulfur is released into the atmosphere when 1 ton of coal is completely burned? Assume that the coal contains 1% sulfur.

(*b*) Suppose that all of the sulfur went into SO_2. Calculate the weight of SO_2 released to the atmosphere. Use the fact that an oxygen atom has nearly half the mass of a sulfur atom.

5 Isooctane is a particularly useful component of automotive fuel. The complete combustion of isooctane yields only H_2O and CO_2. Complete the following reaction by balancing the number of carbon, hydrogen, and oxygen atoms on each side.

$$2C_8H_{18} + 25O_2 \longrightarrow$$

6 What are some differences between the kind of smog found in London and that found in Los Angeles?

7 (*a*) Can you think of any reason why it might be misleading to express ozone concentrations in terms of 24-hour averages?

(*b*) Estimate the 24-hour average ozone concentration in Los Angeles on July 19, 1965 [Fig. 8-16(b)]. How does it compare to the peak ozone value for that same day?

8 Suppose that the mass density of particulates in a room with dimensions 3 m × 3 m × 3 m is 20 $\mu g/m^3$. What is the total mass of particulates in this room?

9 Suppose that the resultant force acting on a particle of mass, *m*, is zero. According to Newton's second law of motion, what are the two possible states of motion that the particle may be in?

10 If a freely falling particle is released from rest at a height of 64 feet above the ground, it takes 2 seconds to fall to the ground (without air resistance). What will its speed be when it strikes the ground? To see how fast this really is, convert from feet per second to miles per second by using 60 mi/hr = 88 ft/sec.

11 A lead particle, with a radius of 0.5×10^{-4} cm, is emitted in the exhaust from an automobile. Suppose that the air is still and that the tailpipe is 16 inches above the ground. How long will it take for the particle to settle to the ground? The density of lead is 11.3 g/cm^3. *Hint*: Note that the terminal velocity is proportional to the density of the particle.

12 Looking at Table 8-1, what percentage do sulfur oxides contribute to the total emissions from stationary fuel combustion?

13 How long would it take a particle with a radius of 0.05 cm to fall to the ground from a jet airplane's exhaust 5 miles above the ground? Assume that the air is still and that the particle has a density of 1 g/cm^3.

Chapter 9

The effects of air pollution and the national air quality standards

Introduction

In April of 1971, William D. Ruckelshaus, then Administrator of the Environmental Protection Agency (EPA), announced final publication of national air quality standards for six common classes of pollutants: sulfur oxides, particulate matter, carbon monoxide, photochemical oxidants, nitrogen oxides, and hydrocarbons.

The Clean Air Act Ammendments of 1970 required the states to devise implementation plans to attain the national air quality standards within three years after approval of these plans by the EPA.* The implementation plans are a strategy for reducing the emission of air pollutants from both stationary and mobile sources such that the air quality standards are attained and maintained.

The information on which the air quality standards were based is found in the Air Quality Criteria, published by the National Air Pollution Control Administration and the Environmental Protection Agency. In the development of criteria many factors, including the chemical and physical characteristics of the pollutants, must be considered along with the exposure time and conditions of the environment. The criteria must also include consideration of the contribution of all such variables to the effects of air pollution on human health, agriculture, materials, visibility, and climate.

The two basic sources of information on the effects of air pollutants on human health are toxicological studies on animals and man, and epidemiological studies. Before we present findings from these types of studies we discuss the general strategy and limitations of each of these approaches to determine the effects of air pollutants on human health.

Toxicology

Toxicology is defined as the study of poisons. However, a toxicologist often works with substances we don't normally think of as poisons, such as food additives or air pollutants. Ideally, a toxicologist would like to know all of the effects of a known dose of chemical on man (Fig. 9-1). He tries to obtain this information by administering the substance in known amounts to animals or to human volunteers. The data he obtains can be summarized in a dose-effect curve [also called a dose-response curve (Fig. 9-2)].

Some of the questions of interest to the toxicologist are:

1 Is there a threshold dose for the effect of interest?

2 Is the substance toxic only when large doses are given in a short time (days or weeks)? (This is called acute toxicity.)

3 Is the substance toxic when small doses are given over a long time (years)? (This is called chronic toxicity.)

4 Which tests on animals can be extrapolated to man?

*The final attainment date for each of the six air quality standards depends on the particular pollutant and region of the U.S. under consideration.

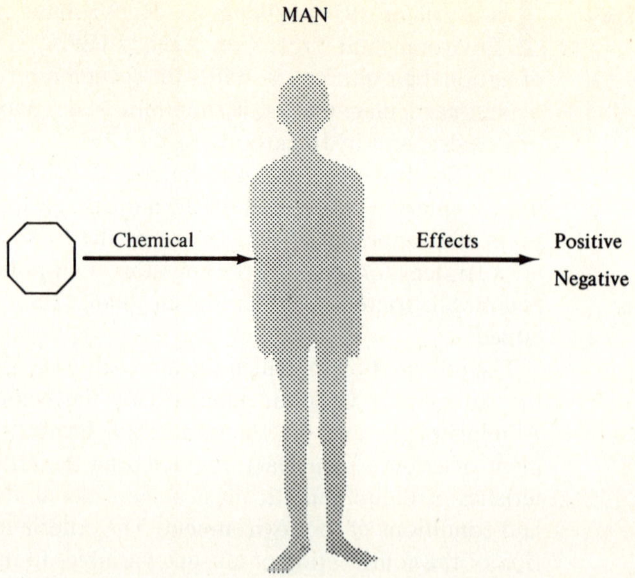

Figure 9-1 A toxicologist studies the adverse effects of chemicals on man.

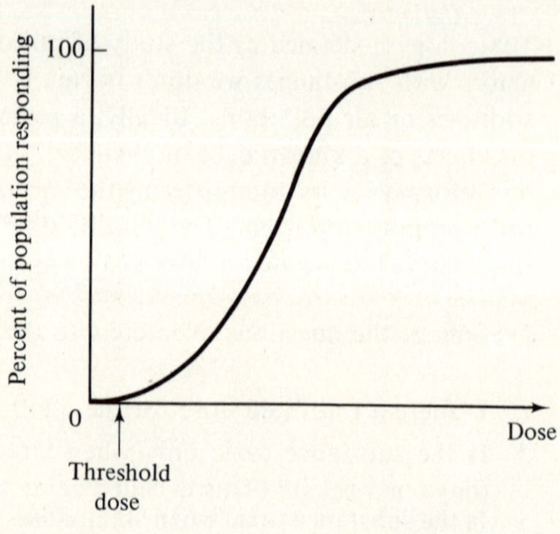

Figure 9-2 A dose-response curve tells us what percent of a population will have a particular response to a known dose of a physical, chemical, or biological agent. The dose might be in milligrams of lead and the response might be some symptom of lead poisoning. Few people respond to small doses, but everyone responds to large enough doses.

The short-term responses a toxicologist looks for include changes in respiratory rate, visual acuity, psychological responses, diseases of various kinds, or death. Some of the long-term effects which are usually looked for are cancer, birth defects, and alteration of genetic material.

Environmental epidemiology

Epidemiological studies, as distinguished from toxicological or experimental studies, analyze the effects of pollution, from existing exposures, on groups of people living in a community. Such studies have the advantage of examining illness where it occurs naturally, rather than in a laboratory, but they have the disadvantage of not being able to control precisely all the factors of possible importance. Nevertheless, the main conclusions in air quality criteria documents prepared in the U.S. have ultimately been based upon epidemiological studies.

Environmental epidemiology is the study of the incidence, distribution, and control of diseases which are caused by environmental contaminants (such as air and water pollutants). Most epidemiological research deals with tests of association and a single study can almost never be interpreted with respect to causation. For example, an epidemiological study might show that there is a strong association between cigarette smoking and lung cancer, but it cannot prove that cigarette smoking causes lung cancer. A conclusion from a hypothetical study might read as follows: The probability of a smoker getting lung cancer is twice as great as that of a nonsmoker. Epidemiological studies are statistical in nature and always deal with probabilities. If many epidemiological studies reach similar conclusions, or if experimental studies reach the same conclusions as epidemiological studies, then the causation of a particular disease or response to some physical or chemical agent becomes more likely.

Epidemiological studies are costly and very time consuming. Often they are aimed at specific groups in the population which may be particularly sensitive to some environmental pollutant. For example, persons with disease states requiring increased oxygen demand may be very sensitive to increased levels of carbon monoxide in the air.

Another approach often used is to study groups of people who are exposed to large doses of an environmental contaminant (for CO this could be traffic policemen). If the environmental contaminant has some adverse effects on the general population, then they should certainly show up in these high-risk groups. Some of the possible effects of air pollutants on people are:

1 acute sickness or death
2 insidious or chronic disease
3 unfavorable symptoms
4 alteration of important physiological functions

5 discomfort from air pollution sufficient to lead the individual to change his residence or place of employment.

The general aim of the national air pollution standards is to protect the whole population from adverse health effects caused by air pollutants. If the most sensitive people in the population are protected, then everyone should be protected from any ill effects caused by air pollutants. For this reason, and also because it is easier to detect adverse health effects in these groups, epidemiological studies of sensitive groups in the population are very useful and plentiful.

Other studies have investigated the relationship between air pollutants in the community as a whole and human health. When the epidemiologist is looking for a very small effect (such as the possible increased mortality from air pollution), a large population is needed since these small sensitive groups do not provide adequate numbers of people to study. Some other requirements of an epidemiological study involving the general population are good mortality and morbidity* statistics, adequate reporting of air pollutant concentrations and meteorological conditions in different parts of a city, and social and economic data. It is desirable to know air pollutant concentrations in various parts of a city or a metropolitan area so that high and low air pollution regions can be compared with each other.

Since there are many other factors (other than air pollution) which affect death rates from specific causes or the incidence of specific diseases, epidemiologists usually try to eliminate the effect of as many of these other factors as possible. They do this by choosing two or more population groups which are similar in terms of median age, sex ratio, social status, racial composition, cigarette smoking, etc., but different in terms of exposure to some air pollutant. The length of such a list of factors affecting health status is staggering and presents the epidemiologist with an almost impossible task. Therefore, it is always possible to criticize epidemiological studies by saying: The association which is claimed to exist (for example between SO_2 concentration and death rates from respiratory diseases) is really due to some variable that was not considered in the study.

Another difficult problem faced by epidemiologists is the determination of individual exposure to a pollutant. The concentration of air pollutants in different parts of a large metropolitan area can usually be obtained as a function of time of day. However, people often move around from high to low or low to high pollution areas. They may receive high exposures of certain pollutants on their jobs, or none at all. Thus, it is almost impossible to obtain

**Morbidity* is the occurence of a disease state.

accurate individual dose estimates for community air pollution studies.* All members of a particular section of a city are lumped together as far as exposure is concerned.

Even with all these difficulties, environmental epidemiological studies are still useful, if done properly, to establish associations between specific pollutants and certain responses. They are particularly useful if taken in conjunction with the other major source of information used in establishing the air quality standards, that is, toxicological data.

We now review briefly the effects of air pollutants on public health, vegetation, materials, and visibility. The effects of particulates and carbon dioxide on climate are discussed later.

The effects of carbon monoxide on human health†

At present the most dangerous results of exposures to carbon monoxide are the deaths that occur from CO production in closed areas such as automobile garages. Several hundred Americans die from CO poisoning each year; many are suicides.

Carbon monoxide enters the body through the respiratory system and reacts primarily with the hemoglobin of the circulating blood. Hemoglobin (Hb) is the iron-containing protein (found in red blood cells) that carries oxygen to the body's cells. The affinity of hemoglobin for CO is over 200 times that for oxygen, indicating that carboxyhemoglobin (COHb) is a more stable compound than oxyhemoglobin (OHb). The absorption of CO is associated with a reduction in the oxygen-carrying capacity of blood and also with a reduction of the ease with which the blood gives up its available oxygen to the tissues. The latter effect can lead to oxygen starvation even though the blood is carrying more oxygen than the body needs.

Toxicological appraisal of atmospheric carbon monoxide Long-term exposures of animals to sufficiently high CO concentrations can produce structural changes in the heart and brain. However, it has not been shown that typical urban concentrations will produce these effects. The lowest exposure that has produced any such changes has been 50 ppm continuously for six weeks. Daily averages in urban areas are more like a few parts per million.

The normal or background concentration of COHb in nonsmokers‡ is

*An exception is CO. The concentration of CO in expired breath can be measured, giving an indication of CO exposure. It would be very time consuming to do this for large numbers of people.

†The publication, "Air Quality Criteria for Carbon Monoxide," NAPCA Publication No. AP-62 was used as source material for this section. The other five Air Quality Criteria publications were used as source material for their respective sections. They are listed at the end of the chapter.

‡Smokers have higher COHb levels in their blood since they are exposed to high CO concentrations in cigarette smoke.

about 0.5% ($\frac{1}{2}$ of 1%) and is attributed to internal sources such as hemoglobin breakdown. The body's intake of atmospheric CO increases blood COHb according to the concentration and length of exposure to CO as well as the respiratory rate of the individual.

In human exposure studies, continuous exposure for about 8–12 hours to various CO concentrations led to the COHb levels listed in Table 9-1.

Table 9-1 LEVELS OF COHb IN BLOOD VERSUS CO CONCENTRATION IN AIR

CO concentration (ppm)	*COHb in blood*
10	2%
20	3.7%
30	5.0%

The effects of CO exposure on humans may be summarized as:

1 Experimental exposure of nonsmokers to 50 ppm for 90 minutes has been associated with impairment in time-interval discrimination. This exposure is likely to lead to a COHb level in the blood of about 2.5%. This same COHb level will occur with continuous exposure to 10–15 ppm CO for 8 or more hours.

2 Experimental exposure to CO concentrations sufficient to produce blood COHb levels of about 5% has provided in some instances evidence of impaired performance on certain psychomotor tests and an impairment in visual discrimination.

3 Experimental exposure to CO concentrations sufficient to produce blood COHb levels above 5% has produced evidence of physiologic stress in patients with heart disease.

Epidemiological appraisal of carbon monoxide Epidemiological studies of the effects of CO on human health have mostly concentrated on occupationally exposed groups. These studies cannot be extrapolated quantitatively to the general population because persons with severe anemia, cardiovascular insufficiency, or other debilitating conditions are unlikely to be working in places like traffic tunnels, steel mills, or parking garages. Since air quality criteria must consider all of the population, epidemiological studies must especially consider the impact of air pollutants on the most sensitive groups in the population.

Hospitalized patients with myocardial infarction* have been studied in

*Myocardial infarction refers to the changes that take place in heart muscle when an artery is suddenly blocked, for example, by a blood clot.

Los Angeles County. The evidence to date, though inconclusive, suggests that weekly average CO values in excess of from 8–14 ppm may be associated with an increase in mortality in these patients.

Automobile drivers are a large group of people who are exposed to relatively high concentrations of carbon monoxide. In two studies, persons driving motor vehicles that were involved in accidents had higher COHb levels than "control" populations. Possible mechanisms by which CO might affect the ability to drive a motor vehicle are suggested in the available data on the effects of CO upon visual sensitivity, psychological test performance, and accurate estimation of time intervals.

Aside from these studies on hospital patients and drivers, epidemiological studies have been made to determine if there might be an association between community exposure to CO and mortality or morbidity. The results of these studies are inconclusive.

Evidence from other studies of the effects of CO does not currently demonstrate an association between existing levels of CO and adverse effects on vegetation, materials, or other aspects of human welfare.

Now we are ready to look at the air quality standard for CO (Table 9-2). The maximum allowable 8-hour concentration of CO is 9 ppm, and the maximum 1-hour concentration is 35 ppm, each not to be exceeded more than

Table 9-2 NATIONAL AIR QUALITY STANDARDS

Sulfur oxides	80 $\mu g/m^3$ (0.03 ppm) annual arithmetic mean; 365 $\mu g/m^3$ (0.14 ppm) as a maximum 24-hour concentration not to be exceeded more than once a year.
Particulate matter	75 $\mu g/m^3$ annual geometric mean; 260 $\mu g/m^3$ as a maximum 24-hour concentration not to be exceeded more than once a year.
Carbon monoxide	0.01 g/m^3 (9 ppm) as a maximum 8-hour concentration not to be exceeded more than once a year; 0.04 g/m^3 (35 ppm) as a maximum 1-hour concentration not to be exceeded more than once a year.
Photochemical oxidants	160 $\mu g/m^3$ (0.08 ppm) as a maximum 1-hour concentration not to be exceeded more than once a year.
Hydrocarbons	160 $\mu g/m^3$ (0.24 ppm) as a maximum 3-hour concentration (6 to 9 A.M.) not to be exceeded more than once a year.
Nitrogen oxides	160 $\mu g/m^3$ (0.05 ppm) annual arithmetic mean.

Note: The concentration of gaseous pollutants is now usually given in terms of a mass density instead of parts per million. Therefore, we have expressed the standards in both ways.

once per year. From the presently available data it appears that a concentration of 9 ppm is low enough to prevent the occurrence of any detectable physiological effects, such as impairment of time-interval discrimination. Myocardial infarction patients should also be protected by these standards since they may be adversely affected by *weekly* average CO concentrations of 8 to 14 ppm. Thus it appears that the CO standards are quite stringent and should protect even the most sensitive groups in the population.

The respiratory system

Air pollutants enter the human body mainly via the respiratory system. Damage to the respiratory organs may follow directly, or the pollutant may be transported by some mechanism to a remote susceptible organ. Thus, when we study the effects of particulates (including sulfuric acid droplets)

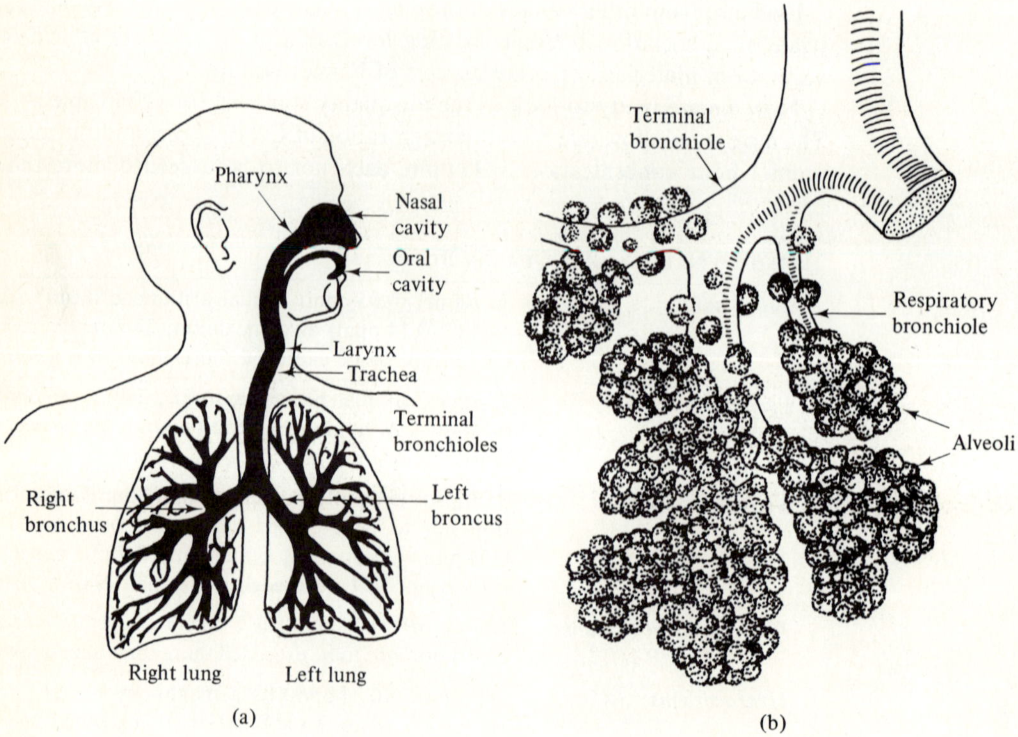

Figure 9-3 (a) The major anatomical features of the human respiratory system; (b) The terminal bronchial and alveolar structure of the human lung. (From "Air Quality Criteria for Particulate Matter," NAPCA Publication No. AP-49)

on health, it will be useful to have some understanding of the mechanisms of deposition and clearance of particles from the respiratory system.

The respiratory system (Fig. 9-3) can be broken down into a conducting division whose walls are too thick to allow exchange of gases between the air in the tube and the blood stream, and a respiratory division whose walls are thin enough to permit exchange of gases between the tubes and the blood capillaries surrounding them. The respiratory division begins with the respiratory bronchioles which are small branches off of the terminal bronchioles, and ends with the little air sacs called *alveoli*.

Deposition Since the respiratory system transports gases from the atmosphere to the lungs, a wide variety of potentially dangerous microorganisms, antigens, and particulate material may be inhaled. Some of the inhaled particulate material is deposited and retained in the various parts of the respiratory system. Two important factors determining the likelihood of particle deposition are particle size and density. Figure 9-4 shows the fraction of particles deposited in the alveolar region versus particle radius. Large particles (5×10^{-4} cm or greater) may be deposited in the conducting division and never reach the alveoli, which are in the most sensitive region of the lungs.

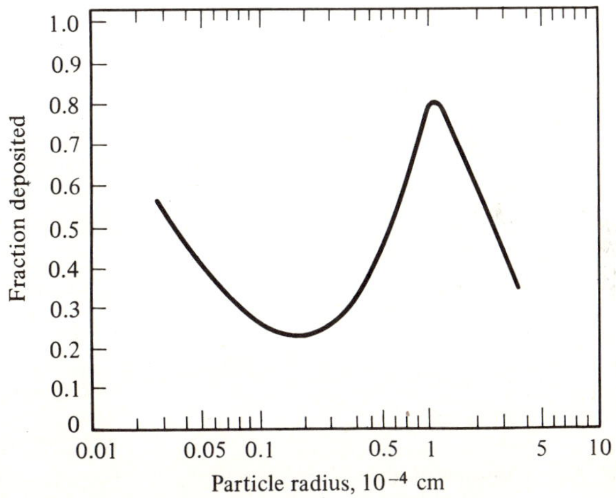

Figure 9-4 Calculated fraction of particles deposited in the alveolar region versus particle radius. The particles are assumed to be spherical of density 1 g/cm^3. (Adapted from "Air Quality Criteria for Particulate Matter," NAPCA Publication No. AP-49)

Oxygen and carbon dioxide exchange takes place in the alveoli. Small particles (3×10^{-4} cm or less) have a 30% probability or greater of being deposited in the alveoli, depending on their size. We can see now why it is important to know the particle size distribution of aerosols in the atmosphere.

Clearance mechanisms The respiratory system can protect itself from foreign matter by:

1 Filtering of large particles through hairs in the nose.

2 Reflex mechanisms such as sneezing and coughing to force foreign particles from the conducting system.

3 Trapping particles on the mucous membrane that lines all of the surfaces of the conducting system. The particles are then moved upward along the mucous layer to the throat by ciliary action (microscopic hairs which beat the mucus upwards). Swallowing pollutants that are cleared in this manner can adversely affect the digestive system. Perhaps the high incidence of stomach cancer in areas of high air pollution is associated with this mechanism.

4 Destruction of bacteria, particles of dust, or other harmful substances by special cells (phagocytic cells) in the respiratory division.

5 Secretion of an antibody called immunoglobulin in all parts of the respiratory system.

The effects of particulates

Particulate matter may exert a toxic effect because:

1 The particle may be intrinsically toxic due to its inherent chemical or physical characteristics (for example, lead and asbestos).

2 The particle may interfere with one or more of the clearance mechanisms in the respiratory system.

3 The particle may act as a carrier of an adsorbed* toxic substance.

Carbon or soot is an efficient adsorber of many organic and inorganic compounds and can carry toxic gases such as SO_2 into the lungs. This can lead to a potentiating effect in the human body in which particles that contain an adsorbed toxic substance increase a person's physiological response to that toxic substance to a level above what it would be if the substance were present without the adsorbant particle.

Epidemiological studies indicate that particulates are associated with health effects of varying severity. Respiratory illness, especially from chronic

*Adsorption is the phenomenon by which gases or liquids are attracted and retained on a surface.

Figure 9-5 Particulates in the air can reduce visibility greatly as shown in these two photographs of the same view, taken on consecutive October days. (a) The Golden Gate Bridge and San Francisco Bay at a distance of about 20 miles; (b) Trapped air pollution has diminished visibility.

diseases such as bronchitis* and emphysema† show the strongest positive association with levels of particulate matter. Adverse health effects have been observed for annual mean levels of as little as 80 $\mu g/m^3$. However, these particulate levels are usually associated with high levels of SO_2 and the effects of the two have not been separated.

Nonhealth effects Particles suspended in the air reduce visibility by scattering and absorbing light that comes from an object and its background, thereby reducing the contrast between them. Moreover, suspended particles scatter light into the line of sight, illuminating the air between; this further degrades the contrast between an object and its background (Fig. 9-5). The scattering of light by particles in the air is evident when we send a beam of light into a fog and see it spread out in all directions (Fig. 9-6). In addition to aesthetic degradation of the environment, reduced visibility has serious implications for safe operation of aircraft and motor vehicles.

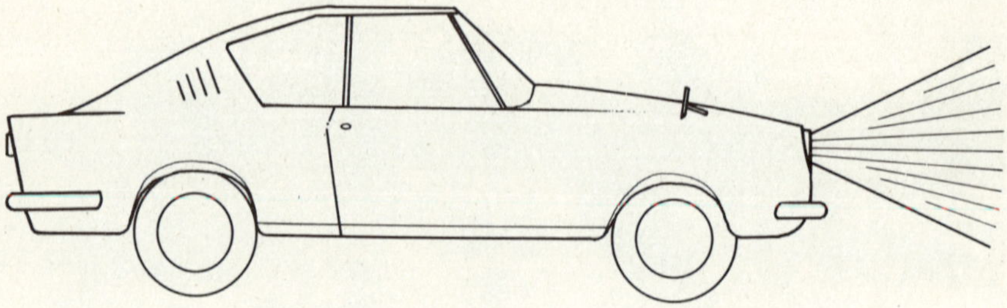

Figure 9-6 A beam of light starts off narrow from automobile headlights, but spreads out rapidly because of scattering by water droplets in a fog.

Particulate air pollution causes a wide variety of damage to materials. Particulates can soil and damage buildings, textiles, and other surfaces. They can chemically attack materials through their own corrosivity, or through the corrosivity of substances adsorbed on them. Merely by soiling materials (Fig. 9-7), and thereby causing their more frequent cleaning, particulates can accelerate deterioration. At concentrations ranging from 60 $\mu g/m^3$ to 180 $\mu g/m^3$ (annual mean), in the presence of SO_2 and moisture, corrosion of steel and zinc occur at an accelerated rate.

*Bronchitis is an inflammation of the bronchial tubes.

†Emphysema is a condition in which there is a swelling of the alveoli and destruction of lung tissue.

Figure 9-7 District Building, Washington, DC, taken Oct. 27, 1960, shows soiling effects of air pollution. (Courtesy EPA—Documerica)

The effects of sulfur oxides

From 5% to 20% of the total suspended particulate matter in urban air consists of sulfuric acid and other sulfates. They are mostly of the right size (less than 10^{-4} cm) to scatter visible light efficiently and can contribute significantly to reductions in visibility.

Atmospheres polluted with oxides of sulfur attack and damage a wide range of materials and property. Much of this damage is due to the conversion of sulfur oxides to highly reactive sulfuric acid. Corrosion rates of most metals, especially iron, steel, and zinc, are accelerated by sulfur dioxide polluted environments. Particulate matter, high humidity, and high temperatures also play an important synergistic* role in this corrosion reaction,

*Synergism is the action of two or more substances to achieve an effect greater than the sum of their effects when acting separately.

although SO_2 concentration has the greatest influence on the degree of corrosion that takes place.

Atmospheres containing oxides of sulfur also attack and damage building materials such as limestone, marble, roofing slate, and mortar as well as statuary and other works of art, causing discoloration and actual physical deterioration. It appears that SO_2 is an important factor in the deterioration of marble statues and buildings in Venice, Italy. These objects, which lasted for hundreds of years before the days of air pollution, seem to be suffering under the onslaught of modern technology. In addition to building materials, certain textile fibers, such as cotton, rayon, and nylon, are also harmed by atmospheric sulfur oxides.

Figure 9-8 Lead-based house paint damaged by sulfur gases from burning coal-refuse pile in Johnstown, PA, Sept. 20, 1962. (Courtesy EPA—Documerica)

Vegetation The absorption of SO_2 by plants can produce leaf injury. At a concentration of 0.03 ppm of SO_2 (annual mean) a gradual yellowing of leaves and excessive leaf drop may occur. Short-term exposures, such as 0.3 ppm for 8 hours, can cause injury in some species of trees (for example, pine needle tips turn brown) and shrubs. Sulfur dioxide also reacts synergistically with either ozone or nitrogen dioxide in short-term exposures (about 4 hours) to produce moderate to severe injury to sensitive plants.

Toxicological effects of sulfur oxides on man Various animal species, including man, respond to SO_2 by a constriction of their bronchial tubes, which increases the resistance to the flow of air through the respiratory system. Laboratory observations suggest that most individuals exhibit respiratory irritation at concentrations of 5 ppm and above, while concentrations of 1 ppm to 2 ppm can elicit responses from more sensitive individuals.

Sulfuric acid is a much more potent irritant to man than is SO_2, and its effects are highly dependent on particle size. Sulfuric acid droplets form when sulfur dioxide reacts with water vapor in the atmosphere. The probability of a sulfuric acid droplet reaching the lungs depends on its size. More than 80% of sulfuric acid and sulfate particles found in urban air are smaller than 10^{-4} cm in radius. As can be seen in Fig. 9-4, many of these particles will be deposited in the alveolar region of the respiratory system.

The potentiation by particulate matter of toxic responses to sulfur dioxide has been observed under conditions which would promote the conversion of SO_2 to sulfuric acid. The degree of potentiation is related to the concentration and size of the particulate matter.

Epidemiological studies on SO_2 Epidemiological studies of acute air pollution episodes in Europe and the United States have shown a clear association between simultaneous high sulfur dioxide and smoke concentrations and increased mortality and morbidity. It is extremely difficult to separate the effects of SO_2 and particulates since they mostly occur together in urban air. British studies suggest that a small rise in the daily death rate is detectable in large populations if the concentration of SO_2 rises abruptly to about 0.25 ppm in the presence of smoke at 750 $\mu g/m^3$. Most of the excess deaths were among the aged population and patients with chronic lung or heart disease. Studies in Rotterdam, Holland, indicate that there is an association between SO_2 levels as low as 0.19 ppm (24-hour average) for a few days, and mortality and morbidity. These studies are especially significant since the particulate levels in Rotterdam are generally low. Another British study indicates that an annual SO_2 concentration of about 0.04 ppm, accompanied by a smoke concentration of about 160 $\mu g/m$, may cause an increase in mortality from bronchitis and lung cancer.

The air quality standards for sulfur oxides and particulates in Table 9-2 seem adequate to protect the most sensitive people in a population—the aged and those who already have respiratory diseases such as bronchitis or emphysema.

The effects of photochemical oxidants

Nonhealth effects Photochemical oxidants result from a complex series of atmospheric reactions initiated by sunlight. Some of the compounds found in photochemical smog are ozone, peroxyacyl nitrates (PAN), acroleins, and formaldehyde.

Adverse effects on sensitive vegetation have been observed from exposure to photochemical oxidant concentrations of about 0.05 ppm for 4 hours. A common injury is called silverleaf, a shiny, oily effect on a leaf's lower surface that develops into a glaze, then turns silver. Adverse effects on materials from exposure to photochemical oxidants have not been precisely quantified, but have been observed at the levels presently occurring in many urban areas. In particular, it is known that ozone attacks and cracks rubber.

Health effects In the general community, eye irritation is the major effect of photochemical air pollution. In southern California it has affected 75% of the population on some days. Eye irritation is likely to occur in a large fraction of the population when instantaneous oxidant concentrations reach 0.10 ppm.

Ozone, which is one of the major photochemical oxidants, is not an eye irritant by itself. However, oxidant concentrations correlate well with eye irritation, indicating that oxidant levels are probably a measure of the photochemical activity which produces the eye irritants. Eye irritation has been reported in places other than southern California, but the irritants have not been identified.

Epidemiological studies have shown no association between mortality and morbidity rates and urban oxidant levels. However, maximum daily oxidant values of 0.13 ppm have been associated with *aggravation* of asthma.* High school athletes were studied in southern California to see if there might be any impairment of performance during severe smog. It was found that when photochemical oxidants ranged from 0.03 to 0.3 ppm for 1 hour immediately prior to the start of activities, there was impairment of performance. More automobile accidents have occurred on days of high oxidant concentration, but no threshold level could be determined from the analysis.

It appears that the standard for photochemical oxidants, 0.08 ppm as a maximum 1-hour concentration, is the only one of the six standards in Table 9-2 that is questionable in terms of its protection of public health. Health effects have been associated with peak oxidant concentrations of 0.10 ppm and 0.13 ppm. During an hour when the average oxidant concentration is 0.08 ppm, it will be above that average part of the time and below it part of the time. Experience indicates (Fig. 9-9) that there will be days when the maximum 1-hour concentration of oxidants is 0.08 ppm or slightly less, and the peak concentration will be greater than 0.10 ppm. A small percentage of the population will experience eye irritation on these days.

When the standard for oxidants was first proposed it was to be 0.06 ppm.

*Asthma is a respiratory disease involving constriction of the bronchial tube muscles, resulting in labored breathing.

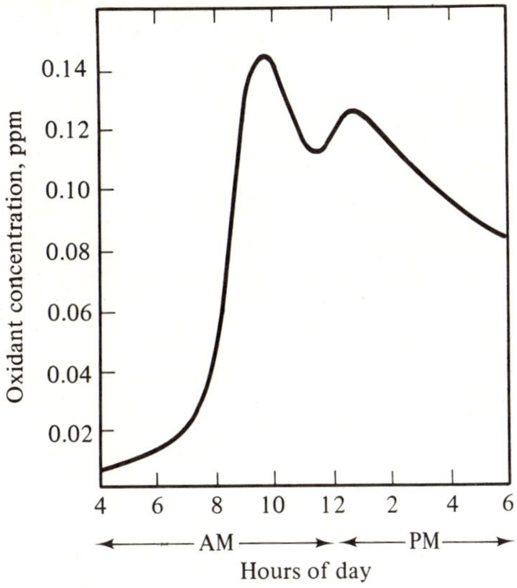

Figure 9-9 Hourly variation of total oxidant in Philadelphia, July 18, 1967. The oxidant concentration typically rises very steeply in the morning, and the peak concentration during any hour in this part of the day can be quite a bit higher than the average concentration for that same hour. (Adapted from "Air Quality Criteria for Photochemical Oxidants," NAPCA/Publication No. AP-63)

Evidence was produced that such a limit is reached naturally in some areas, and questions were raised about some of the above studies used as a basis for the standard. As a result, the standard was raised to 0.08 ppm.

The effects of hydrocarbons

The air quality standard for hydrocarbons is based on the important role they play in the formation of photochemical smog. No direct health effects have been noted from the concentration of gaseous hydrocarbons found in urban air. Injury to sensitive plants, however, has been reported in association with ethylene concentrations of 0.001 to 0.5 ppm over a time period of 8 to 24 hours.

Studies were made in Washington, DC, Denver, and Philadelphia to determine the relationship between early morning hydrocarbon concentrations and the subsequent maximum daily oxidant concentration. It was found that an early morning (6–9 A.M.) concentration of 0.3 ppm of nonmethane hydrocarbon can be expected to produce a maximum hourly average oxidant concentration of up to 0.1 ppm. Methane is excluded since it is not involved in photochemical smog formation. The air quality standard in Table 9-2 reflects the result of these studies.

Nitrogen oxides

Nonhealth effects Damage to textiles and metals from exposure to nitrogen oxides has been observed in urban atmospheres, but the precise concentrations producing these effects have not been determined. Leaf injury has been observed in crops and ornamental plants at concentrations as low as 0.25 ppm of NO_2 for 8 months. A decreased yield of navel oranges has resulted at this exposure.

Health effects Nitrogen oxides, like hydrocarbons, play an important role in photochemical smog formation. A 6–9 A.M. concentration of NO_x of 0.04 to 0.16 ppm, plus a nonmethane hydrocarbon concentration of 0.3 ppm could lead to a 0.1-ppm maximum daily 1-hour average oxidant level. This level of oxidant causes eye irritation.

Nitrogen dioxide, the only oxide of nitrogen examined in epidemiological studies, can be significantly correlated with increased respiratory disease at mean concentrations between 0.06 and 0.11 ppm. In the late 1960s, four residential areas in Chattanooga, Tennessee, were studied to determine if there was any association between NO_2 concentration and health effects. One residential area had high NO_2 levels and low particulates, one had low NO_2 and high particulate levels, and the other two areas had both low particulates and low NO_2. Among families living in the high NO_2 area there was 18% more respiratory illness than in the two control areas during a 24-week study. This increased incidence of acute respiratory disease was observed when the mean 24-hour NO_2 concentration measured over a 6-month period was between 0.06 and 0.11 ppm and the mean suspended-nitrate level was 3.8 $\mu g/m^3$ or greater.

In a later study of the same Chattanooga areas, an increased frequency of acute bronchitis was observed when the mean 24-hour NO_2 concentration, measured over a 6-month period, was between 0.06 and 0.08 ppm and the mean suspended nitrate level was 2.6 $\mu g/m^3$ or greater.

The results of the Chattanooga studies can be extended to other cities and, since NO_2 does not exhibit marked seasonal variations, it is possible to compare yearly averages with the lower limit at which health effects were observed in Chattanooga. Ten percent of the cities in the United States with populations less than 50,000 have a yearly average NO_2 concentration equal to or exceeding 0.06 ppm. In the range from 50,000 to 500,000, the yearly average NO_2 concentration equals or exceeds 0.06 ppm in 54% of the cities. The situation is even worse in the over 500,000 population class; 85% of these cities equal or exceed a yearly average NO_2 concentration of 0.06 ppm.

The air quality standard for nitrogen oxides (0.05 ppm yearly average) seems adequate to protect the public health. However, as the above figures indicate, there are many cities where the standard is surpassed, and most of them have large populations. Automobiles account for about half of the nitrogen oxides emitted into the atmosphere each year. The new controls on automobile emissions should lower nitrogen oxide levels in urban areas. We discuss automobile emission and controls more fully in the next chapter.

In order to see how one metropolitan area was approaching some of the federal air quality standards as far back as 1972, let us look at the San Francisco Bay area (population greater than 4.5 million). The area embraces a large, shallow basin ringed by hills (Fig. 9-10). This topography gives it great potential for trapping and accumulating air pollutants. Within this basin, air contaminants are emitted at a fairly constant rate throughout the year. However, because of weather changes (winds and inversions) the amount of air pollution actually present in the air fluctuates widely from day to day.

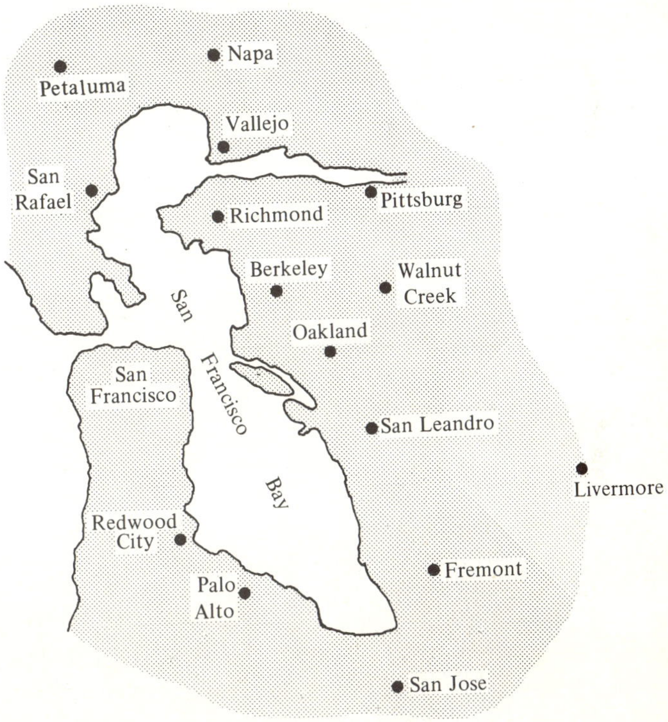

Figure 9-10 The San Francisco Bay Area. There is an air-pollution measuring station at each city on this map, except in Berkeley.

Table 9-3 gives some idea of how closely this region of California was approaching the air quality standards during 1972. As can be seen, the daily CO levels were already low enough to meet the 8-hour standard on almost every day in 1972. The 1-hour standard was never exceeded in 1972. The air quality standards for nitrogen oxides and sulfur oxides were also met almost 100% of the time. The amount of oxidants and particulates was variable with location, but generally needed improvement. The California air quality standard for particulates is stricter than the federal standard, so it will not be as difficult as it appears to satisfy the federal standard for particulates. The

Table 9-3 AIR POLLUTION IN THE BAY AREA—1972 BY STATION

For oxidant and nitrogen dioxide, "max" is the highest hourly average value expressed in parts per million. For carbon monoxide "max" is highest 12-hour average value in parts per million. (The one-hour standard for CO was never exceeded during the year.) For sulfur dioxide, "max" is highest 24-hour average value expressed in parts per million. For suspended particulates, "mean" is the annual geometric mean in micrograms per cubic meter.

Stations	*Oxidant max**		*Carbon monoxide max**		*Nitrogen dioxide max**		*Sulfur dioxide max†*		*Suspended particulates mean‡*	
San Francisco	.08	0	11.7	1	.33	2	.070	0.8	46	1.6
San Rafael	.17	5	7.7	0	.14	0	.016	0	55	3.8
Richmond	.12	7	9.1	0	.16	0	.028	0	41	5.5
Pittsburg	.19	25	5.1	0	.07	0	.042	0.4	43	2.3
Walnut Creek	.17	30	—	—	—	—	—	—	—	—
Oakland	.12	1	7.2	0	.20	0	—	—	—	—
San Leandro	.17	15	—	—	—	—	—	—	—	—
Fremont	.34	44	6.5	0	—	—	.005	0	56	7.6
Livermore	.22	27	6.5	0	.16	0	.005	0	67	16.8
San Jose	.20	19	13.8	11	.23	0	.006	0	66	17.4
Los Gatos	.21	15	—	—	—	—	—	—	—	—
Mountain View	.19	10	—	—	—	—	—	—	—	—
Redwood City	.28	17	9.2	0	.14	0	.002	0	44	4.5
Burlingame	.14	8	9.9	0	—	—	.020	0	—	8.1
Petaluma	.07	0	—	—	—	—	—	—	—	—
Napa	.18	20	7.4	0	.08	30	.007	0	—	13.9
Vallejo	.26	15	12.1	5	.12	0	.009	0	—	1.2
Fairfield	.13	4	—	—	—	—	—	—	—	—

*Number of days ambient air quality standard was exceeded.

†Percent of observed days when ambient air quality was exceeded.

‡Percent of observed days when ambient air quality standard (100 $\mu g/m^3$ for 24 hours) was exceeded.

(From "Air Pollution and the San Francisco Bay Area" by the Bay Area Pollution Control District, Dec. 1973)

standard for photochemical oxidants will probably be the most difficult to meet in California because of the great dependence on the automobile.

The effects of carbon dioxide and particulates on climate

Man may be changing his climate on a global scale. It is possible that man-made alterations in the physical and chemical composition of the atmosphere might bring about an ice age or a melting of the ice caps. At this time, we do not know the probability of occurrence of either of these events.

Geological and historical records clearly show that major climatic fluctuations have taken place throughout history. The last ice age ended about 10,000 to 15,000 years ago. From A.D. 800 to A.D. 1000 there was a warming trend, and from 1650 to 1840 there was a cooling trend. One of the variables used to describe the world climate is the average annual temperature at the earth's surface. Beginning in 1890, this average temperature started an irregular climb (Fig. 9-11). By 1940 it was 1.1°F higher than it had been during the decade from 1880 to 1890. However, in the 30 years from 1940 to 1970 the average temperature fell about 0.5°F. The recent lowering of the average temperature has been accompanied by the shifting of frost and ice boundaries to the south and a marked increase in rainfall in parts of previously arid continental regions. In recent winters ice coverage in the North Atlantic has

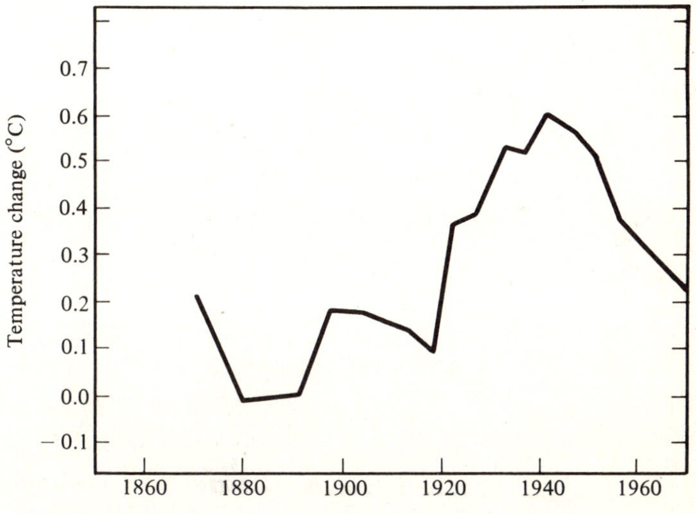

Figure 9-11 Trend of mean annual temperature between 0° and 80°N latitude. Temperature change is relative to time period from 1880–1890. (From *Global Effects of Environmental Pollution*, edited by S. Fred Singer, Springer-Verlag, Inc., New York, 1970)

been extensive enough to cause great loss to Icelandic fisherman. In contrast, the rains in India and East Africa contributed to high wheat yields. These experiences illustrate the significant impact of small shifts in climate on human activities.

Carbon dioxide There is reason to believe that the increasing concentration of pollutants in the atmosphere may be affecting both local and global climates. It is thought that increasing levels of carbon dioxide tend to increase average temperatures while increasing particulate levels tend to decrease average temperature.

There is a seasonal cycle for carbon dioxide concentration in the atmosphere. Its concentration drops in the spring and summer, as plants and trees increase their intake of carbon dioxide, and rises in the fall and winter when vegetable matter decays and carbon dioxide is liberated. Superimposed on

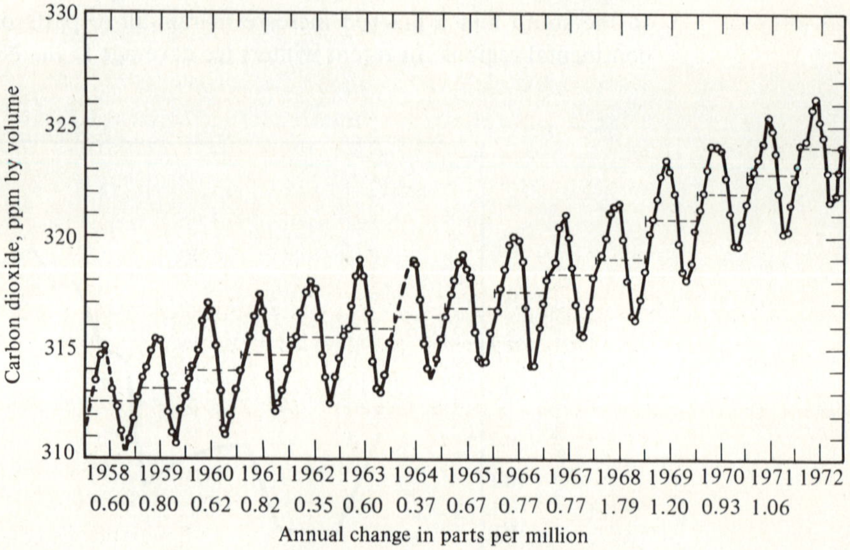

Figure 9-12 Increase in carbon dioxide concentrations at Mauna Loa Observatory. Change based on less than 12 monthly measurements per year. (From Department of Commerce, National Oceanic and Atmospheric Administration, based on Figure 5 of "Is Carbon Dioxide from Fossil Fuels Changing Man's Environment?" Charles D. Keeling *Proceedings of the American Philosophical Society,* Vol. 114, No. 1 (1970))

this cycle is a long-term increase. Carbon dioxide concentration in the atmosphere has increased from 290 ppm to 325 ppm within the last 100 years. Figure 9-12 shows an increase in carbon dioxide concentration (at Mauna Loa Observatory) which has taken place during the time period from 1958 to 1972.

Carbon dioxide concentrations at monitoring stations over the globe have grown faster since about 1968 than in the previous decade. The rate of CO_2 increase is about 0.7 ppm per year. Man is adding enough CO_2 to the atmosphere each year to increase the CO_2 concentration by more than 1 ppm annually. The excess CO_2 is probably mainly dissolving in the oceans. So as a result of man's activities the concentration of carbon dioxide is increasing in both the oceans and the atmosphere.

The increase of carbon dioxide in the atmosphere affects the solar radiation reaching the earth only slightly since carbon dioxide is essentially transparent to incoming solar radiation. Solar radiation is mostly in the ultraviolet and visible parts of the electromagnetic spectrum, and carbon dioxide is a strong absorber of infrared radiation. However, the radiation emanating from the earth (blackbody radiation) is all in the infrared region of the electromagnetic spectrum and the carbon dioxide in the atmosphere absorbs some of the infrared radiation coming from the earth. The captured radiation is reradiated in all directions; some will reach the ground (Fig. 9-13). Therefore, an increasing carbon dioxide concentration tends to increase the temperature of the lower troposphere by trapping some radiation that would otherwise escape into outer space. This process is called the greenhouse effect since glass in a greenhouse has the same properties as carbon dioxide; that is, it absorbs radiation in the infrared part of the spectrum and is essentially transparent to visible radiation. Actually, the analogy is inexact since a greenhouse achieves its result less from the fact that the glass absorbs infrared radiation than from the fact that it cuts down the convective transfer of heat. This same trapping of heat energy takes place in a car which is in the sunlight with the windows closed.

Many people have carried out theoretical calculations to determine the effect of increasing the CO_2 concentration in the atmosphere. Unfortunately, the results show a tremendous variation. Depending on whose calculation you trust, you can find that a doubling of the CO_2 concentration would result in a temperature increase of 0.8 to 2.9°C in surface temperature. One calculation* predicts a much smaller effect. In this case if you increase the CO_2 concentration by as much as a factor of ten, the temperature would increase only by about 2.5°C. At the present rate of increase, CO_2 concentration will

*Rasool, S. I., and S. H. Schneider, "Atmospheric dust increase could lower Earth's temperature," *Physics Today* (October 1971), 19–20.

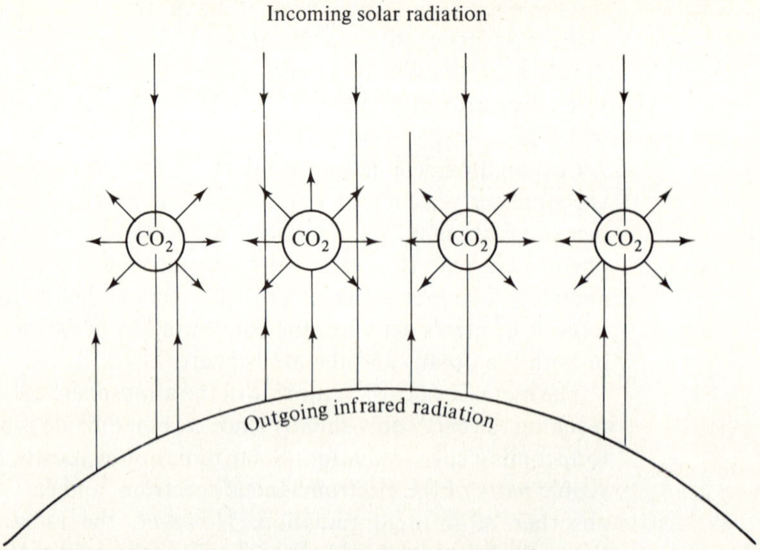

Figure 9-13 Carbon dioxide is transparant to incoming solar radiation, but opaque to much of the outgoing blackbody radiation emitted by the earth. The infrared radiation that is captured by molecules of CO_2 in the atmosphere is reradiated in all directions. Some of this radiation reaches the earth, thereby causing it to be warmer than if the CO_2 were not present.

double before the year 2000, so if the former calculations are correct, the average global temperatures may rise by 1 to 3 degrees Celsius by 2000 if CO_2 were the only factor affecting average global temperatures.

Particulates Since 1940 there has been a rather steep decline in average global temperature (see Fig. 9-11). This cannot be explained by the CO_2 increase in the atmosphere. It is probable that the increasing load of particulates in the atmosphere is responsible for this decrease in temperature. Particles are released into the air by volcanic eruptions, combustion of fossil fuels, industrial activities such as stone crushing, and by soil movement. Part of the increase in particulates is thought to be due to man's activities and part to volcanic eruptions. An increase in particle concentration in the atmosphere reduces the amount of solar radiation reaching the earth's surface by a corresponding increase in the absorption and back-scattering of incident radiation. This would cause a lowering of terrestrial temperatures.

There are no direct measurements to support this hypothesis, but there is evidence that volcanic eruptions in the past spewed enough fine ash into the atmosphere to significantly lower temperatures on a worldwide basis. The 1815 eruption of Tambora in the Dutch East Indies was the largest known volcano eruption. A rough idea of the magnitude of the temperature drop after this eruption can be obtained from old agricultural and weather records. The two years after the 1815 eruption were the coldest in English history. Furthermore, 1816 was one of the four coldest years in the period 1698 to 1957, and July of 1816 was the coldest July in that same 260-year time span. Crop failures occurred throughout Europe and America, and in England the price of a sack of flour doubled in two years.

Dust from volcanic eruptions can cool the earth for as long as a few years since some of the dust is sent into the stratosphere, from where it slowly descends back to earth. It is generally believed that volcanic dust in the stratosphere promotes cooling at the earth's surface. Some scientists believe that most of the temperature drop since 1940 can be explained by natural volcanic activity. Others feel that air pollution has caused a significant portion of this temperature decrease. It is more difficult to calculate the effect of particulates than of CO_2 on global temperatures since particulates can also act to increase temperature. Dust which settles on snow and ice reduces the reflection and increases the absorption of solar radiation, thereby aiding melting, and the presence of particulates in the troposphere may lead to increased cloud formation which tends to enhance the greenhouse effect.

In spite of these difficulties, theoretical calculations have been made to estimate the effect of increased particulates in the atmosphere on climate. Excluding the effect of clouds, it has been determined* that if the dust concentration increased by a factor of four (and assuming a particle size of 0.5×10^{-4} cm), the surface temperature would decrease by 3.5°C. Some scientists feel that this temperature decrease could trigger another ice age. The doubling time for particulates in the atmosphere has been estimated to be between 15 and 20 years, so a factor of four increase in particulate loading may take only 30 to 40 years.

It is clear that carbon dioxide and particulates have opposing effects on climate, but it is not known which pollutant has the stronger effect. Since worldwide temperatures have been decreasing, there is real concern that man may be precipitating an ice age. Geologic studies indicate that, through natural processes (perhaps volcano eruptions), ice ages have occurred many times in the past. Even though particulate air pollution may not be the main cause of the temperature decrease in the last 30 years, it certainly is con-

*See footnote on page 243.

tributing to this decrease and may tip the balance in the direction of a continuing temperature decrease.

Local climatic effects The residence time of particles in the air depends on their size, density, and region of the atmosphere in which they are located. Particles in the air may act as *condensation centers*; that is, particles upon which water vapor condenses into liquid water or into solid ice particles. The cloud cover and frequency of fog have been observed to be greater in cities than in the surrounding countryside. This is probably a result of the greater particulate concentrations over cities. The large airborne particles generated in metropolitan areas serve as a base for the condensation of moisture and lead to the rapid formation of rain droplets or ice crystals. Rainfall patterns over cities and the surrounding suburbs show a correlation with particle concentration in the atmosphere.

Atmospheric particles scatter and absorb light from the sun, thereby reducing the solar radiation available to cities. The visible light reaching the ground in some cities may be reduced by as much as 33% of its value in surrounding rural areas. In general, cities receive 15 to 20% less total (direct plus scattered) solar radiation than do rural environments. Most of the energy reduction occurs in the ultraviolet region of the solar spectrum.

The effect of SSTs on the atmosphere

Supersonic transports (SSTs) will cruise in the lower part of the stratosphere, at an altitude between about 60,000 and 65,000 feet. When jet fuel is burned the usual fossil fuel combustion products are released into the atmosphere. Scientists are particularly concerned about the possible effects of water vapor and nitric oxide (NO) in the stratosphere. Water vapor, like CO_2, absorbs infrared radiation, so it can enhance the warming of the earth through the greenhouse effect. There is very little water vapor in the stratosphere, so a fleet of SSTs could alter the natural composition of the stratosphere and increase temperatures on earth.

The addition of nitric oxide to the atmosphere could initiate a series of ozone-depleting chemical reactions. Ozone is a very efficient absorber of ultraviolet radiation and protects life on earth from the ionizing effects of short wavelength ultraviolet radiation. The following reaction takes place when ozone absorbs ultraviolet radiation:

$$O_3 \underset{}{\overset{\text{u.v. rad.}}{\rightleftarrows}} O_2 + O$$

This process and its reverse take place continually in the atmosphere. If NO were introduced into the stratosphere by combustion of jet fuel, then the following reaction could take place:

$$NO + O_3 \longrightarrow NO_2 + O_2$$

NO can be regenerated by a different reaction:

$$NO_2 + O \longrightarrow NO + O_2$$

As a result of these two reactions, the concentration of ozone (O_3) is reduced and the concentration of NO remains the same (NO acts as a catalyst). Therefore, a small amount of NO could deplete a large amount of ozone. There is some experimental evidence that an increase in NO concentration in the stratosphere causes a decrease in ozone concentration. It has been observed that cosmic rays produce nitrogen oxides in the stratosphere, and that when cosmic ray intensity increases, ozone concentration decreases.

Dr. Harold Johnston, a leading authority on atmospheric chemistry at the University of California, Berkeley, campus, has calculated that 500 American SSTs operating for about 5 years would decrease the concentration of ozone in the stratosphere by somewhere between 15 and 30%. It would take an ozone reduction of only several percent to bring about adverse effects on plants, animals, and people. The increase in ultraviolet radiation reaching the earth would increase the number of cases of skin cancer and would also cause a temporary blindness of those people staying outdoors without eye protection. The possible adverse effects of the sonic boom and the reduction of the ozone layer were the main considerations in the U.S. government's decision not to build the SST.

Suggested reading

1 GOLDSMITH, JOHN R., and SEYMOUR I. COHEN, "Epidemiological Bases for Possible Air Quality Criteria for CO," *Journal of the Air Pollution Control Association*, **19**: 9 (Sept. 1969), 704–713.

2 HEXTER, ALFRED C., and JOHN R. GOLDSMITH, "Carbon Monoxide: Association of Community Air Pollution with Mortality," *Science*, **172** (April 16, 1971), 265–267.

3 LOOMIS, T.A., *Essentials of Toxicology*. Philadelphia: Lea & Febiger, 1968.

4 "Air Quality Criteria For CO," NAPCA Publication No. AP-62.

5 "Air Quality Criteria For Hydrocarbons," NAPCA Publication No. AP-64.

6 "Air Quality Criteria For Photochemical Oxidants," NAPCA Publication No. AP-63.

7 "Air Quality Criteria For Particulate Matter," NAPCA Publication No. AP-49.

8 "Air Quality Criteria For Sulfur Oxides," NAPCA Publication No. AP-50.

9 "Air Quality Criteria For Nitrogen Oxides," Environmental Protection Agency, Air Pollution Control Office, Publication No. AP-84.

10 "E.P.A. Sets National Air Quality Standards," *Journal of the Air Pollution Control Association*, **21** (June 1971), 352–353.

11 "Air Pollution Injury to Vegetation," NAPCA Publications No. AP-71.

12 KEELING, C.D., "Is CO_2 from Fossil Fuels Changing Man's Environment?" *Proc. Amer. Phil. Society*, **114** (1970), 10–17.

13 LANDSBERG, H.E., "Man Made Climatic Changes," *Science*, **170** (1970), 1265–1274.

14 NUESSLE, V.D., and R. W. HOLCOMB, "Will the SST Pollute the Stratosphere?" *Science*, **168** (1970), 1562.

15 SINGER, S.F., ed., *Global Effects of Environmental Pollution*. New York: Springer-Verlag New York Inc., 1970.

Questions

1 Suppose that you are going to do an epidemiological study of the effect of nitrogen dioxide (NO_2) on the rate of respiratory disease occurrence in your home city. Make a list of as many factors as you can think of, other than NO_2 concentration, that might affect the morbidity rate. These factors will have to be as similar as possible in the two or more population groups that you study.

2 What percentage of inhaled sulfuric acid droplets of 10^{-4}-cm radius will reach the alveoli in the lungs? (See Fig. 9-4.)

3 (*a*) Eye irritation is a common affliction of photochemical smog. According to the data in Fig. 9-9, for about how many hours was eye irritation experienced in Philadelphia on July 18, 1967?

(*b*) During what 1-hour time period that same day might the average oxidant concentration have been below 0.08 ppm with a peak concentration of 0.10 ppm during that same hour?

4 In the 30-year period from 1940 to 1970 the average annual temperature at the earth's surface fell by about 0.5°F. How would you attempt to explain this drop?

5 (*a*) In the time period from 1958–1967, what was the percent increase in carbon dioxide concentration in the atmosphere?

(*b*) Carbon dioxide absorbs infrared radiation. Both the incident solar radiation and the blackbody radiation coming from the earth contain infrared radiation. Why, then, does adding CO_2 to the atmosphere tend to increase the mean global temperature?

6 Does ozone concentration in the stratosphere influence the greenhouse effect? Explain why or why not.

Chapter 10
Transportation

Introduction

Transportation of all types accounts for about two-thirds of the carbon monoxide and about half of the hydrocarbons and nitrogen oxides emitted into the atmosphere each year (Table 8-1, Chapter 8). The effects of these pollutants on vegetation, materials, visibility, and human health has already been discussed (Chapter 9). We know that the hydrocarbons and nitrogen oxides emitted from automobiles play a major role in the formation of photochemical smog with its eye irritants. Although transportation, as it is considered in Table 8-1, includes motor vehicles, aircraft, railroads, etc., it is the automobile which accounts for almost all of the air pollution listed under the category transportation.

The number of registered motor vehicles (cars and buses) in the United States has grown from only 10,000 in 1900 to well over 100 million. There is now about one car for every two people in the population, including children. Figure 10-1 shows the growth in U.S. population and the growth in motor vehicle registrations. Percentagewise, the number of motor vehicles is grow-

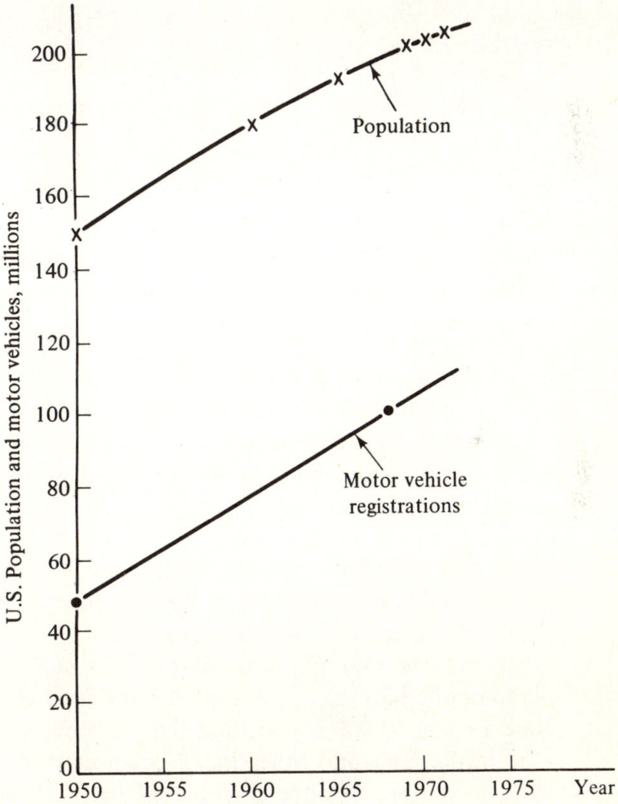

Figure 10-1 The number of motor vehicle registrations has doubled in the last 20 years, whereas the U. S. population has increased only by about 30%.

Figure 10-2 A professional football game caused this traffic jam on a freeway near the Berkeley-Oakland border in California.

ing faster than the number of people, and by the year 2000 there may even be more automobiles than people to drive them.

Aside from their contribution to the total air pollution we are subject to, automobiles affect our lives in many other ways. Everywhere we look we see the ubiquitous auto and its support structure. The U.S., and many other nations, are now crisscrossed by so many highways that it is extremely difficult to escape from motor vehicle noise. Until very recently, every cent of Federal taxes on gasoline went into the construction of even more highways. A small percentage of this money now goes towards the construction of mass transit. Our physical environment suffers from the proliferation of automobile junkyards, gas stations, and used car lots. Traffic fatalities average around 55,000 a year, and there are even more injuries. Even with all the traffic jams and other inconveniences (for example, parking difficulties and increasing gasoline prices) that the auto commuter is subject to, his love affair with the auto does not appear to have lessened.

In the last few years it has become evident that the U.S. will not be able to produce and refine enough crude oil to supply the increasing number of automobiles with the amount of gasoline that they now require. We will need to import more and more crude oil (both for transportation and heating) and thus become even more dependent on foreign nations (in the Middle East especially) for our petroleum sources. This importation of huge quantities of oil will be bad for our balance of payments problem and there will also be the added hazard to marine life from possible oil spills by ocean-going tankers.

So we see there are two different types of problems associated with the large number of automobiles presently on the road. There is the air pollution, which is being diminished by various control devices and engine modifications, and then there are all the problems mentioned above which will not go away even after air pollutant emissions are controlled. An increase in the number of cars can even negate the effects of pollution control devices on total pollution reduction. In order to solve most of the other problems that we have mentioned it would be necessary to induce more people to use mass transit instead of their autos in urban areas. Mass transit systems are very expensive to build, and federal or state assistance will probably be needed to encourage their construction.

In addition to providing for the promulgation of air quality standards by the Environmental Protection Agency, the Clean Air Act of 1970 also called for the cleaning up of the internal combustion engine (ICE). The Act calls for a 90% reduction of hydrocarbon and carbon monoxide levels, as measured on 1970 cars, for all 1975 model year cars. It also provides for a 90% reduction in the emissions of nitrogen oxides, as measured in 1971 model year cars, for all 1976 model year cars. However, auto manufacturers have been granted a one-year delay on these emission standards.

Automobile manufacturers have been working on several types of emission control systems for the internal combustion engine and also on several different types of power sources (other than the ICE). According to the EPA, the engine systems which have the greatest potential for becoming viable alternatives to the conventional internal combustion engine are the gas turbine, the heat-electric hybrid, and the steam engine. It may surprise you to see two old engine types (electric and steam) making a comeback attempt at this late date, but their low pollutant emissions now look very attractive.

We begin this chapter with a description of the basic operation of an internal combustion engine, the methods of pollution control, and then some alternative engines. Then, we consider some operational and proposed mass transit systems.

The internal combustion engine

An internal combustion engine is able to convert the potential energy of a mixture of air and gasoline vapor into the kinetic energy of a piston. The basic energy conversion process begins in one of the six or eight cylinders found in most automobile engines. The operating cycles of most gasoline engines are basically similar. A piston moves up and down in a cylinder, transmitting its straight-line motion through a connecting rod to the crankshaft. The crankshaft is thus made to rotate, the rotary motion is carried through the power train* to the car wheels so that they rotate, and the car moves. The four strokes of the conventional piston engine are shown in Fig. 10-3.

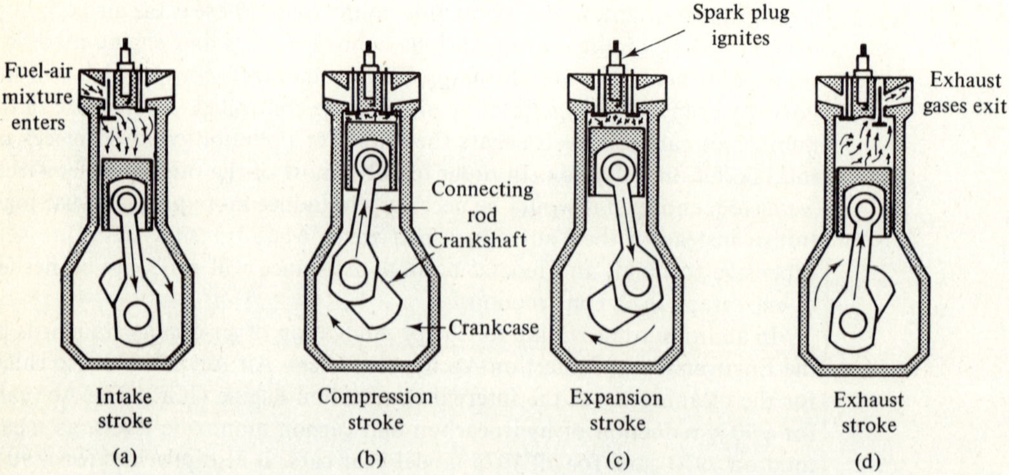

Figure 10-3 Schematic diagram of four-stroke-cycle engine. (Adapted from "Control Techniques for Carbon Monoxide, Nitrogen Oxide, and Hydrocarbon Emissions from Mobile Sources," NAPCA Publication No. AP-66)

Intake On the intake stroke, the intake valve has opened and the piston is moving down. As the piston moves down, the volume of air in the cylinder increases and the air pressure decreases, allowing outside air at higher pressure to move into the cylinder. This outside air first passes through the carburetor where it picks up some gasoline vapor and then through the intake valve.

Compression On the compression stroke, the piston moves upward and both the intake and exhaust valves close, so the cylinder is now sealed. As the piston moves upward, the air-fuel mixture is compressed and the pressure rises inside the cylinder. On a molecular basis, we can say that the molecules

*The power train contains several mechanisms, such as the transmission and propeller shaft, that carry the engine power to the rear wheels.

of gasoline vapor and air are pushed closer together and bump into the cylinder walls and piston more often. The increased frequency of bumps means that a stronger push is registered on the walls and piston, or that the pressure is higher. As the piston reaches the top of its compression stroke, an electric spark is produced at the spark plug, starting combustion of the air-fuel mixture.

Expansion This is the one stroke of the four that delivers power. The burning mixture expands, forcing the piston down. The pressure and temperature increase inside the cylinder to as much as 600 pounds per square inch and 4,500°F, respectively.

Exhaust As the piston moves back up again on the exhaust stroke, the exhaust valve opens and the rising piston forces the products of combustion

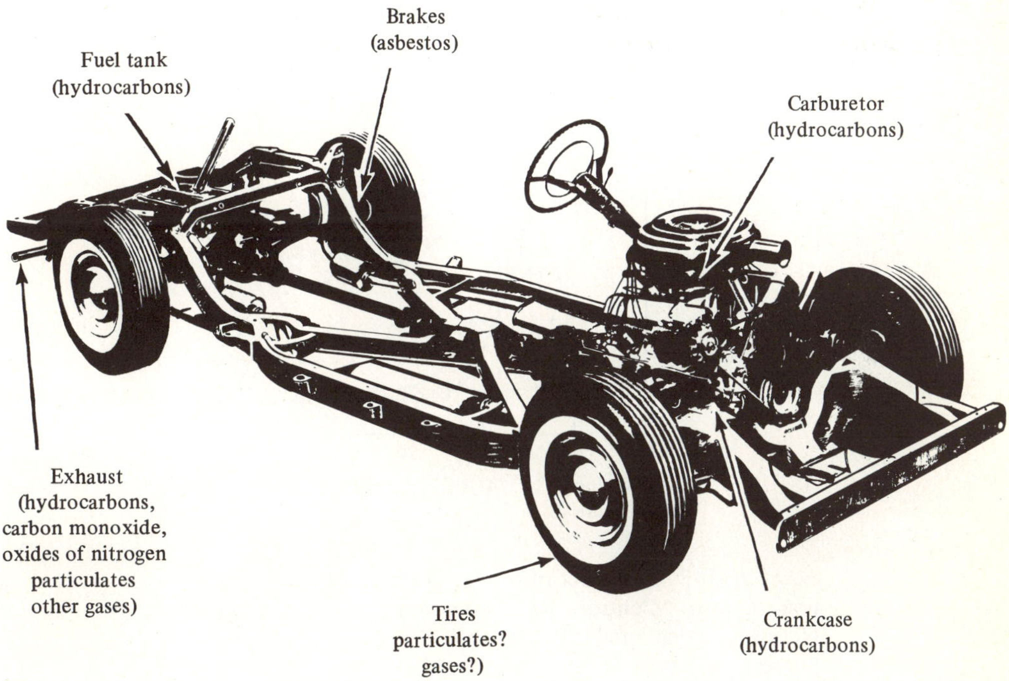

Figure 10-4 Sources of automotive air pollution

out of the cylinder. As the piston reaches its highest position the exhaust valve closes and the intake valve opens. The cycle starts all over again as the piston moves down and a fresh mixture of air and fuel is brought into the cylinder.

The exhaust system of the engine is made up of a manifold* connected to each cylinder exhaust port, and piping to carry the gases to the rear of the vehicle, where they are discharged to the atmosphere. At some point between the manifold and the discharge, a muffler is used to quiet the exhaust noises.

In Chapter 8 we discussed the origin of the various products of combustion. Complete combustion at a high temperature results in only carbon

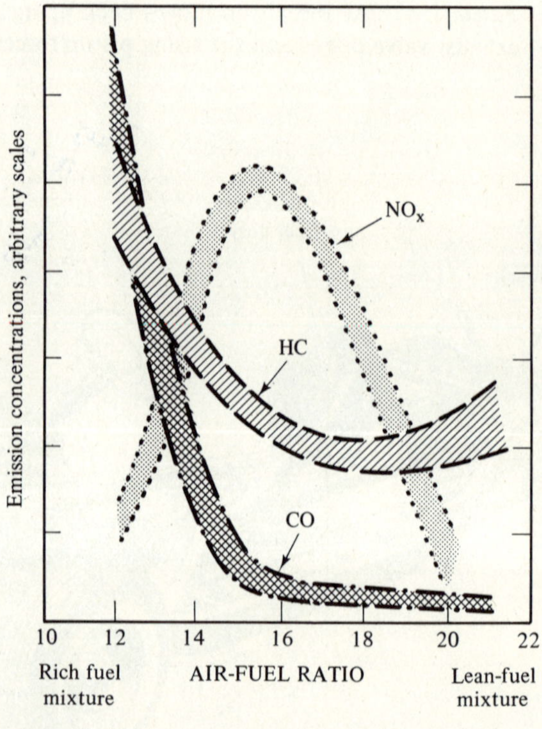

Figure 10-5 The effect of air-fuel ratio on exhaust composition. (From "Control Techniques for Carbon Monoxide, Nitrogen Oxide, and Hydrocarbon Emissions from Mobile Sources," NAPCA Publication No. AP-66)

*A manifold is a chamber that delivers exhaust gases from the engine cylinders to the exhaust system.

dioxide, nitrogen oxides, and water vapor being emitted. Incomplete combustion (which is the situation in auto engines) results in carbon monoxide and various hydrocarbons also being emitted along with carbon dioxide, water vapor, and nitrogen oxides.

The predominant source of pollutants from gasoline-fueled engines is the exhaust gas, although there is some emission of hydrocarbons due to gasoline evaporation from the fuel tank and carburetor (Fig. 10-4). A graphic illustration of a basic difficulty in the attempt to eliminate exhaust emissions is found in Fig. 10-5. In this graph the emission concentrations of the three main pollutants from auto exhaust are plotted as a function of the ratio of air to fuel in the engine cylinders. The only region where all three pollutants have relatively low concentrations is at very high air-to-fuel ratios such as 20 or 21. With such a lean fuel mixture, operating difficulties such as misfire and stalling would be encountered with most commercially available, gasoline-fueled, internal combustion engines, causing poor performance and high emissions of CO and hydrocarbons. Therefore, the choice is to operate the engine either at an air-fuel ratio where nitrogen oxides are low and CO and hydrocarbons high in concentration (rich fuel mixture), or where CO and hydrocarbons are relatively low and nitrogen oxides relatively high in concentration (air-to-fuel ratio of around 15 or 16). Aside from the air-to-fuel ratio, some of the other factors affecting pollutant emission are spark timing, engine speed, compression ratio, and combustion chamber deposits.

Automobile emission controls

Early controls Earlier we mentioned that a small percentage of the hydrocarbons emitted by automobiles comes from carburetor and fuel tank evaporative losses. Prior to 1971, gasoline tanks and carburetors had been vented directly to the atmosphere when an automobile was not in use. Beginning in 1971, these emissions were controlled by collecting gasoline vapors from the fuel tank and carburetor either in the crankcase* or in activated carbon canisters, from which they are sent to the engine when it is started again.

In order to produce a better seal between the piston and cylinder walls, metal rings are placed around the pistons. Some combustion gases still manage to escape from the cylinders by slipping in between the piston rings and cylinder walls, and down into the crankcase. Starting in 1968, new autos contained a positive crankcase ventilation (PCV) system, designed to prevent gases that blow by the piston rings from escaping into the atmosphere. The gases escaping from the cylinder are recycled back to the engine intake manifold instead of being vented to the atmosphere.

*The crankcase holds oil for lubricating the crankshaft, connecting rods, and other parts of the engine.

Exhaust emission control systems Automobile manufacturers have taken two approaches toward meeting the Federal emission standards for 1976 and 1977. They have tried to reduce the production of pollutants in the cylinders by modification of the gasoline engine, and they have also tried to remove pollutants from the exhaust.

Thermal reactors offer a partial solution for cleaning automobile exhaust. They are simply enlarged exhaust manifolds which provide a longer detention time for the exhaust gases while they are still at high temperatures (Fig. 10-7). An air pump supplies atmospheric air to mix with the exhaust gas. The greater detention time and extra oxygen allows more complete oxidation of carbon monoxide to CO_2, and unburned hydrocarbons to CO_2 and water.

'It may keep the pollutants out of the air, but I dont think you've licked the whole problem !'

Figure 10-6 Editorial cartoon by Pat Oliphant. Copyright 1966, *The Denver Post*. (Reprinted by permission of Los Angeles Times Syndicate.)

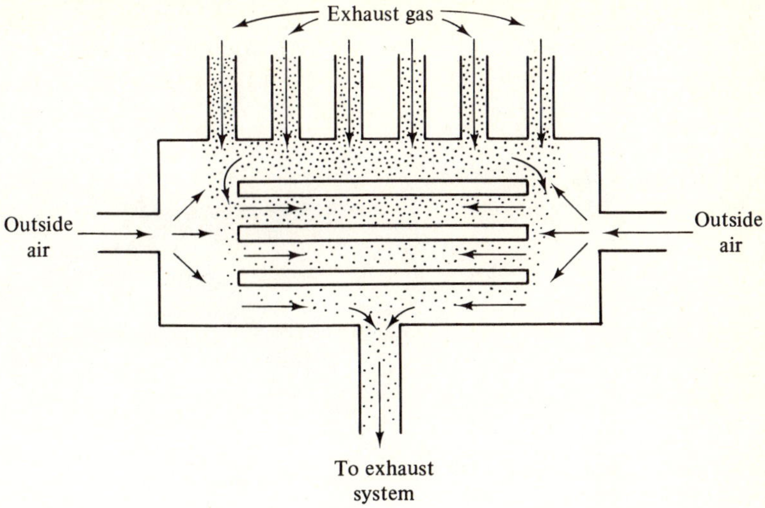

Figure 10-7 In an exhaust manifold thermal reactor, outside air is mixed with engine exhaust gases and oxidation is promoted by the longer retention time of the hot exhaust gases and by the presence of excess oxygen.

The main problem with the thermal reactor is that the high temperatures and excess oxygen, which promote more complete oxidation of CO and hydrocarbons, tend to increase nitrogen oxides formation.

The other exhaust control device which has been receiving great attention is the catalytic converter. A catalyst is a substance that modifies (usually increases) the rate of a chemical reaction without being consumed in the process. The catalytic converter, which generally consists of a catalyst deposited on a support matrix, is positioned in the exhaust stream at the tailpipe or in the exhaust manifold near the engine (Fig. 10-8). There are two types of catalysts: oxidation catalysts which promote the conversion of CO and hydrocarbons to CO_2 and water, and reduction catalysts which convert nitrogen oxides into nitrogen and oxygen gas. Both types will have to be used to reduce all three pollutants.

Materials which are being used as catalysts include rare metals, such as platinum and palladium, and base metals, such as cobalt, copper, and molybdenum, in various combinations. The exact chemical makeup of the catalysts used in catalytic converters is secret. Even the automobile manufacturers have promised not to analyze the catalysts supplied to them.

It appears that some catalytic converters built thus far (for example, Universal Oil Products) can meet the 1976 emission standards (for CO and

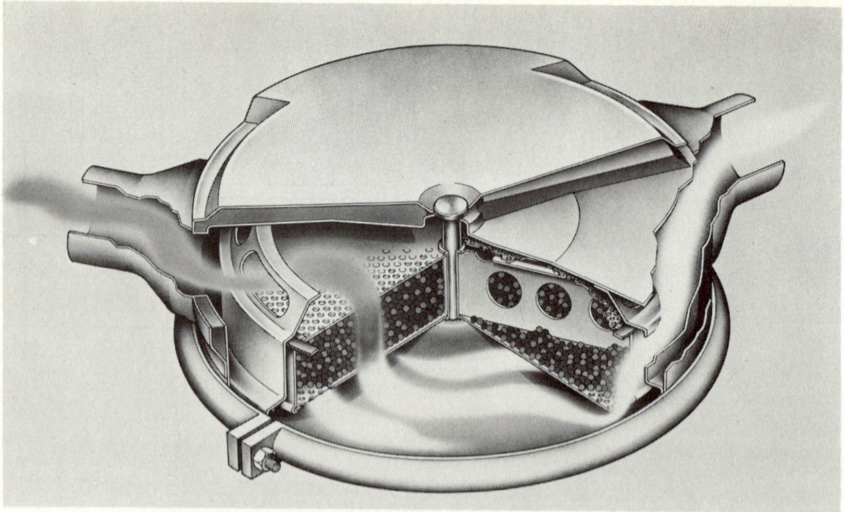

(a)

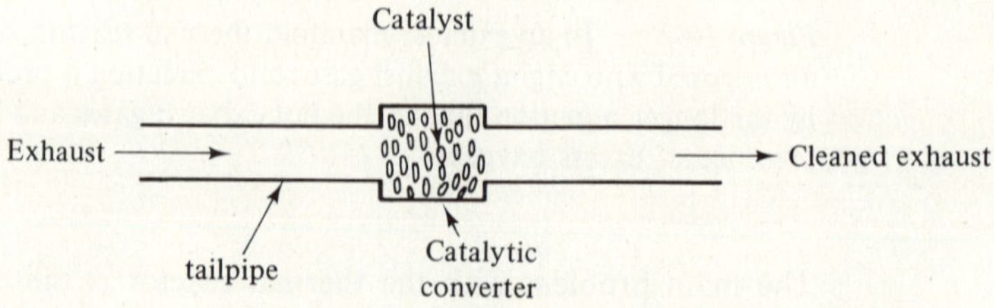

(b)

Figure 10-8 (a) In this experimental catalytic converter the catalyst is coated on tiny beads, so there is a great deal of catalytic substance exposed to the exhaust stream. (Courtesy of General Motors Corp.) (b) The concentration of pollutants in the exhaust gas is reduced after passage through the catalytic converter.

hydrocarbons) for the 50,000 miles required by the EPA, but then the catalyst has to be replaced. Unleaded gasoline must be used in cars containing catalytic converters; otherwise, the catalyst becomes poisoned by the lead and rapidly loses its effectiveness.

It is much more difficult to achieve NO_x reduction than CO and hydrocarbon reduction. The technology of reduction catalysts is less developed than for oxidation catalysts, and it is not certain that the 1977 standards can be met. Another method that is being used to reduce NO_x emission is recirculation of some of the exhaust gases back into the engine to lower the combus-

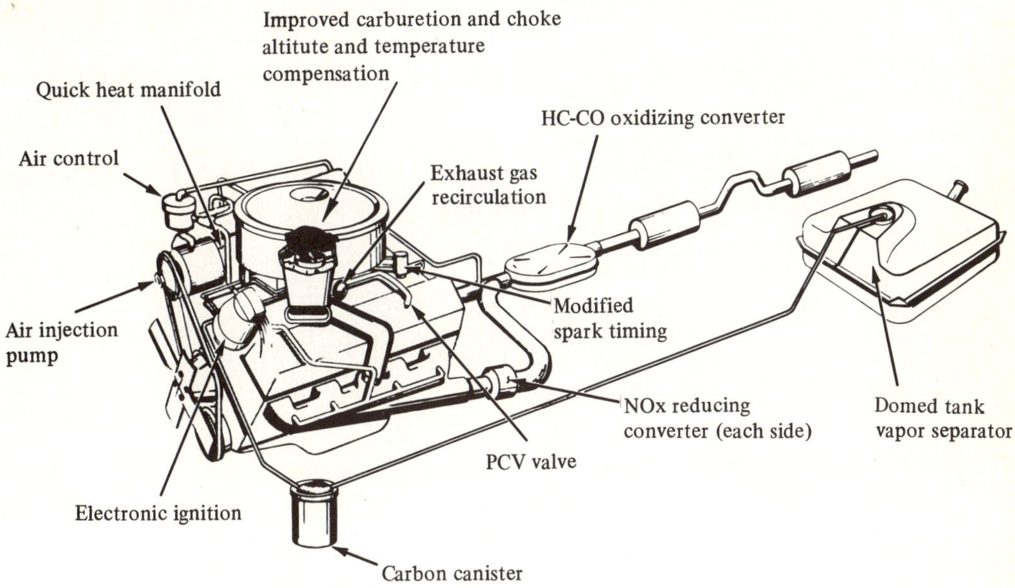

Figure 10-9 This complete emission control system, designed by General Motors, is expected to meet the 1977 emission control standards for NO_x, CO, and hydrocarbons. (Courtesy General Motors Corp.)

tion temperature and, therefore, the amount of NO_x produced. There is a chance that the nitrogen oxides standard may be eased since the EPA discovered that some of the original studies conducted to set these standards were in error. Thus, the car manufacturers may be able to meet the 1977 NO_x standard after all.

Internal combustion engine modifications

The Honda Motor Co., Ltd., of Japan has made a simple modification of the internal combustion engine which apparently allows it to meet the 1976–1977 emission standards. The Honda engine is a version of the "stratified charge" engine (*charge* in this case refers to the mixture of gasoline vapor and air), which decreases exhaust emissions by burning an exceptionally lean mixture of gasoline and air. (See Fig. 10-5.) Since such a mixture cannot be ignited by a spark plug, the problem facing engineers was how to form a small pocket of richer mixture (a stratified charge) around the plug to initiate combustion.

In the Honda engine the spark plugs are isolated in a little chamber of their own above the main cylinders. Two separate carburetors are used to supply the different mixtures of fuel to the two chambers: the rich mixture to the spark plug chamber and the lean mixture to the main chamber (Fig. 10-10). After the spark plug ignites the fuel-rich mixture, flame travels down into the cylinder and explodes the lean mixture to drive the piston. The National Academy of Sciences has found this engine to be the most promising candidate to fulfill the provisions of the Clean Air Act. Ford Motor Co. is also working on a "stratified charge" engine.

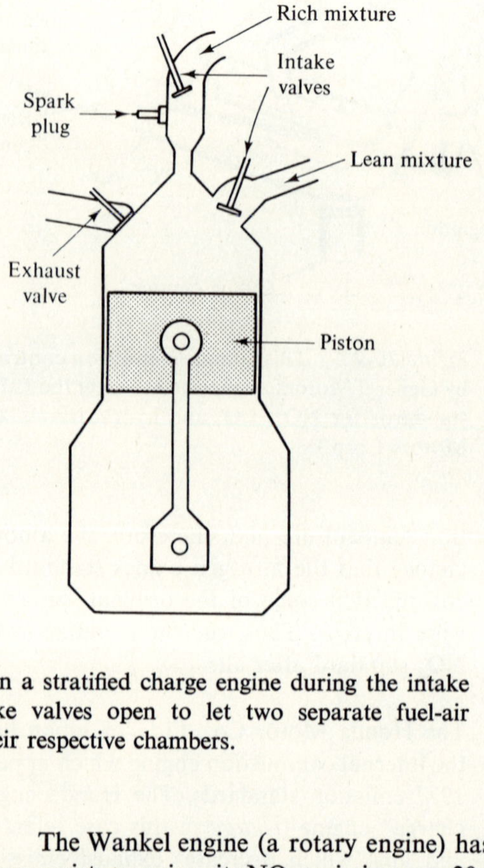

Figure 10-10 In a stratified charge engine during the intake stroke, two intake valves open to let two separate fuel-air mixtures enter their respective chambers.

Wankel engine The Wankel engine (a rotary engine) has a very important advantage over a piston engine: its NO_x emissions are 20-60% lower. Its CO and hydrocarbon emissions are higher than for a piston engine, but these are easier to control than NO_x emissions. These differences between the Wankel and piston engines with regard to their pollutant emissions can be explained by the lower compression ratio and, therefore, lower pressure and temperature in the rotary engine. Other advantages of the Wankel engine are its fewer parts, smaller size, and greater power per chamber. Since the Wankel engine is about one-third the size of a piston engine of equivalent horsepower, there is more room for pollution control devices.

The Wankel engine, which is a gasoline-fueled internal combustion engine, employs a rotor in its combustion chamber rather than a piston; hence, there is no need to convert linear motion to rotational motion as is done in a piston engine. The rotor is roughly triangular in shape with seals on each of three apexes (Fig. 10-11). Power to the wheels is provided by a gear arrangement in the center of the rotor which turns the driveshaft.

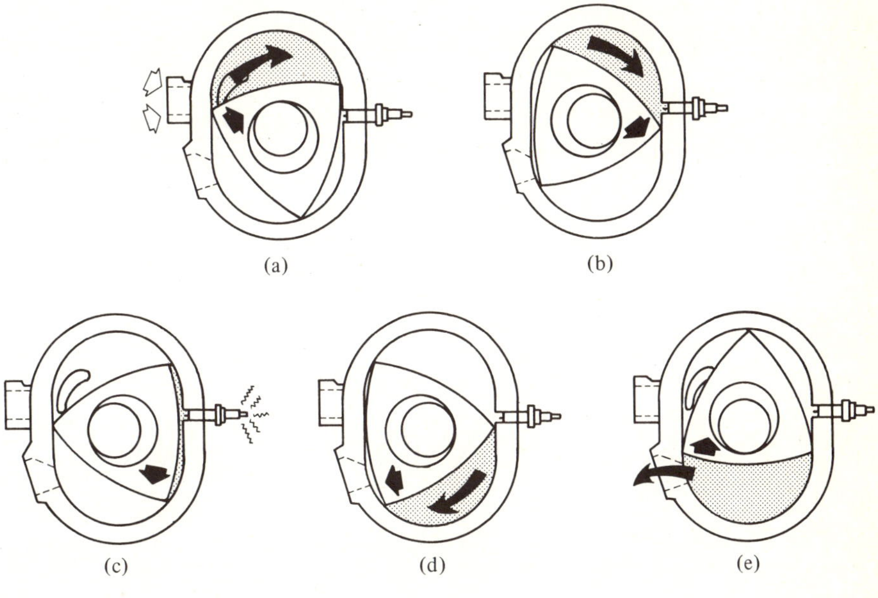

Figure 10-11 Sequence of Wankel rotary engine cycle events: (a) Intake; (b) Compression; (c) Ignition; (d) Expansion; (e) Exhaust

The rotor divides the chamber into intake, combustion, and exhaust zones. The air-fuel mixture entering from the intake port is compressed in the combustion zone where it is ignited by a spark plug. The explosion forces the rotor to spin, ejecting combustion products from the exhaust port, and the cycle is repeated. Rotary engines are presently used in the Mazda, manufactured by Japan's Toyo Kogyo Co., Ltd. Both General Motors and Ford have acquired rights to make the Wankel engine.

Energy and power requirements of an automotive power source

Although the internal combustion engine (with modifications) will probably be the main power plant found in American cars in the near future, by the 1980s we may see one or more other power sources being used. Two very important considerations in the selection of a vehicle power source are vehicle speed and range between stops to refuel or recharge. The speed and range characteristics of several power sources are shown graphically in Fig. 10-12. A vehicle of high specific power is one that has a high ratio of power to weight, and can move at high speeds. A vehicle of high specific energy is one that has a high ratio of stored energy to weight and can travel long distances. According to the data in Fig. 10-12, gas turbines, internal combustion

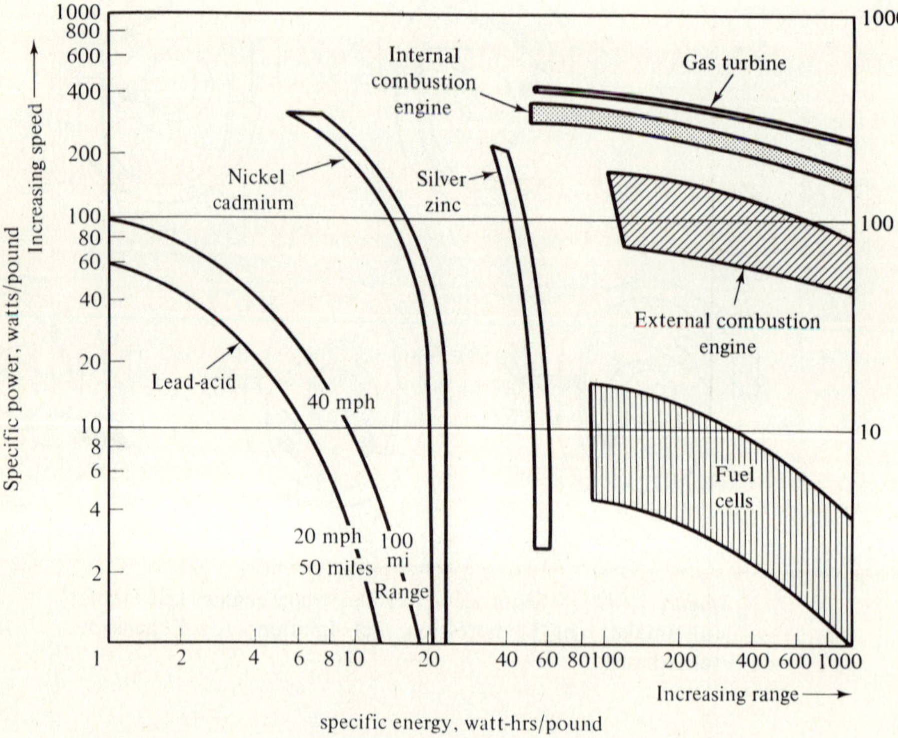

Figure 10-12 Specific power and specific energy of various power sources suitable for motor vehicle propulsion. *Note:* Assumes 2,000-lb vehicle, 500-lb motive source, and steady driving. (Adapted from "Control Techniques for Carbon Monoxide, Nitrogen Oxide, and Hydrocarbon Emissions from Mobile Sources," NAPCA Publication No. AP-66)

engines, and external combustion engines (steam engines) have both high specific energy (range) and high specific power (speed). The rest of the power sources in Fig. 10-12 are mostly batteries of different types.

We now use Newton's laws of motion to derive some equations which show how the required specific power and specific energy of a vehicle depend on the desired speed and range capabilities.

Newton's third law of motion states that for every action there is an equal and opposite reaction. If we press down on the floor with our feet, then the floor pushes back on our feet with a force of equal magnitude but opposite direction. A rocket ejects gases from its rear which, in turn, propel the rocket forward. The rocket is pushed forward by the reaction force of the gas molecules on the engine (Fig. 10-13). It is important to keep in mind that action and reaction forces never act on the same body, so they cannot cancel each other to produce zero resultant force on a body.

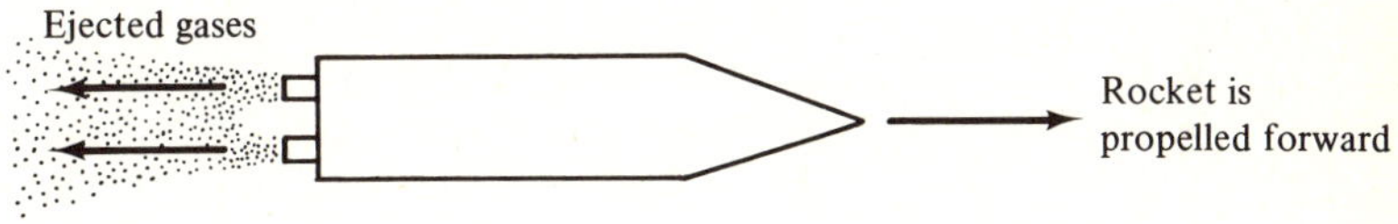

Figure 10-13 Newton's third law can be used to explain forward motion of a rocket when gases are ejected from the rear. The engine pushes on the hot gas, and the molecules react with an equal and opposite force on the engine.

We can use Newton's third law to help us understand how an automobile moves. The engine provides power to turn the wheels which push against the road and the road pushes back against the tires, providing the force which moves the car (Fig. 10-14). There are other forces which act on a car in motion, such as rolling and sliding friction, air resistance, gravity, and the upward push of the road against the car. Since the car pushes down on the road with a force equal to its weight, the road pushes up against the car with a force equal in magnitude to the weight but in the opposite direction. Therefore, if the car is not on an incline, the two vertical forces on the car (the force of gravity and the force of the road) are equal in magnitude but opposite in direction and cancel each other. When a car is on an incline this is not true. (See Question 8 at end of this chapter.)

It turns out that problems involving forces in two or three dimensions can be broken down into two or three parts. In treating the motion of a vehicle we can separate the horizontal and vertical forces and apply Newton's second

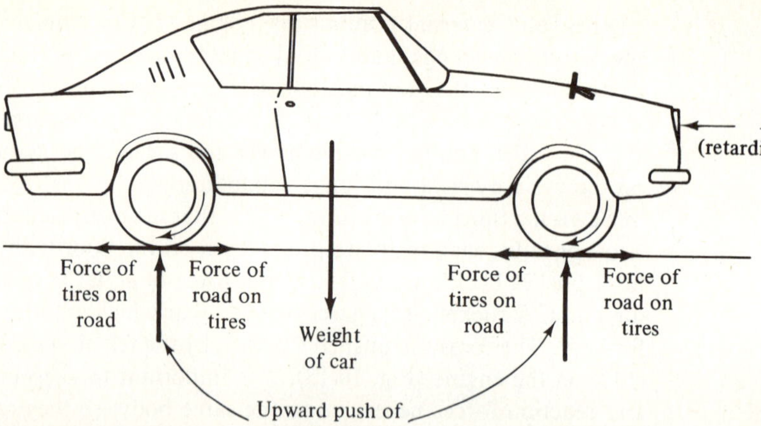

Figure 10-14 Forces acting on and forces produced by an automobile moving on level ground. The force of gravity and the upward push of the road on the car add up to zero. The resultant horizontal force is equal to $F_{\text{thrust}} - F_r$, where F_{thrust} is the horizontal force exerted by the road on the tires.

law for each direction. We have already done this for the vertical direction and found that the resultant force in this direction is zero and, not surprisingly, there is no motion in this direction.

Experimentally, it is found that the resistive force, F_r, is usually proportional to the mass of the car and its speed:

$$F_r = kmv$$

where k is a constant, m is the mass of the car, and v is its speed. From Newton's second law, we obtain the following equation for motion in the horizontal direction:

$$F_{\text{thrust}} - F_r = ma$$

where F_{thrust} is the force of the road on the car (equal in magnitude to the force of the car on the road). Subsituting for F_r,

$$F_{\text{thrust}} = ma + mkv$$

This means that the car must provide a force which accelerates the car and also overcomes resistive forces.

Once the car finishes accelerating and reaches a constant speed, the force needed to keep it moving at that speed is just equal to the resistive force, mkv. The car must do work* against this resistive force. The work, W, done by any constant force, F, is the product of the force and the distance, d, through which it acts:

$$W = F \times d$$

In the metric system, if force is in newtons and d is in meters, then work is in joules. Work and energy have the same units. In this example, the initial work done by the car goes into increasing its kinetic energy and overcoming frictional losses (heat), and when a constant speed is reached, the work done is all dissipated as heat.

The power, P, or rate of doing work is equal to the work done divided by the time it takes to do it, t:

$$P = W/t = F \times d/t$$

Since $d/t = v$ equals speed,

$$P = F \times v$$

The units of power are joules/sec or watts, where 1 watt equals 1 joule per second.

The rate at which work is done by the car, or its power, is given by

$$P = F_{\text{thrust}} \times v = mav + mkv^2$$

Power is needed to accelerate the car and to keep it moving at constant speed after the acceleration. Once the car stops accelerating ($a = 0$), the power which must be developed by the car is:

$$P = mkv^2$$

The greater the mass and the greater the speed of the car, the greater the power required to maintain the speed. It takes four times as much power to maintain a speed of 50 miles per hour as it does for a speed of 25 miles per hour. When automobiles of different mass (and, therefore, weight) are being compared, it is convenient to talk about specific power which is power divided by mass (or sometimes it is defined as power divided by weight as

*Work, energy, and power are discussed more fully in Chapter 11.

in Fig. 10-12):

$$\text{Specific power} = P/m = kv^2$$

This relationship tells us that if we want an automobile to be able to travel at a certain speed, v, then we need a source of specific power of magnitude kv^2. The units of specific power, as it is defined in the above equation, are watts/kg.

If we also want the automobile to be able to travel a certain distance, d, then it will have to be able to store a certain minimum amount of energy. This energy may be stored in the form of chemical energy as found in gasoline or other petroleum derivatives, or in the form of electric energy as in a battery. We can find the energy needed to travel a distance d (once the auto has accelerated to speed v) by calculating the work done by the car in moving against an opposing force equal to mkv:

$$W = F \times d = mkv \times d$$

So the energy required is proportional to the mass of the car, its speed, and the distance traveled. Not surprisingly, it takes more energy (or fuel) to travel a certain distance when it is done at a higher speed. The specific energy is obtained by dividing W by m:

$$\text{Specific energy} = W/m = kvd$$

The units of specific energy are joules/kg or watt-hr/kg.

In summary, the minimum specific power needed by an automobile depends on the square of the desired cruising speed. Doubling the desired speed quadruples the required specific power. The minimum specific energy needed to travel a certain distance depends on both the speed at which that distance is covered and the distance itself. If we would like an automobile to have a range of 200 miles instead of its present 100-mile range (while keeping its speed constant), then we would have to double its specific energy. Likewise, if we want to double the speed of an automobile while it is covering the same range, then we must also double the specific energy. In more practical terms, driving at slower speeds saves gasoline.

Alternatives to the internal combustion engine

The electric automobile By 1900 there were more electric automobiles than internal combustion cars on the roads of the United States. By the late 1920s, the electric car had all but disappeared. It could not compete with the internal combustion engine in speed, range, or convenience.

An electric drive system consists essentially of an electric power source, an electric motor, and a means of controlling energy flow from the power source to the motor. Electric motors can convert electric energy to mechanical energy and turn the wheels. The battery problem has been the principal obstacle to serious consideration of the electric auto as a replacement for the ICE. The conventional lead-acid battery, which is now used in most cars to operate the headlights, windshield wipers, etc., has a low specific energy (Fig. 10-12) and would give a car a range of about 80 miles under the most favorable driving conditions and less than 50 miles in city traffic. Therefore, there is a lot of research on alternative batteries of higher specific energy, but they are presently too expensive to be used in quantities sufficient to operate current automobiles. In time we can expect cheaper mass-produced versions of these more exotic batteries to become available, thus giving the electric auto a greater range.

Electric autos presently cannot compete with the ICE in terms of range, but they may be very useful in city traffic where only short distances are driven daily. They have the great advantage of being essentially nonpolluting, and in urban areas this is a particularly desirable property of an alternative power source. Electric autos have been and still are used in England for delivery purposes in large cities. They have been found to be reliable, quiet, and low in cost to operate.

Even though electric autos do not directly pollute the atmosphere, they must obtain electric energy from power plants to recharge their batteries. Power plants do pollute the atmosphere (see next two chapters) and there may not be enough electric energy available to recharge millions of batteries, even if it is mostly done at night. If electric cars were substituted for ICE autos, then the nature and distribution of air pollutants would change drastically. With our present mixture of power plants (mostly fossil fuel), it is not at all certain that electric autos would reduce overall air pollution. There would be a dramatic improvement, however, in the air of cities which did not have coal- or oil-burning power plants nearby.

The lead-acid battery The lead-acid battery found in most cars supplies electrical current* for operation of the cranking motor and the ignition system when the engine is being started. It also supplies current for lights, radio, and other electrical accessories at times when the generator is not operating fast enough to handle the electrical load.

*In Appendix B, some basic electricity, including the concepts of current and voltage, is discussed.

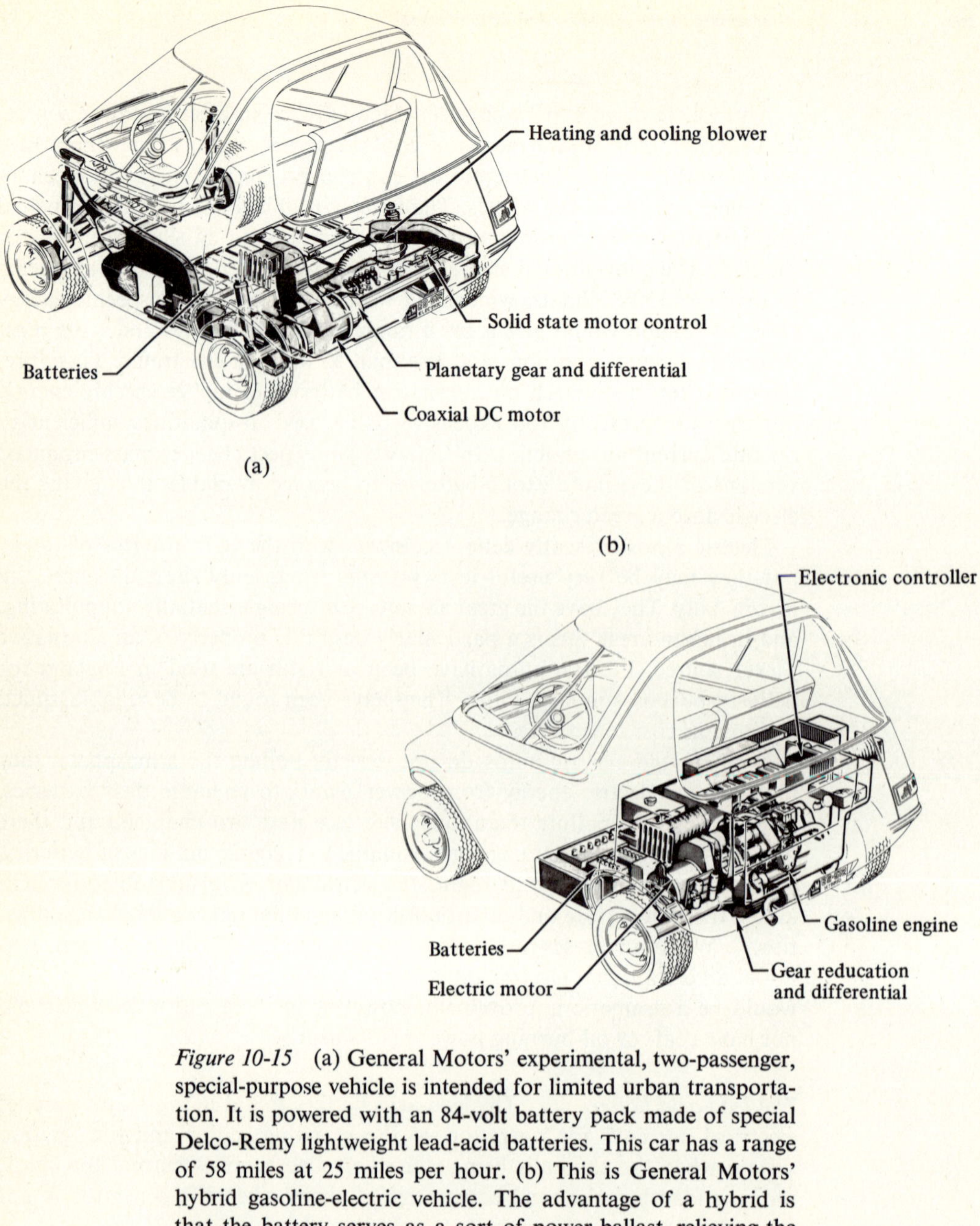

Figure 10-15 (a) General Motors' experimental, two-passenger, special-purpose vehicle is intended for limited urban transportation. It is powered with an 84-volt battery pack made of special Delco-Remy lightweight lead-acid batteries. This car has a range of 58 miles at 25 miles per hour. (b) This is General Motors' hybrid gasoline-electric vehicle. The advantage of a hybrid is that the battery serves as a sort of power ballast, relieving the engine of the requirement to respond to very rapid transient demands. Thus, the engine can operate more nearly at its optimum condition for low emissions. (Courtesy General Motors Corp.)

The chemicals used in the automobile battery are lead (Pb), lead peroxide (PbO_2), and sulfuric acid (H_2SO_4). These three substances are brought together in such a way that they react chemically to produce a flow of electrons (or what is called an electric current). The lead and lead peroxide are put on metal plates, and a 12-volt battery may have as many as 48 separate plates inside it. These plates are immersed in a solution of approximately equal parts of water and sulfuric acid (Fig.10-16).

As the battery is discharged, both types of plates become coated with lead sulfate ($PbSO_4$) and the solution becomes less dense (since the acid which is denser than water is used up). The action in the lead storage battery is reversible and the charge and discharge reactions can be combined into the single equation:

$$Pb + PbO_2 + 2H_2SO_4 \underset{\text{charge}}{\overset{\text{discharge}}{\rightleftharpoons}} 2PbSO_4 + 2H_2O$$

During the discharge, electrons are released at the negative terminal (called the negative electrode) and they can travel through an electrical circuit to the positive electrode. After a certain amount of time, both types of plates become coated with lead sulfate to the point where no charges will flow (the battery is run down or dead). When the battery has reached this state it may sometimes be recharged. This can be done by supplying it with a flow of current from some external source, such as a generator or other battery, which forces current through the battery in a reverse direction. This reverses the chemical activities in the battery and it can then be used again.

Automotive gas turbines The gas turbine* has many inherent advantages over the ICE. It has less than half as many parts, it needs no cooling system, needs no warmup, and provides vibrationless operation. It can burn a variety of fuels, such as kerosene, gasoline, and diesel fuel. Of all the fuel-burning engines, gas turbines are considered to be among those that have the lowest emissions. There is good reason to believe that they can easily meet the 1976 emission standards for CO and hydrocarbons, and also the 1977 NO_x standards with some exhaust controls.

*A turbine is a machine in which the kinetic energy of a moving fluid is converted to useful rotational energy by the interaction of the fluid with a series of blades arranged around the circumference of a wheel. See the next chapter and Fig. 11-4.

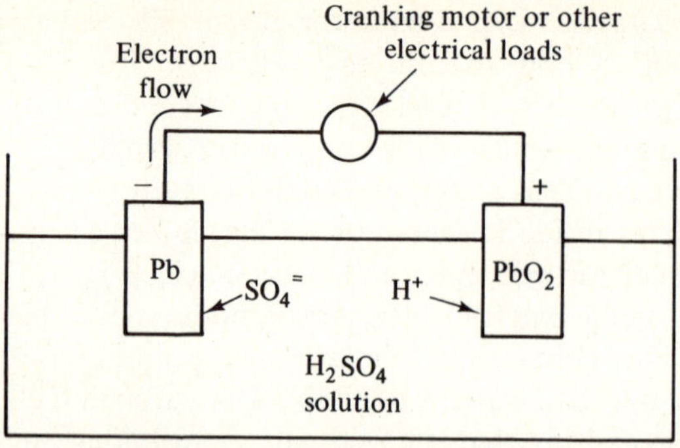

Figure 10-16 The basic action in a lead-acid storage battery involves two dissimilar electrodes and an acid solution.

Figure 10-17 is a schematic diagram of a gas turbine which is a member of the internal combustion family of heat engines. Atmospheric air is drawn into the compressor, squeezed to several times atmospheric pressure, and sent to the combustor. Fuel is sprayed into the combustor where it is burned, raising the temperature of the air-fuel mixture to as high as 1,700°F. The high-temperature, high-pressure gas expands through the turbine and then exhausts to the atmosphere. While the hot gases are passing through the

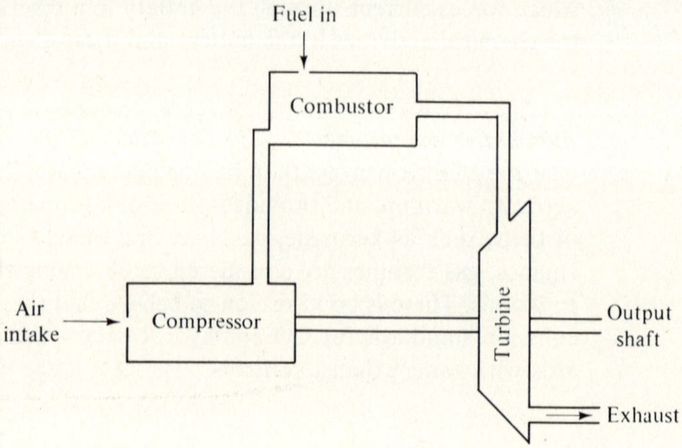

Figure 10-17 Schematic diagram of a simple gas-turbine engine.

turbine they spin it, providing power to the rear wheels of the vehicle. Part of the power developed by the turbine is used to drive the compressor.

As can be seen in Fig. 10-12, the turbine has both very high specific power and specific energy. It develops significant power, however, only at high speeds. This means that expensive variable-speed mechanisms are needed between the turbine and drive wheels to allow for speed changes in stop-and-go traffic. The turbine is well suited to aircraft which travel long distances at

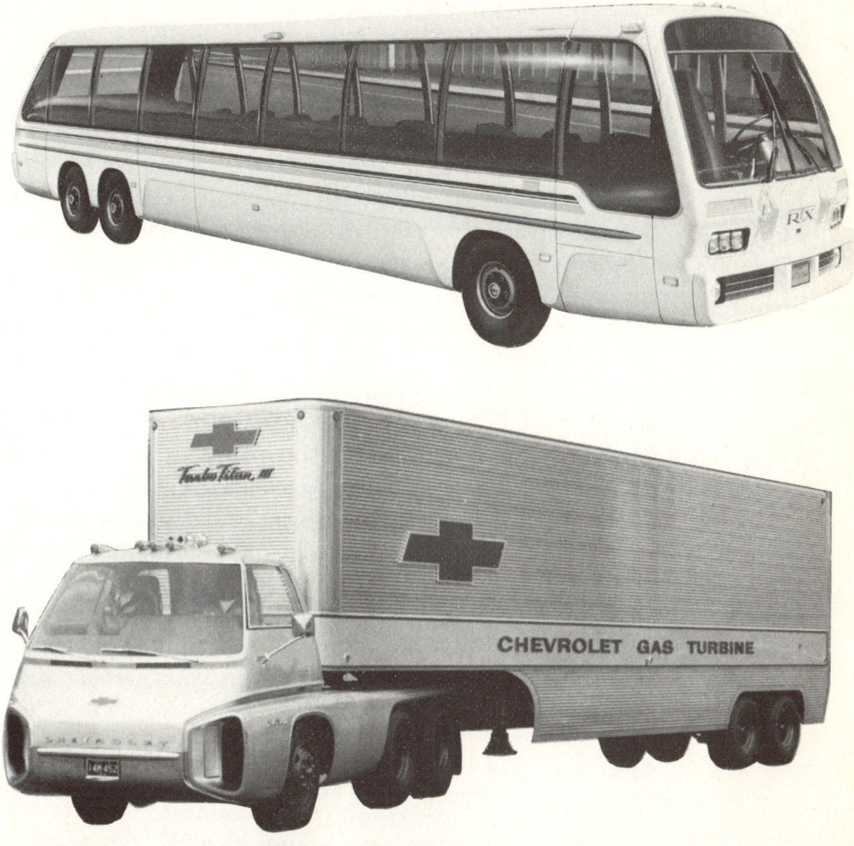

Figure 10-18 Gas turbine buses and trucks have been built and promise to become quite common in the near future. The poor acceleration and initial high cost of the gas turbine engine are not so crucial in the heavy-duty commercial market. (Courtesy General Motors Corp.)

high, almost constant speeds. Experience with turbine engines in aircraft has also shown that they can operate for very long periods of time between overhauls.

Turbine engines will be more expensive than ICEs until they are mass produced and maintenance personnel are trained to maintain this new type of engine. General Motors, Chrysler, and Ford have ongoing turbine programs. If there is going to be an alternative power source for the automobile in the 1980s, the turbine may be it.

Steam engines Steam engines were used to power coaches as early as 1827. The Stanley Steamer was the most popular of the autos run with steam engines in the early 1900s. The automotive steam engine is an external combustion engine in which high-pressure steam or some other working fluid vapor (such as Freon, a fluorocarbon) is used either to spin a turbine or drive a piston. It is called an external combustion engine because the fuel combustion takes place external to the piston or turbine.

Several inventors have developed prototype vehicles which use steam engines. Will P. Lear's (Lear Motor Corp.) steam bus uses water vapor to turn a turbine, and Edward Pritchard's car uses water vapor to move a piston. Many of the faults of earlier steam engines have been corrected. Special boiler designs allow *small* volumes of water or other fluid to be quickly heated to high temperatures and pressures, thus minimizing the explosion hazard and shortening the warmup time to about 20 seconds. The use of water as the working substance makes operation impossible at subfreezing temperatures, but other liquids with lower freezing points could be used.

The advantages of the steam engine over the ICE are low pollutant emissions, quietness, durability, and simplicity. Presently available steam engines are costly and bulky; however, further development could change both of these factors. Table 10-1 compares the pollutant emissions of some selected alternative engines to the 1976-1977 emission standards.

Table 10-1

	Hydrocarbons (grams/mile)	*CO (grams/mile)*	*NO_x (grams/mile)*
Federal standard 1976–1977	0.41	3.4	0.4
Lear's steam turbine, a 220-hp diesel bus engine	0.10	2.0	0.35
Williams Research Corp. 80-hp gas turbine	0.34	4.5	2.15

Mass transportation

Interestingly enough, traffic jams existed even before the automobile left its experimental stage. In 1900, New York City was planning its subway which, it was hoped, would relieve the spectacular traffic jams created by horses, streetcars, carts, and pedestrians. By this date, Boston, London, and Budapest had already built their subways for the same reason. Once again, city planners and transportation managers are turning to mass transit as a means of coping with unmanageable urban traffic; only this time it is automobile traffic. It is also hoped that the use of mass transit will reduce air pollution in urban areas.

This new concern for public transportation comes at a time when patronage is at its lowest point since early in the century. Since 1945 there has been a steady decrease in the use of public transportation (a factor of three decrease from 1945-1970). City officials hope to reverse this trend by building new transportation systems in Washington, DC, and the San Francisco Bay Area, and by modernizing and extending existing systems in cities such as New York, Boston, Chicago, and Cleveland.

The construction of subways is extremely expensive. The minimum excavation cost alone is about $5 million per lane mile, and the total cost can be about $30 million per lane mile. Therefore, Washington, DC's new 100-mile system (expected completion around 1980) may cost 6 billion dollars or more for two lanes of traffic. Federal legislation passed in 1970 provided 10 billion dollars of aid for public transportation in cities, to be spent before 1982. Even this large sum will be able to cover only part of the cost of building and improving many mass transportation systems. The rest must be borne by the taxpayers of the region in which the system is being built.

Some city planners are now contesting the assumption that building more mass transportation will relieve congestion. Cities such as New York and Boston, which have extensive mass transit systems, also have some of the worst traffic congestion. What can happen is that as soon as new transit systems are planned, new commercial and residential development takes place along its proposed route. In a short period of time this can result in more automobiles being used in a particular area of a city than was the case before the new system was constructed.

Instead of making it easier to move around, many community planners are attempting to reduce the need to do so. This is being done on a small city-within-a-city level, and also in planned communities of as many as 100,000 people (Columbia, MD, and Reston, VA). The idea is to bring housing, employment, and community facilities (shopping, medical care, etc.) close together in order to reduce daily transportation needs and to provide the inhabitants with a feeling of community. Other features which are usually included in this type of community are preservation of open space around medium-density housing, reduction in the amount of urban areas devoted to streets, and the encouragement of walking and bicycling. Perhaps by combining community planning and transportation planning, our cities may become more livable in the future.

Many older cities are already set up in such a way that large numbers of people must commute from residential areas in the suburbs to the central business section. In order to attract people to mass transit (buses or trains) in these situations, it will be necessary to offer them some of the comfort and convenience afforded by their automobiles. It may be necessary to make mass transit cheaper than automobile use or even free. Seattle has tried a one-year experiment of offering free bus service to all passengers. Bus use increased about 50%, but the effects on vehicular traffic were not determined as of this writing. Many cities are creating special lanes on highways for buses in order to increase their average speed and thus make them more attractive to commuters. Rising gasoline prices may also convince more people to use mass transit. Inducements of various kinds will be necessary to draw people away from their cars.

Some transportation systems (present and future?) There are many factors which can be considered when comparing various means of transportation, such as cost, speed, safety, land use, comfort, air pollutant emissions, and efficiency. Because of the 1974 fuel shortages, efficiency became particularly important. It also effects operating costs and pollutant emissions. One measure of the efficiency of a mode of transportation is the distance a given number of passengers can be moved with a set amount of fuel. These variables are usually combined into a quantity called passenger efficiency (passenger-miles per gallon of fuel), which is obtained by multiplying the number of passengers carried by the number of miles per gallon obtained by the type of transportation being considered. A 4-cylinder car typically averages around 23 miles per gallon and carries one or two people. Therefore, the number of passenger-miles per gallon of gasoline for a small 4-cylinder car is:

$$1.5 \text{ passengers} \times 23 \text{ miles/gallon} = 34.5 \text{ passenger-miles/gallon}$$

Table 10-2 lists the number of passenger-miles per gallon for several modes of transportation. We can see that the American automobile does not fare well under this criterion. Cars in urban situations normally carry one or two people. If this could be improved to an average of four passengers per car, then a typical American car would be capable of 50 passenger-miles per gallon, approaching that of a diesel train but far from the efficiency of a commuter bus.

We have not included an electric train in this table since it does not use a fossil fuel directly; however, we can estimate its passenger efficiency. First

Table 10-2 PASSENGER EFFICIENCIES OF SEVERAL MODES OF TRANSPORTATION

	Passengers	*Miles/gallon*	*Passenger efficiency (passenger-miles/gallon)*
Typical American car	1.5	12.5	18.75
Small 4-cylinder car	1.5	23.0	34.5
Diesel train	350	0.17	60
Boeing 747	200	0.20	40
Commuter bus	50	3.50	175

we calculate its efficiency in terms of energy conversion. The generating efficiency for a fossil-fuel steam power plant is around 40%. That means that about 40% of the input energy (chemical energy in a fossil-fuel plant) to the power plant is converted to electric energy. This electric energy must be converted into mechanical energy by an electric motor in order to propel the train (Fig. 10-19). The efficiency of a large electric motor is around 90%; it

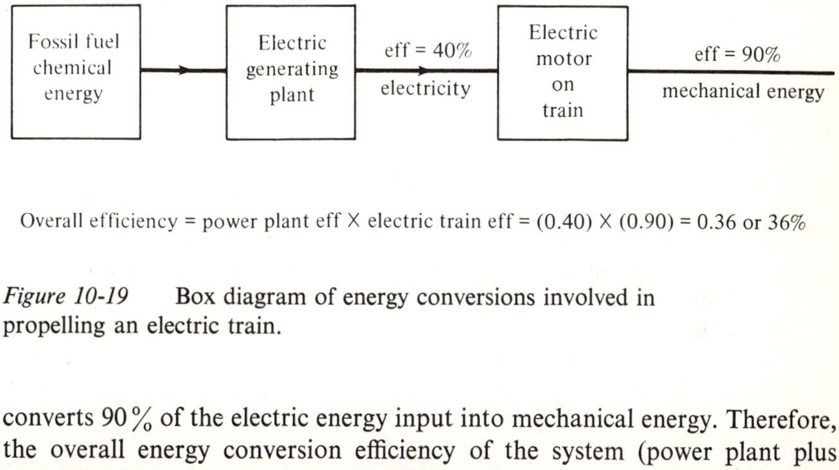

Figure 10-19 Box diagram of energy conversions involved in propelling an electric train.

converts 90% of the electric energy input into mechanical energy. Therefore, the overall energy conversion efficiency of the system (power plant plus electric train) is $0.90 \cdot 0.40 = 0.36$. Thirty-six percent of the original input energy to the system is converted to mechanical energy in the electric train. In this calculation we have neglected transmission losses between the power plant and the train; these are small, however.

A diesel engine has an energy conversion efficiency of around 38%; it converts 38% of the fuel energy in diesel oil into mechanical energy. So it appears that diesel and electric trains have about the same overall conversion efficiency, and since they can carry about the same number of passengers, their passenger efficiencies are also about equal. There are many other factors to take into account before attempting to choose between two such alternative transportation modes. For example, electric trains produce much less air pollution along their route, but require electric energy which is often in short supply.

BART In 1962 voters in the San Francisco region decided to tax themselves in order to build a 75-mile Bay Area Rapid Transit system (called BART for short). By 1974 the system had cost about \$1.5 billion (only \$200 million was from Federal funds). This was close to double what the voters were told in 1962 that it would be. Limited service began in late 1972, was expanded in 1973, and finally completed in 1974. There have been many

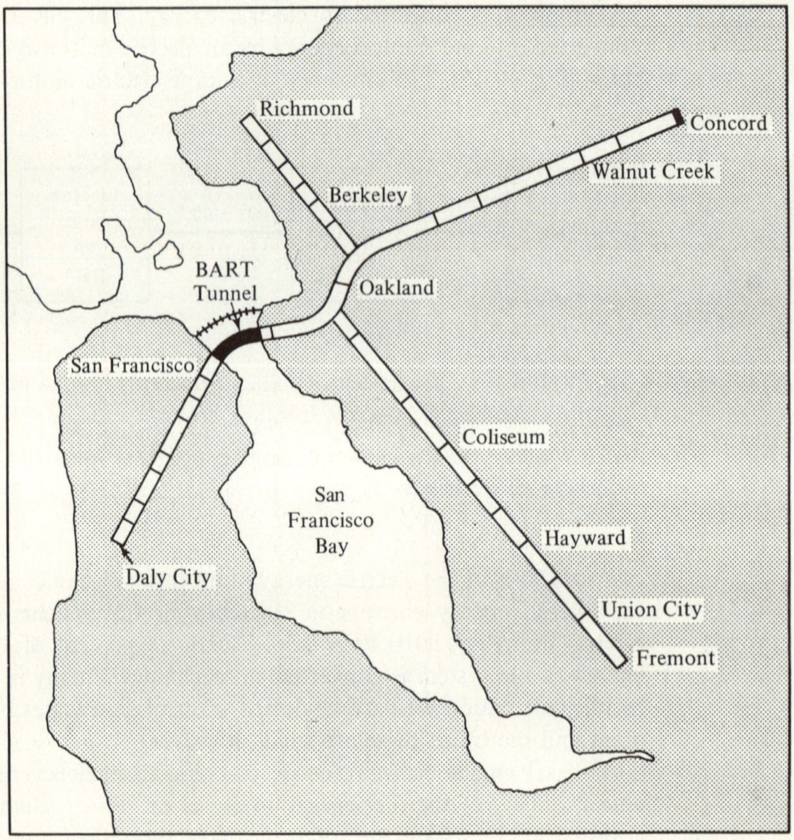

Figure 10-20 The Bay Area Rapid Transit System (BART)

serious technical difficulties such as failure of automatic braking controls and opening of doors between stations. Patronage has been well below the expected rates thus far, but this may change as people become familiar with the system and gain confidence in it.

The vehicles for BART, the nation's first all-new transit system in more than half a century, have wall-to-wall carpeting, air conditioning, ceiling-high picture windows, and extremely comfortable seats. The top speed of these cars, which are run with electric motors, is 80 miles per hour, and the average speed, including stops, is about 45 to 50 miles per hour.

Other transportation concepts There is considerable research effort to develop high-speed land transportation between cities such as Boston and New York and Philadelphia and Washington. It is felt that trains traveling at speeds of 200 to 300 miles per hour can be competitive (in terms of time and cost) with jet aircraft (and cars) when the cost and time for transport to and from the airport are taken into account. In addition, since electric trains will probably be used, air pollution can be directed away from the urban area to centralized power plants where it may be easier to control.

At speeds above 200 miles per hour, conventional means of suspending high-speed ground vehicles lose their reliability. Therefore, alternative, noncontact means of suspending and guiding such vehicles have been tested. In one method of support, the train rides on a cushion of compressed air. This same principle is used in air tracks which are often used in physics experiments (Fig. 10-22). Another method employs magnetic levitation and is being investigated in the United States, Japan, and Germany. The last two countries are actually testing prototype vehicles.

The source of magnetic levitation is the magnetic force generated by moving a magnet over a conducting (aluminum) guideway (Fig. 10-23). The situation is analagous to that created by aligning two magnets in such a way that they repel each other. In this case one magnet is carried on board the train and the other can be thought of as being induced in the aluminum guideway. The train gets its propulsive power from electric energy.

Magnets are already available which can lift a train 1 foot above the guideway. They are made of superconducting* material; that is, material in which an electric current persists for very long periods of time (if very low temperatures are maintained), even after all input energy has been withdrawn. It is the internal electric current which produces the magnetic field. Magnetic levitation vehicles should offer a smooth and quiet ride at high speeds.

*Superconducting material can conduct an electric current with much less resistance than the best metal conductors. At very low temperatures its electrical resistance approaches zero.

Figure 10-21 Most of the BART tracks were built beside freeways. This minimized disruption of communities, both during and after construction of the system. (Official BART photograph)

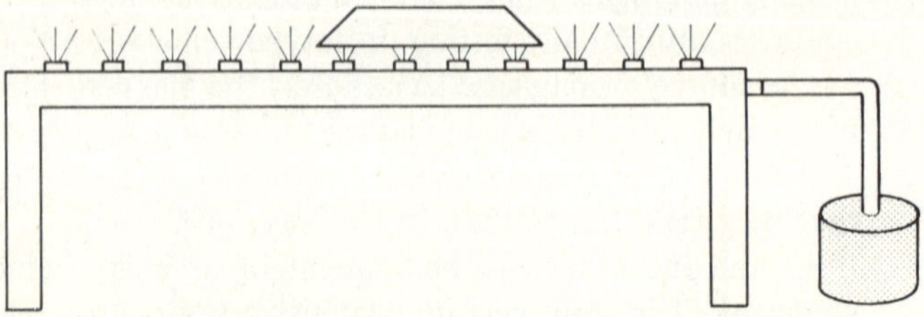

Figure 10-22 In an air track, air is forced through small openings on the top of the track in order to support a moving vehicle. Experiments can be performed with negligible friction between the vehicle and the track.

High-speed tube transportation A very imaginative method of transporting people at high speeds involves the use of gravity and pressure differences to propel a train. The trains would run in a closed and sloped underground tube from which air has been partially evacuated. As a train begins to descend from one station on its way to the next (Fig. 10-24), air at normal atmospheric pressure would be let in behind it. The pressure difference between the front and back of the train produces a force tending to propel it forward. Therefore, the train gains speed from two sources, the force due to the pressure difference and the force due to gravity. The magnitude of the gravitational force which accelerates the train depends on how steeply the curved

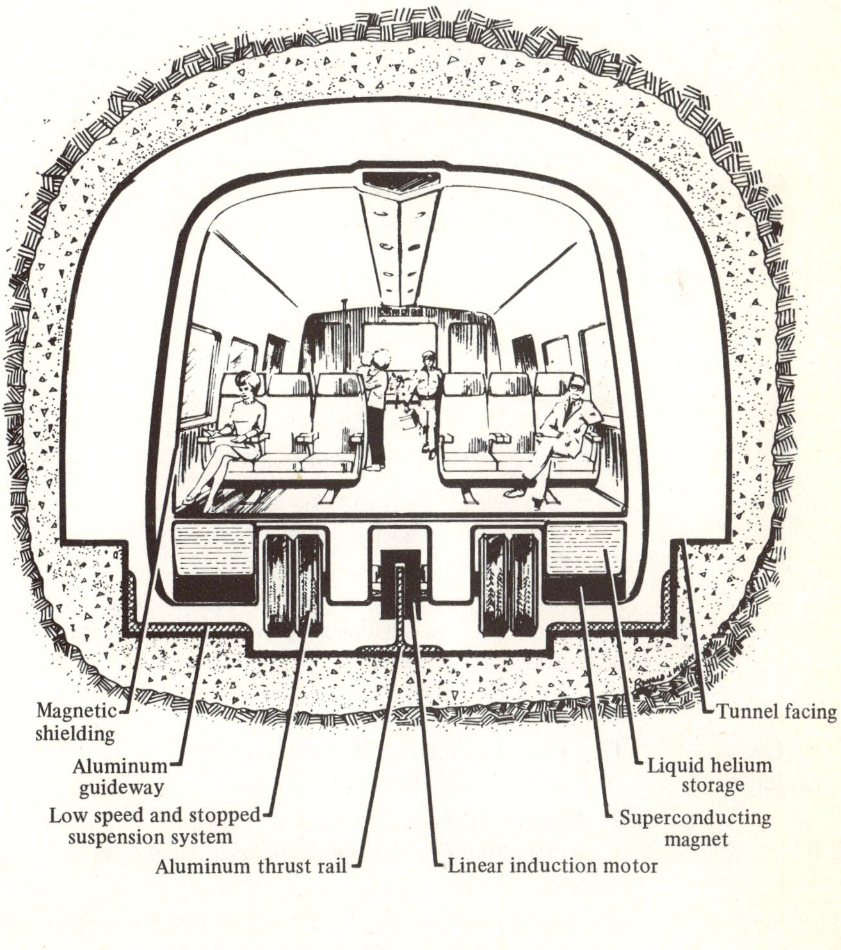

Figure 10-23 Stanford Research Institute Maglev vehicle. Superconducting magnets will lift the magnetic levitation (Maglev) vehicle off its track in order to propel it to high speeds. (Courtesy Howard Coffey, Stanford Research Institute)

tube descends from the station. The construction of such a tunnel, between Boston and Washington, say, would be very expensive but not much more than the cost of a surface transportation system.

This type of system (called a gravity-vacuum system) operates partly in the way a pneumatic delivery system does in a library and partly like a pendulum. Neglecting the effect of the pressure difference for the moment, we can look at the analogy with the pendulum. At station 1 the train starts from rest. As it descends along the tube, it picks up speed until finally it reaches its maximum speed and, therefore, maximum kinetic energy at the lowest point of its path (point 3). This kinetic energy which it possesses at the bottom is enough to allow it to reach the other station if there are no energy losses due to friction. Since in reality there are losses due to friction, these can be offset by the additional energy increase obtained from the pressure difference. This system has the advantage of being relatively nonpolluting, as does the magnetic levitation system, since the only energy sources are gravity and electricity to run some air compressors. At the present time this system is only theoretical and no prototypes have been built.

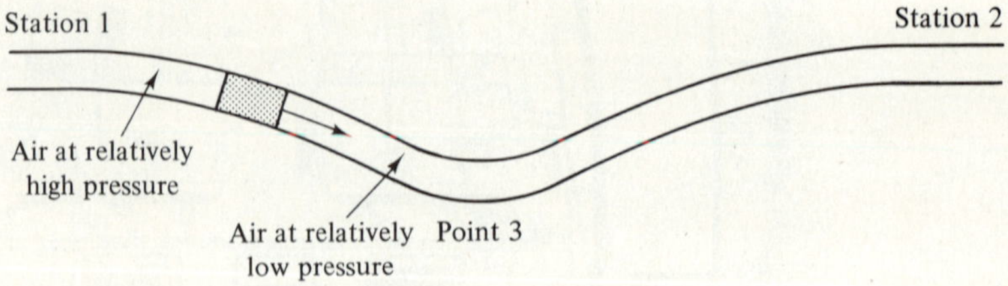

Figure 10-24 A train in a sloped evacuated tube accelerates due to two forces: an unbalanced gravitational force and a force due to pressure difference.

Suggested reading

1 Crouse, William H., Automotive Mechanics, 5th ed. New York: McGraw-Hill Book Co., 1965.

2 Malin, Marty, and Carol Lewicke, "Pollution-free Power for the Automobile," *Environmental Science and Technology*, **6**: 6 (June 1972), 512.

3 Douglas, John H., "What Next for Auto Emissions Control? (or) Is There a Honda in Your Future?" *Science News*, **103**: (April 28, 1973), 276.

4 "Control Techniques for Carbon Monoxide, Nitrogen Oxide, and Hydrocarbon Emissions from Mobile Sources," NAPCA Publication No. AP-66.

5 CHINITZ, WALLACE, "Rotary Engines," *Scientific American*, **220**: (Feb. 1969), 90.

6 HOFFMAN, GEORGE A., "The Electric Automobile," *Scientific American*, **215**: (Oct. 1966), 34.

7 WALKER, GRAHAM, "The Stirling Engine," *Scientific American*, **229**: (Aug. 1973), 80.

8 CHILTON, FRANK, HOWARD T. COFFEY, and L.O. HOPPIE, "Magnetic Levitation: Tomorrow's Transportation," *Bulletin of the Atomic Scientists*, **28**: 3 (March 1972), 39.

9 EDWARDS, L.K., "High Speed Tube Transportation," *Scientific American*, **213**: 2 (Aug. 1965), 30.

10 BUEL, RONALD, *Dead End: The Automobile in Mass Transportation*. Englewood Cliffs, NJ: Prentice-Hall, Inc., 1972.

11 STONE, TABOR, *Beyond the Automobile, Reshaping the Transportation Environment*. Englewood Cliffs, NJ: Prentice-Hall, Inc., 1971.

12 OWEN, WILFRED, *The Accessible City*, Washington, DC: The Brookings Institution, 1972.

13 AYRES, ROBERT V., and RICHARD P. MCKENNA, *Alternatives to the Internal Combustion Engine*. Baltimore: Johns Hopkins University Press, 1972.

14 CROSSLAND, JANICE, "Cars, Fuel, and Pollution," *Environment*, **16**: 2, March 1974.

Questions

1 (a) By what percent did the U.S. population increase from 1950 to 1970?

(*b*) By what percent did automobile registrations increase during the same period?

(*c*) At these rates of growth, what will eventually happen?

2 Do you think that there is any limit to such rapid growth as is occurring in car population? What are the limiting factors?

3 (*a*) Suppose that a mixture of gasoline vapor and air burns inside a closed chamber. As the temperature of the mixture increases, what does this mean in terms of molecular speeds?

(*b*) Explain, on a molecular level, why the gas pressure of this mixture increases with increasing temperature.

4 Could a thermal reactor alone be used to meet the 1976–1977 automobile emission standards for an internal combustion engine? Why?

5 How does the stratified charge engine get around the problem of poor engine performance (misfires and stalling) when a lean fuel mixture is used?

6 Why does a Wankel engine have lower NO_x emissions than a conventional internal combustion engine, but higher CO and hydrocarbon emissions?

7 A ball falls towards the earth. What are the action and reaction forces?

8 When a car is parked on a hill there is an unbalanced force acting on it in the downhill direction. The reaction force of the road on the car acts perpendicular to the road. Explain why there is an unbalanced force in this situation but not when the car is on level ground. Drawing a diagram would help.

9 (*a*) A 4,000-pound Cadillac and a 2,000-pound Volkswagen are moving along a highway at the same constant speed. Is there more resistance to the motion of the Cadillac or the Volkswagen?
(*b*) Which car needs more power in this situation? How much more?
(*c*) If they start off with the same amount of fuel, which car travels a greater distance? How much greater?

10 Do electric generators and electric motors perform the same or opposite functions?

11 When a lead-acid battery discharges, some sulfuric acid molecules are replaced by water molecules. What effect does this have on the density of the solution, and how can this information be useful?

12 What inducements would be effective in convincing you and some of the people you know to use mass transit instead of a private automobile?

13 What is the passenger efficiency of a motorcycle carrying one person and obtaining 70 miles per gallon?

14 In terms of energy usage, which of the following is a more efficient method of transportation: driving a car, bicycling, or walking? Assume that the distance to be traveled is 3 km, that the walking speed is 6 km/hr, the bicycling speed is 10 km/hr and the gasoline mileage of the car is 18 km/gallon. The energy used in walking and bicycling can be found in Table 4-1 in Chapter 4. The energy equivalent of a gallon of gasoline is 3.3×10^7 calories.

15 (*a*) How long does it take a BART train to travel between two stations which are two miles apart when its average speed between stations is 60 miles/hr?
(*b*) A slower train has an average speed of 30 miles/hr between stations. How long does it take to travel the same two miles?

16 Suppose each train in Question 15 stops for 1 minute at each station. How much time is saved by using the faster BART train for the trip?

17 (*a*) If a BART train runs every two minutes during rush hour periods, how many trains will pass a station each hour?

(*b*) If each train is full (6 cars, 72 passengers in a car), how many passengers per hour can be moved past each station?

18 If there are 35 stations in the BART system, each pair 2 miles apart, how many trains would be needed to run a train through each station, in both directions, at two-minute intervals?

Part Six

Energy, power, and man

Chapter 11

Methods of energy conversion

Introduction

This chapter begins with a discussion of the concepts of work, energy, power, and heat. The operation of a heat engine is described and its efficiency calculated. Steam turbines, which are one type of heat engine, are used in all fossil-fuel and nuclear power plants. The various methods of generating electricity are then described.

The universe contains energy in various forms, for example, gravitational, light, heat, and nuclear energy. The predominant form of energy in the universe is gravitational.

The first law of thermodynamics—the law of the conservation of energy—states that energy is neither created nor destroyed, but just changes its form. Therefore, energy is not actually consumed, but its form is changed. When electrical energy passes through a light bulb it is not consumed, it is transformed to thermal energy and light energy. All of the energy is accounted for; about 10% is in the form of light energy and about 90% in the form of thermal energy.

The main energy sources available to us on the earth now are chemical fuels, uranium, and sunlight. In the 21st century we may be able to "burn" deuterium in a controlled way. Converting energy from these present sources pollutes our environment. Some of these sources are being used up rapidly and others appear to be practically limitless in supply. In the future we are going to be faced with still more problems concerning energy use and the decisions we make will, as always, determine the quality of life for man on this planet.

Energy and work

In physics, energy is defined as the capacity for doing work. In the usual sense of the word, work refers to any form of activity that requires the exertion of muscular or mental effort. In the study of physics the term has a very restrictive meaning. Work is done only when a force* is exerted on a body and, at the same time, the body moves in such a way that it has a component of its motion in the direction of the force. Holding a heavy object in your hand involves no work in the technical sense of the word, since there has been no motion of the object (Fig. 11-1). Work is done in raising the object to a greater height (Fig. 11-2). If a constant force, F, is applied to an object of mass, m, and it moves a distance, d, in the direction of the force, then the work, W, done on that object is:

$$W = F \cdot d \tag{11-1}$$

In Fig. 11-2, the force applied is the weight of the object, mg, and the distance traveled is the height, h. Therefore,

$$W = mgh \tag{11-2}$$

*By force we mean that which produces or tends to produce an acceleration of the body to which it is applied.

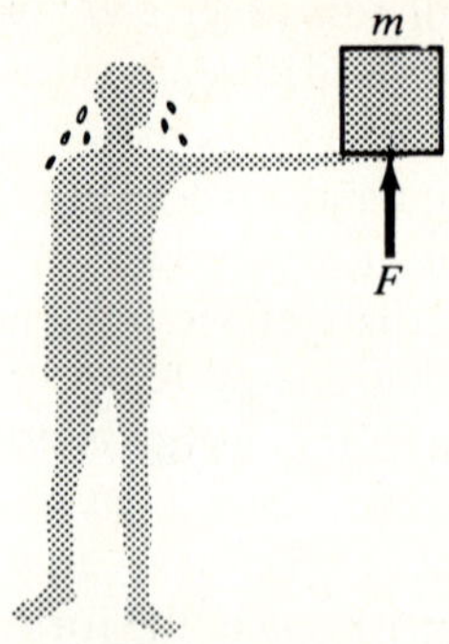

Figure 11-1 A man holding an object of mass, m, kilograms in his hand does no work on the object, although he applies force F to the object. That force F is balanced by the weight of the object and there is no motion, hence no work.

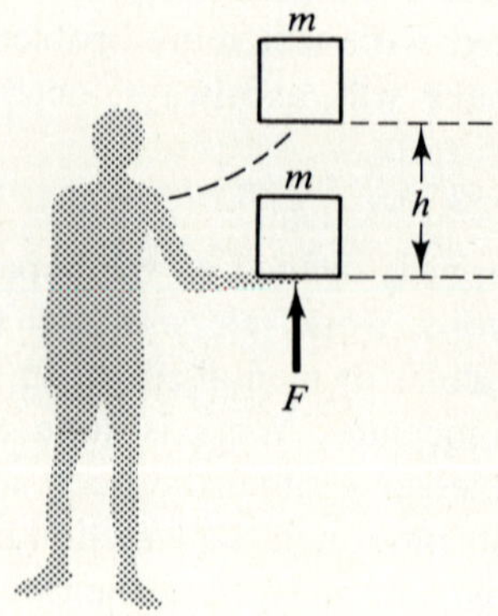

Figure 11-2 In order to raise an object of mass, m, to a height, h, work must be done against gravity. The work done is equal to the force applied, mg times the distance traveled in the vertical direction: $W = F \cdot d = mg \cdot h$. In order to get the object moving from rest, the applied force, F, had to be greater than the weight, mg, of the object for just a short time.

The raised object possesses energy by virtue of its position. This energy is called gravitational *potential energy*. It is a form of energy that is stored and can be released at a later time, just as the chemical energy in a storage battery can be released at a later time. If the object is released and allowed to fall, it has the ability to do work. If the raised object were water in a reservoir, then upon release from the reservoir the potential energy of the water would be converted to kinetic energy* of the moving water (Fig. 11-3).

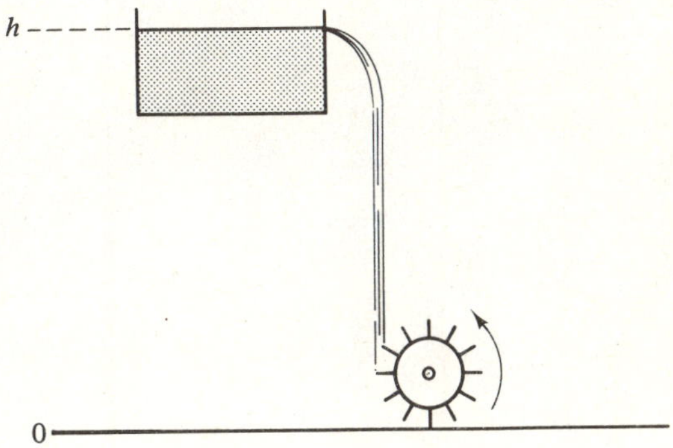

Figure 11-3 Water at a height, *h*, above ground has potential energy which is transformed into kinetic energy when it is allowed to fall. The stream of water does work on the wheel.

The kinetic energy of the moving water could do work in turning a wheel in a flour mill or a turbine in an electric power plant. A *turbine* is a wheel which is made to rotate by the action of flowing water, steam, air, or heated gases, thus forming mechanical energy (kinetic or potential energy) capable of doing work. A steam turbine (Figs. 11-4, 11-5, and 11-6) converts the kinetic energy of steam at high pressures into mechanical energy for the purpose of generating electricity, running machinery, or locomotion. The simplest type of steam turbine consists of a fixed nozzle directing a high pressure jet of steam upon blades arranged at the periphery of a rotating wheel (Fig. 11-6).

In the above examples we speak of work being done as a mechanical aspect of what is in reality either a transfer of energy from one body or system to another, or a conversion of energy from one form to another, or both. We should not think of energy as being converted into work or work into energy, but of energy being converted from one form to another or transferred from one system to another, mechanical work sometimes being done in the process of conversion or transfer.

*Kinetic energy is the energy associated with a body's motion. It is equal to $mv^2/2$, where (in the metric system) m is the mass of the body in kilograms, v is the speed in meters per second, and the energy is in joules.

Figure 11-4 Installation of a turbine at The Geysers geothermal plant, north of San Francisco in California. (Courtesy of Pacific Gas and Electric Company)

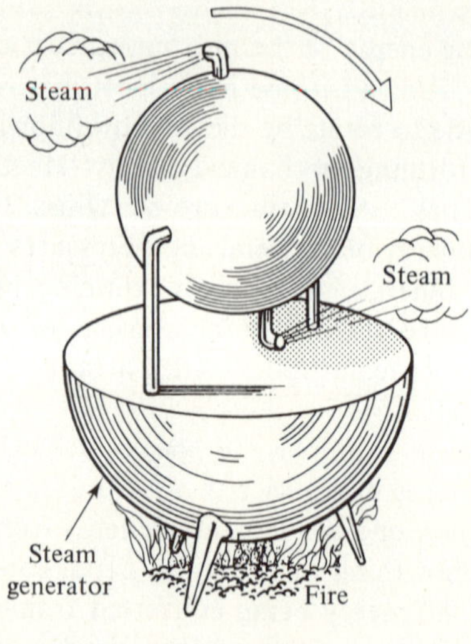

Figure 11-5 Hero's reaction turbine, 2nd century B.C. First recorded use of the ability to extract work from heat was found in the manuscripts of Hero of Alexandria. Steam passes through hollow sphere via support tubes and out through L-shaped pipes. Reaction force on pipes causes the sphere to turn. (From *Collier's Encyclopedia*. New York: Crowell-Collier Educational Corp., 1973. Reprinted by permission of the publisher.)

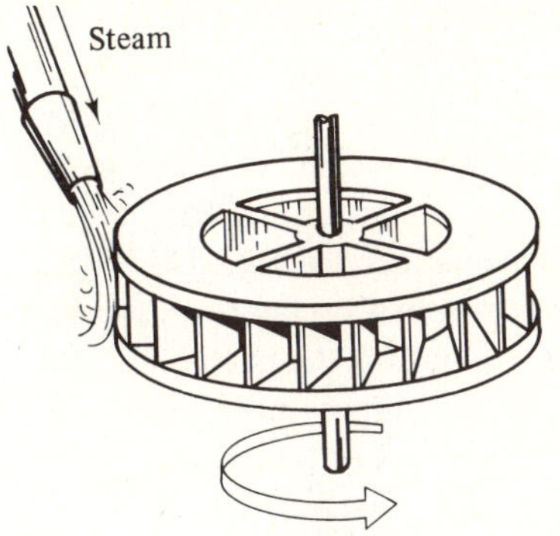

Figure 11-6 Branca's impulse turbine, 1629. Steam issues as a jet from a boiler and strikes the vanes of a horizontal wheel, causing it to turn. (From *Collier's Encyclopedia*. New York: Crowell-Collier Educational Corp., 1973. Reprinted by permission of the publisher.)

If a person lifts an object to a greater height, chemical energy within the tissues of the person is converted to thermal energy. During this energy conversion some work is done in lifting the object. Another way of looking at the same event is as an energy transfer from the person to the object with work being done in the process of transfer. The person's capacity to do further work is reduced since some of his internal chemical energy has been used. The above discussion can be summarized by saying that we need energy to do work and that useful work can be done in the process of energy conversion or transfer. The ability to do further work decreases when energy is converted from one form to another.

Heat and heat engines

Heat Heat is one of the many forms of energy. A body does not contain heat; it contains *internal energy* which is the total energy content of a body. Internal energy consists of both the potential energy within and between molecules, and the kinetic (thermal) energy of the molecules themselves. *Heat* is the thermal energy that transfers from one body to another when there is a temperature difference between them. The greater the temperature difference, the greater the quantity of heat that will flow from the hotter body to the cooler one.

There is a tremendous amount of thermal energy in the oceans, but it is not available for doing work unless there is a region of lower temperature to which the heat can flow. Energy is available when it can be converted into mechanical energy or useful work.

Heat engines A *heat engine* is a device for getting useful work out of the thermal energy available. There are many types of heat engines, such as hydraulic,* steam, gasoline, and diesel. When we talk about heat engines in this chapter we mean steam or hydraulic turbines. In a steam turbine, steam at a high temperature strikes the blades of a turbine, causing it to rotate and gain mechanical energy. This mechanical energy can be transferred to a generator which converts this mechanical energy to electrical energy. The high-temperature steam loses kinetic energy to the turbine and ends up at a much lower temperature, but it is still steam. In order to recycle the original water back to the boiler, this lower-temperature and lower-pressure steam must be condensed to the liquid state. One way of condensing the steam is by pumping cooling water from a nearby waterway through a separate condensing unit, where heat is transferred from the steam to the cooling water. This heat transfer often leads to thermal pollution. (We discuss the environmental effects of power production in the next chapter.)

We can represent a heat engine by the schematic diagram of Fig. 11-7. There is a source of thermal energy which is at a higher temperature than the reservoir. Heat can flow from the source to the reservoir, with useful work

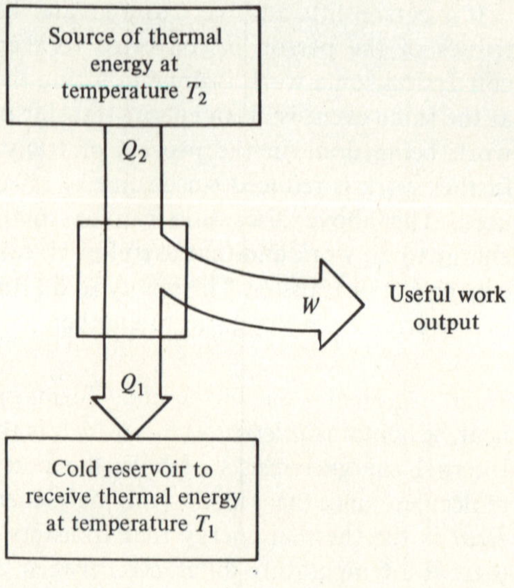

Figure 11-7 Schematic representation of a heat engine

**Hydraulic* means water driven.

being done in the process. All heat engines throw away, in their exhaust, part of the heat supplied to them. The thermal efficiency of the engine is the ratio of work output to heat input. The output is the mechanical work, W, and the heat input is Q_2; therefore, the efficiency is equal to W/Q_2:

$$\text{Efficiency} = \frac{\text{Work output}}{\text{Heat input}} = \frac{W}{Q_2} \tag{11-3}$$

From the conservation of energy, the heat input to the turbine, Q_2, is equal to the mechanical work, W, done by the turbine plus the heat rejected by the turbine, Q_1:

$$Q_2 = W + Q_1$$

or

$$W = Q_2 - Q_1$$

The mechanical work is the difference between the heat input and the heat rejected. The efficiency is thus:

$$\text{Efficiency} = \frac{Q_2 - Q_1}{Q_2} = 1 - \frac{Q_1}{Q_2} \tag{11-4}$$

The second law of thermodynamics can be stated in many different ways, one of which is: It is impossible to construct an engine that, operating in a cycle, will produce no effect other than the extraction of heat from a source and the conversion of this heat completely into work. Therefore, Q_1 is never zero and the efficiency is never 100%. "Waste heat" is always an end product of a heat-engine cycle.

In the 1820s a French engineer, Sadi Carnot, showed that the maximum theoretical efficiency of a heat engine operating between any two temperatures is equal to:

$$\text{Efficiency} = 1 - T_1/T_2 \tag{11-5}$$

where T_1 and T_2 are measured in degrees Kelvin. This is an ideal efficiency which does not include such factors as friction. In order to obtain maximum efficiency it is necessary to make the intake temperature, T_2, as high as possible and the exhaust temperature, T_1, as low as possible. The exhaust temperature cannot be made lower than the lowest temperature available for cooling the exhaust. In the case of a power plant, where the mechanical energy of a turbine is converted into electrical energy in a generator, river water is often used to cool the exhaust which is in the form of steam. The limit on the input temperature T_2 is set by the boiler strength. This is because the vapor pressure of a liquid increases with increasing temperature, and the boiler must be able to contain the water vapor which is produced when the water is boiled.

In modern steam turbines, T_2, which is the temperature of the high pressure steam input to the turbine, is usually about 1,000°F and T_1, the temperature of the exhaust steam, is about 100°F. We can use Carnot's Eq. (11-5) to calculate the maximum theoretical efficiency of such steam turbines:

$$\text{Efficiency} = 1 - T_1/T_2$$

In order to use Eq. (11-5) we must convert the temperatures to degrees Kelvin. The following equations can be used for this conversion:

$$T_{°K} = T_{°C} + 273$$

$$T_{°F} = 9/5T_{°C} + 32$$

For $T_1 = 100°\text{F}$, $T_{°C} = 37.7°\text{C}$ and $T_{°K} = 310.7°\text{K}$
For $T_2 = 1{,}000°\text{F}$, $T_{°C} = 537.7°\text{C}$ and $T_{°K} = 810.7°\text{K}$

$$\text{Efficiency} = 1 - \frac{310.7°\text{K}}{810.7°\text{K}} = 1 - 0.384 = 0.616$$

So the maximum theoretical efficiency of this steam turbine is 61.6%. Since it is not theoretically possible to introduce the steam at a constant upper temperature in a steam cycle, the maximum theoretical efficiency is more like 53% than 61.6%. At present, steam turbines achieve about 89% of the theoretical maximum efficiency of 53%, or a 47% net efficiency.

Modern boilers can convert about 88% of the chemical energy in fuels into heat. Electric generators can convert about 99% of the mechanical energy

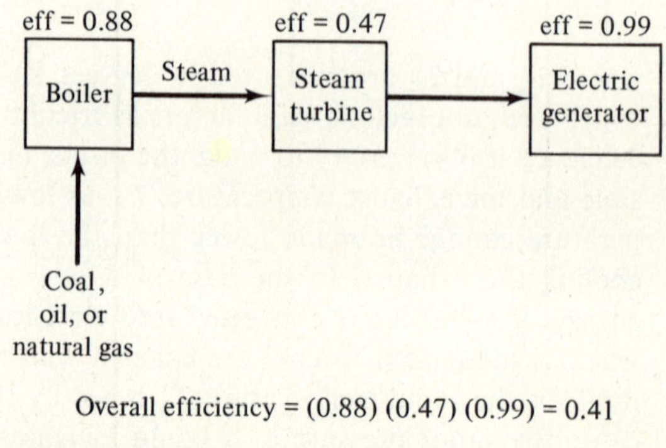

Figure 11-8 Fossil-fuel steam power plant

produced by the steam turbine into electricity. Therefore, the efficiency of a fossil-fuel steam power plant (Fig. 11-8) is about 0.41 or 41%:

$$\text{Efficiency} = (0.99)\,(0.88)\,(0.47) = 0.41$$

Units of energy Energy and work are expressed in the same units. The units of work and energy usually used in discussions about energy and power production are joules and kilowatt hours. Both are derived from the metric system of measurement. If a force of 1 newton is applied to an object that moves 1 meter in the direction of the applied force, then 1 joule of work is done on the object (Fig. 11-9).

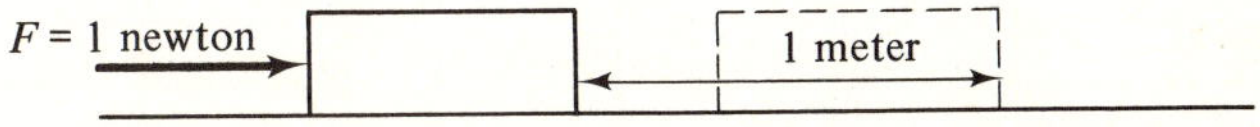

Figure 11-9 $W = F \cdot d = 1$ newton$\cdot$1 meter $= 1$ joule. This object of mass, m, has 1 joule of kinetic energy, $\frac{1}{2}mv_f^2$, after being accelerated from rest to a final velocity, v_f. We can show that the work done is equal to the gain in kinetic energy. We substitute $F = ma$ from Newton's second law, and $d = \bar{v}t$, the distance traveled equals the average velocity, $\bar{v}$, times the time of travel, t.

$$W = ma \cdot \bar{v}t \qquad \bar{v} = \frac{v_{\text{initial}} + v_{\text{final}}}{2}, \quad \text{but} \quad v_{\text{initial}} = 0.$$

So $\bar{v} = v_f/2$. $W = mav_ft/2$. Since $v_f = at$, $W = mv_f^2/2$.

We can get a feel for the size of a joule by considering these examples:

1 A 3,200-pound automobile traveling at 60 miles per hour has an energy due to its motion (kinetic energy) of 5.12×10^5 joules.

2 A 1-kilogram (about 2 pounds) stone moving at a speed of 10 meters per second (about 20 miles/hr) has a kinetic energy of 50 joules.

A joule is not very much energy in terms of the motion of large objects or in terms of energy converted in a power plant.

Power The element of time is not involved in the definition of energy or work. The automobile in the above example will have the same amount of energy when it reaches a speed of 60 miles/hr whether it takes 1 second, 1 hour, or 1 year to do so. In many cases we want to know the rate at which work is done or energy is used as well as the total amount of work done or energy used. The rate at which work is done is called the *power*.

If a quantity of work, W, is done in a time interval, t, then the average power is defined as:

$$\bar{P} = \text{Average power} = \frac{\text{work done}}{\text{time interval}} = \frac{W}{t}$$

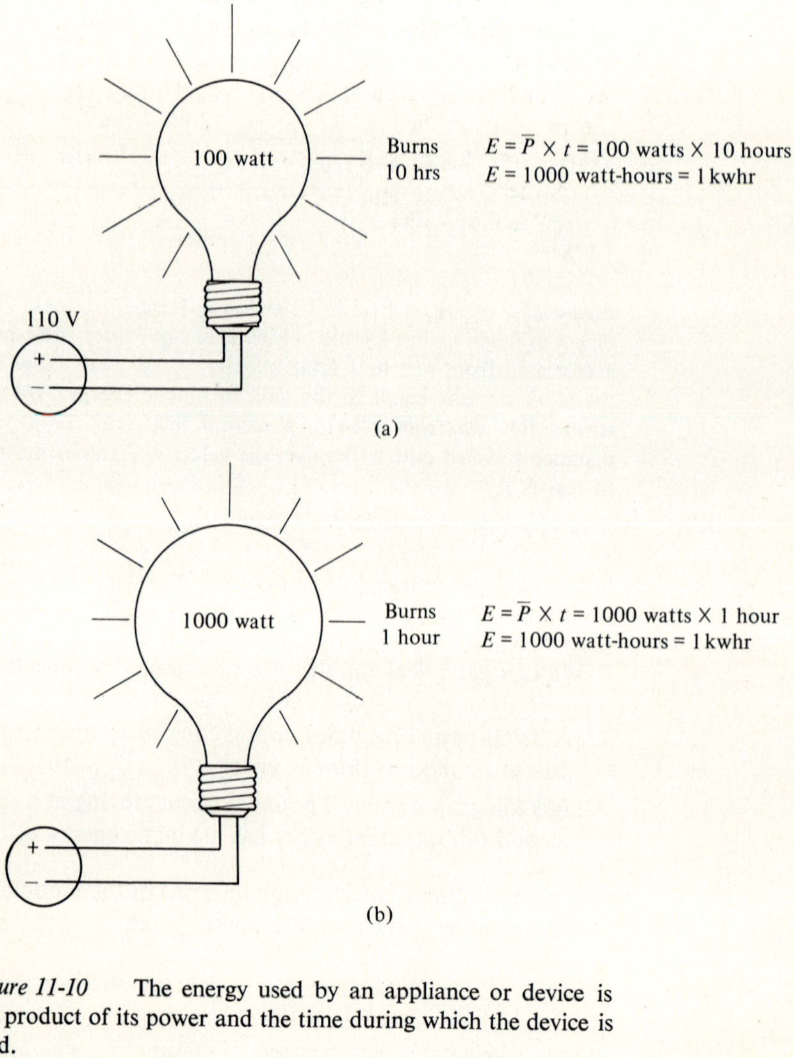

Figure 11-10 The energy used by an appliance or device is the product of its power and the time during which the device is used.

If 1 joule of work is done in 1 second, then the power is 1 *watt*. This is a small unit of power and kilowatts (1,000 watts) or megawatts (10^6 watts) are often used.

From the preceding example (1) we found that 5.12×10^5 joules of work had to be done to accelerate a 3,200-pound automobile from rest to a speed of 60 miles/hr. The work necessary to overcome friction was neglected. The power of this automobile cannot be determined from this information. If we are told that it took 10 sec to reach 60 miles/hr, then the power of this auto is 5.12×10^5 joules/10 sec, which equals 5.12×10^4 watts or 51.2 kilowatts. If it takes 20 sec to accelerate a different 3,200-pound automobile from rest to 60 miles/hr, then the power of this auto is 5.12×10^5 joules/20 sec, which equals 2.56×10^4 watts or 25.6 kilowatts. As might be expected, the more powerful automobile (51.2 kW) reaches a particular speed in a shorter time than the less powerful (25.6 kW) one.

Another unit which is often used in discussing energy is the kilowatt-hour, abbreviated as kWh. It is the energy consumed in 1 hour when the power is 1 kW.

$$1 \text{ kWh} = 1{,}000 \text{ watts-hr} \times 3{,}600 \text{ sec/hr} = 1{,}000 \text{ joules/sec} \times 3{,}600 \text{ sec}$$

$$1 \text{ kWh} = 3.6 \times 10^6 \text{ joules}$$

This same amount of energy (3.6×10^6 joules) could be used over a longer period of time, like 10 hours at 100 watts power. A 100-watt light bulb, if burned for 10 hours, would use 1 kWh of electric energy (Fig. 11-10).

Energy conversion methods and their useful lifetimes

History "A modern industrial society can be viewed as a complex machine for degrading high quality energy into waste heat while extracting the energy needed for creating an enormous catalogue of goods and services."*

"The use of energy has been a key to the supply of food, to physical comfort and to improving the quality of life beyond the rudimentary activities necessary for survival.† The use of energy depends on available resources and the technological skill to convert the resources to useful heat and work. Energy resources have always been generally available, and the heating process is ancient. Power devices able to convert energy into useful work have been a recent historical development. . . .

"Water power for irrigation purposes, exploiting natural differences in elevation, was known in very early times. The horizontal waterwheel

*From *Conversion of Energy* by Claude Summers.

†These paragraphs are from *Energy and Power* by Chauncey Starr.

appeared about the first century B.C. with a power capacity of perhaps 0.3 kilowatts. By the 4th century the vertical waterwheel had been developed to about 2 kilowatts of power. These wheels were primarily used for grinding cereals and similar mechanical tasks. By the 16th century the waterwheel was by far the most important prime mover, providing the foundation for the industrialization of western Europe. The famous Versailles waterworks at Marly-la-Machine is said to have had a power of 56 kilowatts. The windmill first appeared in western Europe in the 12th century. It was variously used for grinding grain, for hoisting materials from mines, and for pumping water. The windmill had a respectable capacity ranging from several kilowatts to as much as 12 kilowatts. The biggest disadvantage was the intermittent nature of its operation. . . .

"Early in the 18th century steam engines using a moving piston were developed as power sources of several kilowatts. The steam engine was the first mechanical prime mover to provide basic mobility. It was some time, however, before this mobility was used. The early Industrial Revolution was based on the waterwheel and the windmill as prime movers: the location of industrial centers, factories, and cities was primarily determined by the availability of those power sources. It was the geographical limitation on the expansion of water power that gave the steam engine an opportunity to continue the growth of manufacturing centers. The first use of the steam engine was as an auxiliary to the waterwheel: to pump water to an elevation sufficient to increase the wheel's power. It was not until the middle of the 19th century that the steam engine became a principal prime mover for the manufacturing industry of the western world."

In the 20th century the manufacturing industry and most households in the U.S. depend on electricity for their energy requirements. Excepting hydroelectric power plants, steam engines (or turbines) are an integral part of all power plants. The transportation sector of our economy requires mechanical energy which is mostly supplied by gasoline, jet, and diesel engines.

Energy Consumption In the United States, energy consumption has recently been increasing at a rate of several percent each year. During the early 1970s, total energy consumption has averaged about 7.3×10^{19} joules (2×10^{13} kWh) per year. From the data in Table 11-1 we can see that the use of energy is almost equally divided among four categories. Household and commercial uses include space heating, air conditioning, cooking, lighting, water heating, and the running of appliances and machines. Motor vehicles are responsible for most of the energy used by the transportation sector of our economy, with trains, aircraft, and ships accountable for the

Table 11-1 CONSUMPTION OF FUEL RESOURCES BY MAJOR CONSUMER GROUPS*

Household and commercial	21%
Industrial	29%
Transportation	25%
Generation of electricity	25%

*Data are for 1971, but the percentages change rather slowly from year to year.
From U. S. Statistical Abstracts 1974

rest. Industrial uses of energy include the running of smelters and blast furnaces, oil refining, mining, manufacturing of chemicals, food, paper, glass, etc. Electricity generation requires about 25% of the total energy, but accounts for only about 10% of the useful work done. Inefficiencies in generation and transmission are largely responsible for this waste.

The greatest growth in energy usage in the U.S. is for electric power generation (about 9% per year). If the household and commercial consumption of electricity continues to grow as rapidly as it has, then the generation of electricity will account for about 50% of the gross energy consumption in the U.S. by the year 2000. The sources of electric energy and their percent of the total energy consumption are listed in Table 11-2. Power generation by nuclear reactors is the fastest growing source of electricity. The U.S. Federal Power Commission estimates that nuclear power plants will account for 22% of the electricity generated in the U.S. by 1980.

Table 11-2 SOURCES OF ELECTRIC ENERGY†

Source	*Percentage of total*
Coal	44.3
Gas	23.3
Hydroelectric	16.5
Oil	13.5
Nuclear	2.4

†Data are for 1971. Except for nuclear power, the percentages are changing slowly from year to year.
From U. S. Statistical Abstracts 1974

About 80% of the electricity is generated by coal, oil, and gas, which are known as fossil fuels. When total energy consumption is considered, we find that about 96% of the total energy is generated by fossil fuels. This is not surprising since the two other main energy uses (transportation and heating) are almost totally dependent on fossil fuels. Even though there is much research in alternative methods of generating electricity and provid-

Figure 11-11 Pacific Gas and Electric Company's 2,000-megawatt Moss Landing power plant, north of Monterey, CA, is the largest fossil-fueled generating plant in the United States. Each of the two units on the right generates 750,000 kilowatts of power, enough to supply a city the size of San Francisco. (Courtesy Pacific Gas and Electric Company)

ing mass transit, fossil fuels will probably be our main energy source through the rest of the 20th century.

Fossil Fuels Fossil fuels can be divided into three main components: coal, oil, and natural gas. It is thought that fossil fuels were formed from decomposing organic matter. It probably took millions of years for our present supply of fossil fuels to form. Although some new formation of fossil fuels takes place continually, the rate is too slow to be of significance. Therefore, fossil fuels can be considered a nonrenewable resource.

In a fossil-fuel steam power plant (Fig. 11-11) the fossil fuel is burned to produce heat which is used to boil water. Steam from the boiler spins a turbine which then turns an electric generator. Figure 11-12 is a schematic diagram of a fossil-fuel steam power plant. The electric generator is essentially

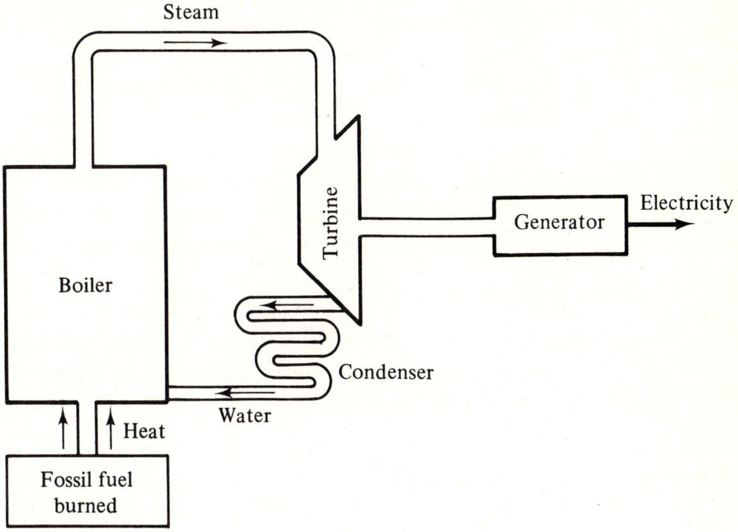

Figure 11-12 A fossil-fuel power plant. In a fossil-fuel power plant one of the fossil fuels is burned to produce heat for boiling water and producing steam. The steam turns a turbine which turns an electric generator.

composed of a number of turns of wire (called a coil) wrapped around an iron core which rotates between the poles of an electromagnet.* A current of electricity is induced in the coil as it cuts through the magnetic field created by the electromagnet. The physical principle which describes this phenomenon is that of electromagnetic induction, discovered by Michael Faraday in 1831.

There are various estimates of the amount of fossil fuels available and also of the length of time the fuels will last. One of the reasons for the variety of estimates is that the reserves of a natural resource often depend on cost of extraction and on present technology. The differences in time estimates stem mostly from varying assumptions concerning production rates of the fuels. Using the U.S. Geological Survey estimates of minable fossil fuels and the assumption that the present production rate of each fuel will only double three more times, the following results are obtained:† for the world as a whole, 90% of the coal will be produced by the year 2300, and 90% of the

*An electrical device that produces a magnetic field when an electrical current is passed through it

†Hubbert, King, "The Energy Resources of the Earth," *Sci. Amer.* (Sept. 1971).

oil will be produced by the year 2020, and for the U.S. 90% of the natural gas will be produced by the year 2015. These estimates can be off quite a bit if new large discoveries of fossil fuels are made. In addition, these estimates do not include such resources as oil shale and tar sands from which oil can be recovered. The amazing aspect of these figures is that the lifetimes of these fuels is measured in terms of hundreds of years or less, even if the estimates are too low by a factor of two. Natural resources that took millions of years to form are being depleted within several centuries.

Hydroelectric power Hydroelectric power is obtained when the potential energy of raised water is converted to kinetic energy as it falls and spins a turbine which turns a generator (Figs. 11-13 and 11-14). The most direct

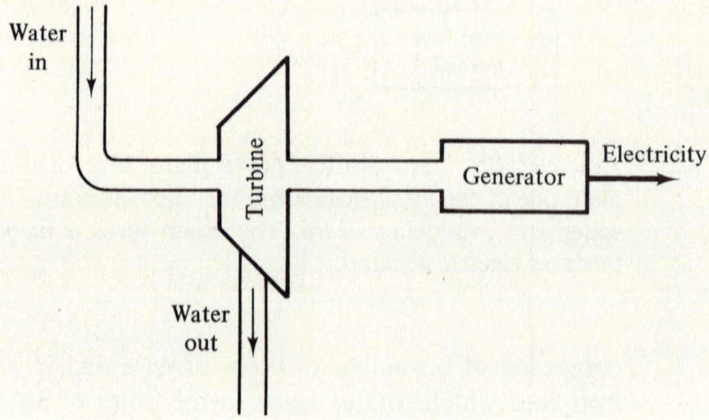

Figure 11-13 A hydroelectric power plant. In a hydroelectric power plant the turbine converts the potential and kinetic energy of falling water into rotational energy. This rotational energy is transferred to the generator where it is converted to electric energy.

and the most efficient energy conversion to electric energy is from falling water. In the U.S. about 16% of the electricity comes from hydropower. It appears that in general the highly industrialized nations cannot substantially increase their hydroelectric power output. First, there are not very many new sites for hydropower left and, second, there is a lot of opposition to building along rivers, for aesthetic and ecological reasons. In these nations (including the U.S.) we can expect the percentage of electricity generated by hydroelectric plants to slowly decrease as energy consumption increases. The greatest potential for water power development is found in Africa, South America, and Southeast Asia, the less industrialized areas of the world.

Figure 11-14 Hydroelectric power plants similar to this project produce about 20% of California's electricity. This plant is one of the largest on the Pit River in the Sierra Nevadas. The large pipes carry water from the reservoir to the hydraulic turbines which are inside the buildings in the lower right-hand portion of the photograph. (Courtesy Pacific Gas and Electric Company)

Nuclear fission In a nuclear power plant (Fig. 11-15) the fission of U^{235} releases the energy needed to make steam which then goes through the same cycle as in a fossil-fuel power plant. The fissioning or splitting of a U^{235} nucleus into two fragments can take place when a neutron is captured by the nucleus. There are many possible reactions. One of them is:

$${}_{92}U^{235} + {}_{0}n^{1} \longrightarrow {}_{54}Xe^{140} + {}_{38}Sr^{94} + 2{}_{0}n^{1} + 200 \text{ MeV}$$

U^{235}, which comprises only 0.7% of natural uranium, is a naturally occurring isotope of uranium which is fissioned by slow neutrons. U^{238}, which accounts for more than 99% of natural uranium, has a very small fission probability as compared to U^{235}; therefore, it cannot be used as fuel in

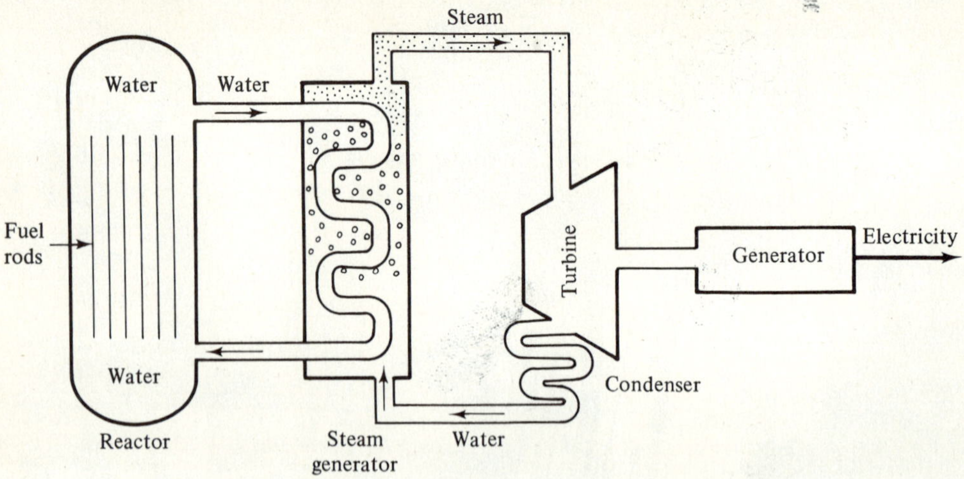

Figure 11-15 A pressurized-water reactor. In a nuclear power plant the fission of U^{235} releases the energy needed to make steam from water. In a pressurized water reactor the water that passes through the reactor is under high pressure and does not turn into steam, even though it is at a very high temperature. In the steam generator, heat is transferred to a separate water system to make steam. The steam goes through the same cycle as in a fossil-fuel power plant.

nuclear reactors. The 200 MeV of energy resulting from each fission reaction is mostly carried away by the kinetic energy of the fission fragments. In the above reaction, the fission fragments are strontium 94 and xenon 140, both of which are radioactive. Some energy is also carried off by the neutrons and by gamma rays.

A self-sustaining chain reaction (Fig. 11-16) is possible because the fissioning of the U^{235} nucleus releases extra neutrons (each fission gives 2.5 neutrons on the average). They more than compensate for the loss of the neutron whose capture initiated the fission event. A certain amount of fissionable material, the critical mass, must be present in a reactor for the chain reaction to be self-sustaining. It is this same chain reaction which is essential to the functioning of an atomic bomb.

In each fission of a U^{235} nucleus, about 0.1% of the original mass of the nucleus is converted to energy. The amount of energy, E, released when some amount of mass, m, is converted into energy is given by Einstein's famous formula:

$$E = mc^2$$

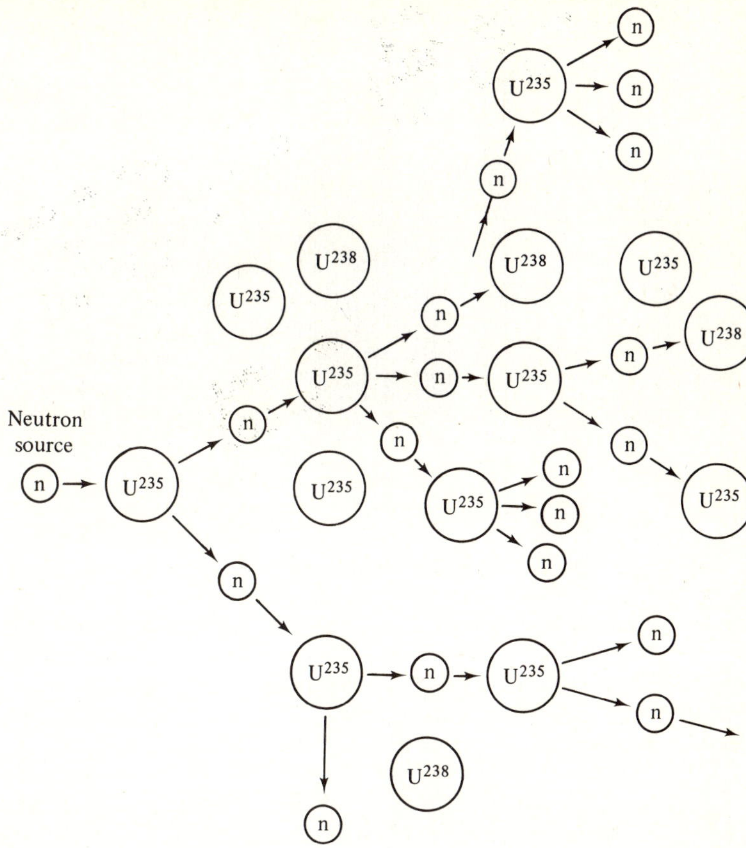

Figure 11-16 In a chain reaction, some of the fission neutrons strike other U^{235} nuclei and cause them to fission while others are absorbed or escape. If the number of neutrons produced per unit time exceeds the number absorbed or lost, then the chain reaction is self-sustaining.

where c is the speed of light in a vacuum ($c = 3 \times 10^8$ meters/second). The conversion can be from energy to mass or mass to energy. If mass is expressed in kilograms in the above equation, then energy will be in joules.

The main components of a nuclear reactor are the fuel, the coolant, the moderator, and the control rods (Fig. 11-17). As a fuel, present-day power reactors use uranium which has been enriched in the proportion of the fissionable isotope (U^{235}) that it contains. The probability of a U^{235} nucleus fissioning increases when the energy of the incident neutron decreases. In order to take advantage of this fact, a moderator is used to slow down the

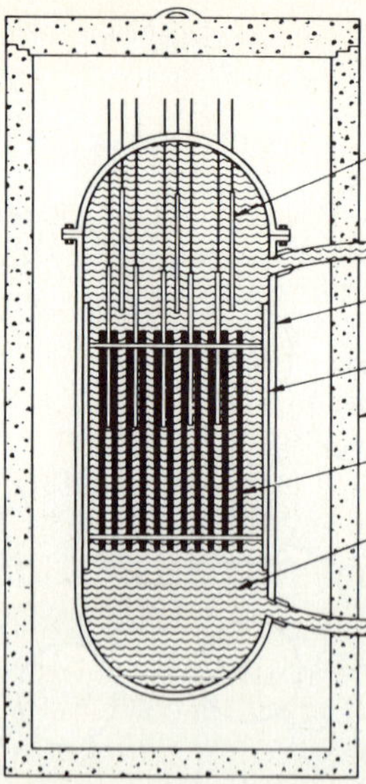

Figure 11-17 Vertical cross section of a pressurized-water reactor. (From Hagerton, J., "Nuclear Reactors," U.S. A.E.C. Understanding the Atom Series)

neutrons from the fast speeds at which they are released in the fission reaction. Water is usually used for this purpose. It also serves as the coolant in most power reactors. Without a coolant, the reactor core (fuel plus moderator) would become hot enough to melt the fuel and supporting structures. The main sources of heat are the fission fragments which transfer their kinetic energy to the fuel rods, causing them to warm up and transfer their thermal energy to the surrounding water.

The heat from the core is carried away either in the form of steam (boiling water reactor—BWR) or in the form of heated pressurized water (pressurized water reactor—PWR). The heated water from a PWR is passed through a heat exchanger (Fig. 11-15) where its thermal energy is used to make steam from water circulating in a separate closed cycle. From this point on, the steam from a nuclear power plant goes through the same cycle as in a fossil-fuel power plant.

The other major component of a nuclear reactor is the control rod. It is made of materials such as boron and cadmium which readily absorb neutrons. The function of the control rods is to control the rate at which the chain reaction proceeds and, therefore, the power output of the reactor. Partially removing the control rods from the reactor core increases the rate of the chain reaction, and moving the rods deeper into the core slows the chain reaction.

In the case of an emergency or for the purpose of a maintenance shutdown, the control rods are pushed all the way into the core to stop the chain reaction by absorbing neutrons which are needed to keep the chain reaction going.

The first self-sustaining chain reaction took place at the University of Chicago in 1942. This reactor was assembled by Enrico Fermi and his associates in a squash court under the stadium of Stagg Field. Unenriched uranium was used as the fuel and graphite as the moderator. This reactor, which proved that a chain reaction could be produced and controlled, was the prototype of all the power reactors that have followed it. However, in 1942 the motivation was to produce an uncontrolled chain reaction (a bomb!), and the Chicago reactor facilitated this end by showing that the conversion of U^{238} into Pu^{239} by neutron absorption was a practical process. Plutonium 239, like U^{235}, is fissionable and could be used for bomb material.

Once the Chicago reactor proved that a chain reaction could be maintained, reactors specifically designed for the manufacture of plutonium were built at Oak Ridge, Tennessee, and Hanford, Washington. The first atomic bomb contained U^{235} and the second contained Pu^{239}. The separation of U^{235} from U^{238} for use as bomb material required the construction of costly separation plants. A book by Nuel Pharr Davis* describes the building of the first atomic bombs in terms of the personalities and interests of those scientists most involved in the decision making.

The supply of U^{235} is very dependent on economic factors. Uranium occurs naturally as uranium oxide (U_3O_8). It sells at $5 to $10 a pound. If nuclear reactors supply 22% of our electric energy by the year 2000, then the cheaper uranium ore will have been used up by that time and the price of uranium oxide will climb to around $30 to $50 a pound. Nuclear fuel is being used at such a fast rate that there will probably be a shortage of enriched uranium by the early 1980s. If we are not willing to pay quite a bit more for uranium fuel, then the lifetime of ordinary nuclear reactors as a power source will be severely limited. It has been estimated that if we are willing to use uranium reserves in the $30- to $500-dollar range, then there would be enough fuel for thousands of years.

Breeder reactors The U.S. government announced in 1971 that the intensive development of the fast breeder reactor would be national policy. There is a commitment to complete the successful demonstration of the liquid metal fast breeder reactor by 1980. The justification for this policy is the expected shortage of U^{235}. In a breeder reactor, excess neutrons from the fissioning of Pu^{239} are used to convert nonfissionable U^{238} and Th^{232} into

*Davis, Nuel Pharr, *Lawrence and Oppenheimer*. New York: Simon and Schuster, 1968.

the fissionable isotopes Pu^{239} and U^{233}, respectively (Fig. 11-18). Plutonium 239 is created in the beta decays which follow the absorption of a neutron by U^{238}:

$${}_{92}U^{238} + {}_{0}n^{1} \longrightarrow {}_{92}U^{239} \longrightarrow {}_{93}Np^{239} + {}_{-1}e^{0}$$ U^{239} has a half-life of 24 min.

$${}_{93}Np^{239} \longrightarrow {}_{94}Pu^{239} + {}_{-1}e^{0}$$ Np^{239} has a half-life of 2.33 days.

By breeding, it is possible to create more fuel than is consumed! This means that for every fission reaction, more than one of the fission neutrons must be absorbed by an atom of U^{238} to create more than one new atom of Pu^{239}. If breeder reactors are successful, the entire supply of natural uranium and thorium would become available as fuel for fission reactors. The use of breeder reactors would extend the fuel supply for fission reactors by more

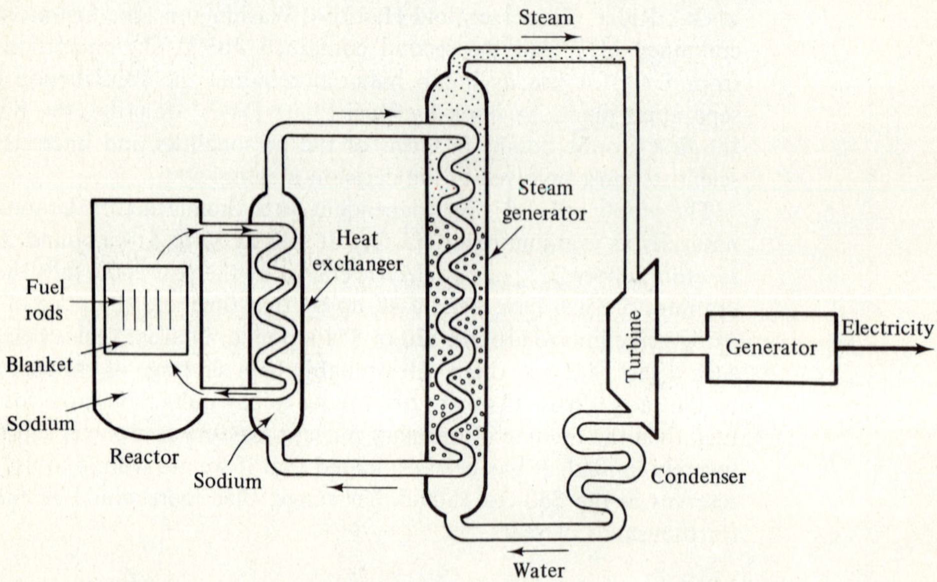

Figure 11-18 A liquid-metal, fast-breeder reactor system. In a breeder reactor, surplus neutrons are captured by a blanket of nonfissioning atoms of U^{238} or Th^{232} which are transformed into fissionable Pu^{239} or U^{233}. The heat of the reactor is removed by liquid sodium. In the steam generator, heat is transferred from the liquid sodium to a separate water system to make steam. The steam goes through the same cycle as in a fossil-fuel plant.

than a factor of 100 compared to the supply if only U^{235} were used in fission reactors.

In addition to the U.S., France, England, and the USSR are all in the race to produce a commercial breeder reactor. In the U.S. there are presently three experimental breeder reactors and two planned for operation in the late 1970s. The two basic kinds of breeder reactors being developed are the liquid-metal fast breeder reactor (LMFBR), which converts U^{238} into Pu^{239}, and the molten salt breeder reactor (MSBR) which converts Th^{232} into U^{233}.

Figure 11-19 The fast flux test facility (FFTF) at A.E.C.'s Hanford works in Washington state is the key project for most of the technical work presently planned in the liquid-metal fast breeder reactor development program. In this photograph, WADCO engineers discuss operation of the simulated core mock-up for the FFTF. (Courtesy U.S. A.E.C. Photo by WADCO)

Most of the U.S. research is on LMFBRs which have a very compact core of highly enriched Pu^{239} and which require fast neutrons for efficient conversion of U^{238} into Pu^{239}. Since fast neutrons are needed in a LMFBR, there is no moderator. The fission energy is generated in a very small space in a fast breeder reactor, and water cannot be used as a coolant since it cannot be circulated through the core fast enough to carry the heat away. Another reason water could not be used as a coolant is that it would also act as a moderator and slow down the neutrons. Liquid sodium, which has a boiling point of 895°C, is used as a coolant since it permits high temperature (without boiling) and low pressure operation. Breeder reactors hold a lot of promise for satisfying our ever-growing energy demands, but they are also more dangerous than ordinary reactors. (See Chapter 12.)

Nuclear fusion In the course of his studies in 1905 on the theory of relativity, Albert Einstein showed that mass and energy were, in a sense, equivalent. Consequently, energy should be liberated in any process associated with a net decrease in mass. By considering the measured masses of nuclei, it became apparent that there were two ways in which nuclear energy could be made available. One is by splitting (or fission) of the heaviest nuclei into two roughly equal parts, and the other is by the combination (or fusion) of some of the lightest nuclei. Actually, there are many other nuclear processes which are accompanied by a liberation of energy, but only with nuclear fission and fusion is there a possibility of producing more energy than is consumed in causing the reaction to occur. In other words, there is some chance that the process, once started, can be self-sustaining like a fire.

Nuclear fusion is an attractive method of energy conversion because of the extremely large and cheap fuel supply, and also because of the low level of radioactivity expected from fusion machines. However, there are tremendously difficult problems to solve before fusion power can be a reality. (See Fig. 11-20.)

We know that nuclear fusion is possible. It is reasonably certain that fusion is the energy source of the sun and other stars. It is thought that, in a series of nuclear reactions, four hydrogen nuclei are fused together to form a helium nucleus, releasing energy which is eventually converted to electromagnetic radiation (mostly visible light). Nuclear fusion is also the source of the large amount of energy released in exploding the so-called H bomb. (The bomb actually contains deuterium and tritium and no ordinary hydrogen.) Finally, fusion reactions have been produced in the laboratory, but on such a small scale that more energy was put into making the reactions occur than was released by the fusion.

Requirements for fusion reactions Fusion reactions with deuterons (nuclei of deuterium atoms; deuterium is an isotope of hydrogen) and tritons

Figure 11-20 Enormous machines and complex equipment are required for nuclear fusion research. One of these is the giant Astron facility at the Lawrence Radiation Laboratory, Livermore, CA, shown here in an overhead view. (Courtesy U.S. A.E.C.)

(nuclei of tritium atoms, also an isotope of hydrogen) take place fast enough to make them possible sources for the release of energy at a useful rate. The possible reactions are:

$$\left.\begin{array}{l} {}_1H^2 + {}_1H^2 \longrightarrow {}_2He^3 + {}_0n^1 + 3.2\text{ MeV} \\ {}_1H^2 + {}_1H^2 \longrightarrow {}_1H^3 + {}_1H^1 + 4.0\text{ MeV} \end{array}\right\} \text{deuteron-deuteron reactions}$$

$${}_1H^2 + {}_1H^3 \longrightarrow {}_2He^4 + {}_0n^1 + 17.6\text{ MeV} \quad \text{deuteron-triton reaction}$$

Since the fusing nuclei have positive charges, and like charges repel each other, these nuclei must be given high energies to overcome this repulsion. The high kinetic energies of the nuclei correspond to temperatures of tens of millions degrees. At these high temperatures the fuel is a gas in which all the atoms have been stripped of their electrons. This state of matter, in which all the particles of a gas are charged, is called a *plasma.*

The plasma must be confined for a long enough time to obtain more energy from fusion reactions than was supplied for heating and confining it. In order to accomplish this, a fuel density which brings about sufficient collisions is required. The plasma cannot be confined by the usual container since it would cool off in collisons with the walls of the container. Even though the plasma is at a very high temperature, the total heat content is small and much of it can be quickly transferred to the walls of a container during collisions.

Complicated configurations of electromagnets called "magnetic bottles" have been used to contain plasmas. Charged particles experience forces upon them in a magnetic field, and the idea of a magnetic bottle is to arrange the magnetic field so that the plasma is confined to a small volume. Thus far, instabilities develop which lead to leakage of the plasma ions from the magnetic field and prevent achievement of the necessary confinement time.

Laser-induced fusion has recently joined magnetic confinement fusion as a prime prospect for generating controlled fusion power. The idea is to use laser light to implode pellets of deuterium and tritium, compressing and heating them enough to induce fusion reactions (Fig. 11-21). The result is essentially a mini H bomb. Calculations indicate that fusion energies 50–100 times larger than laser input energies can be achieved. There is no experimental confirmation of this theoretical prediction, but work on laser-fusion is proceeding at a very rapid rate in both the U.S and the USSR.

Lasers can be built to produce tremendous amounts of energy in very short time periods. Ten-thousand-joule lasers are being built to explore laser-induced fusion. If this amount of energy is emitted in 10^{-9} second, the instantaneous power is

$$\text{Power} = \frac{\text{Energy}}{\text{time}} = \frac{10{,}000\text{ joules}}{10^{-9}\text{ sec}} = 10^{13}\text{ watts}$$

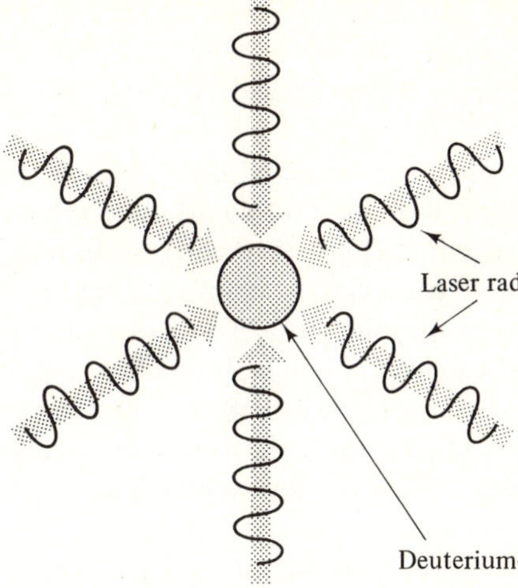

Figure 11-21 Absorption of laser light pulses causes the deuterium and tritium to heat up and become compressed. It is hoped that fusion reactions will be induced by this method, and that more energy will be released than had to be supplied to run the lasers.

The deuterium-deuterium reaction is extremely attractive as an energy source since deuterium is found naturally in sea water in the ratio of 1 deuterium atom to 6,700 atoms of hydrogen. This may not sound like much, but there is an awful lot of water in the ocean. Deuterium can be easily and cheaply extracted from sea water and, if the deuterium-deuterium reaction could be used, there would be enough fuel to supply the world's energy needs for billions of years at the present level of consumption. The problem is that the deuterium-deuterium reaction requires a temperature ten times higher than the deuterium-tritium reaction. Tritium, however, is very scarce and must be obtained from lithium by neutron bombardment.

$$_3L^6 + {}_0n^1 \longrightarrow {}_1H^3 + {}_2He^4 + 4.8 \text{ MeV}$$

Lithium is in shorter supply than deuterium, but there is enough in the seas that can be converted to tritium to supply all the world's energy needs for millions of years at the present rate of consumption. Now that we can see that fusion power would provide an almost unlimited source of energy, we can understand why so much effort is being applied to its achievement.

Solar energy Another possible method of energy conversion would use solar energy. At this time there is still some controversy concerning its technical feasibility in large-scale operations. Solar energy has already been converted to electricity on a small scale. Satellites and other space probes have used panels composed of many small silicon cells to convert solar energy to electric energy for use on board these vehicles. Each cell develops a small electrical potential (like 1/2 a volt), so in order to obtain a large electrical potential it is necessary to connect many of them together. The efficiency of conversion of solar energy to electrical energy is only about 10% and the cost of each cell is high. It has been suggested that a very large solar collector could be placed on a satellite to convert solar energy to electrical energy (Fig. 11-22), and then the electric energy could be converted to electromagnetic radiation (of a wavelength different from visible light), and transmitted to the earth where it would be reconverted to electric power.

Solar energy has been used to heat some specially designed homes and buildings. A firm in California is presently marketing solar panels which are to be used for heating backyard pools. The panels, constructed of plastic,

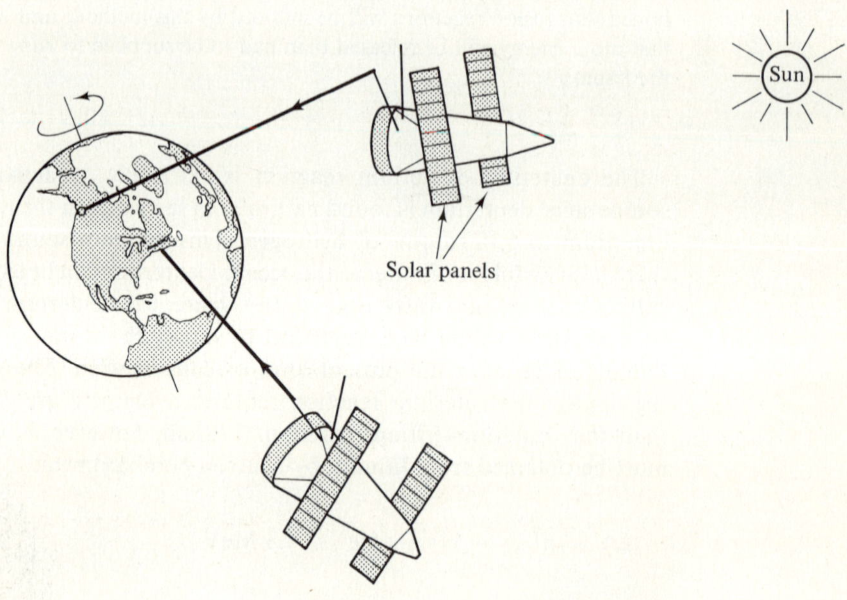

Figure 11-22 Satellites located about 22,000 miles above the earth can remain above a particular city on a continuous basis. The solar panels collect solar energy and convert it to electric energy. This electric energy is converted to electromagnetic energy (microwaves) which is sent down to earth. This energy conversion scheme is being studied.

can be placed on roof tops or fences, and the pool water pumped through pipes running beneath the panels. It is known that solar energy can be used for heating and for producing electricity, but it is not yet clear if this method of energy conversion will be competitive on a large scale with other energy conversion methods.

A solar energy conversion system Even though the sun is 93 million miles from the earth, and even though the radiation emitted spreads as it propagates out from the sun, we still receive as much as an average of 60 watts of power per square foot at the earth's surface (Fig. 11-23). If we could convert

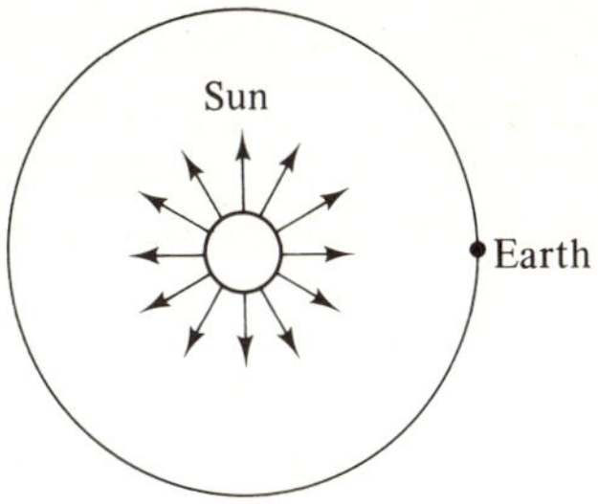

Figure 11-23 Only a very small fraction of the sun's radiation reaches the earth, but it is enough to supply 60 watts of power to a square foot surface area perpendicular to the incident radiation.

this solar radiation to electricity with an efficiency of 100%, then we could light a 60-watt light bulb continuously as long as the sun was out over our 1 square foot of the earth's surface. We can perform a simple calculation to determine how much land we would need to generate enough electricity to supply the total needs of the U.S.

If the sun's radiation is perpendicular to our collecting surface, then the average amount (averaged around the clock) of radiation received at this surface would be 60 watts/ft^2. Since silicon solar cells are too expensive to use on a large scale, we would use specially coated surfaces which absorb a large fraction of the incident solar radiation. These surfaces are heated and their thermal energy is transferred to liquid sodium. A heat exchanger transfers the thermal energy from the liquid sodium to water which changes to steam (Fig. 11-24). The steam goes through the same cycle as in a conventional power plant. A salt reservoir can be used to store enough thermal energy to keep generating steam during the night, or when the sun is hidden by cloud cover. We assume an overall efficiency of 33% for this solar power

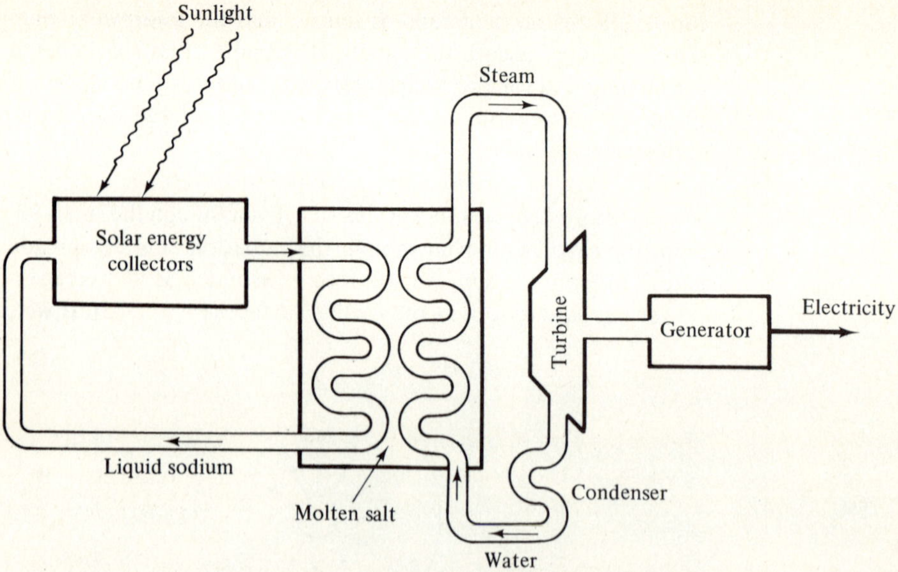

Figure 11-24 A proposed solar power plant. In a solar power plant, specially coated surfaces collect solar energy and their energy is then transferred to a liquid metal such as sodium. Heat is transferred from the sodium to a separate water system to make steam. The steam goes through the same cycle as in a fossil-fuel plant. Some of the heat from the liquid sodium is also transferred to a molten salt where it is stored for use in cloudy weather or at night.

plant and calculate the land area needed to generate 10^{12} watts of power continuously. This is approximately the electric power needs of the U.S. Before we complete this calculation we should reiterate that this is all theoretical and that there is presently no pilot project to determine experimentally if this solar conversion plant is technologically feasible.

We want an electrical power output of 10^{12} watts, but our thermal conversion efficiency is 33%. Therefore, we can calculate the necessary thermal input from

$$\text{efficiency} = \frac{W}{Q}$$

$$0.33 = \frac{10^{12}\ \text{watts}}{Q} \qquad Q = \frac{10^{12}\ \text{watts}}{0.33} = 3 \times 10^{12}\ \text{watts}$$

If we divide this thermal power by the solar input per unit area, then we will have the area necessary to generate 10^{12} watts continuously:

$$A = \frac{3 \times 10^{12} \text{ watts}}{60 \text{ watts/ft}^2} = 5 \times 10^{10} \text{ ft}^2$$

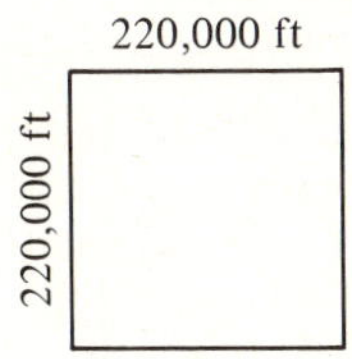

This is the area of a square 2.2×10^5 ft on a side or about 40 miles on a side. But since the sun's radiation is hardly ever perpendicular to the earth's surface at any latitude within the U.S., the average receipt of radiation will be less than 60 watts/ft^2, so we will need a square about 70 miles on a side. Even at equatorial latitudes, the sun's radiation is only perpendicular to the earth's surface for some of the daylight hours and only during certain times of the year.

This large land area would have to be concentrated in those areas of the U.S. that experience maximum sunshine, such as southeastern California and western Arizona. The consequences of generating a large amount of energy in such a small region will be discussed in the next chapter.

Geothermal energy Geothermal energy, which comes from the earth's hot interior, is already being tapped as a cheap source of steam. Extensive geyser steam fields have been discovered in California, 90 miles north of San Francisco, and have already been developed (Fig. 11-25). These energy sources will probably not be important on a national scale, but they do have regional impact. For example, The Geysers field is estimated to have a maximum potential capacity of about 4,000 megawatts, which is enough electrical power to satisfy the needs of the San Francisco Bay Area (4.5 million people). Many other nations, such as Italy, Mexico, and Japan, have developed geothermal energy sources.

The heat of many geothermal reservoirs comes from a large body of molten rock (magma) that has been pushed up into the earth's crust from great depths by geologic forces. This molten rock heats the rocks in the crust near the surface, which in turn heat the water in fissured or porous rocks to a temperature of perhaps 500°F (Fig. 11-26). Since the water may be as much as six miles below the surface, the water is under high pressure and is therefore a liquid. The hot water begins to boil where it escapes through a fissure, and part of it comes off as steam. The geothermal energy can be

Figure 11-25 Steam wells tap thermal energy for the production of power at The Geysers, about 90 miles north of San Francisco. By the end of 1976, power output here is expected to be about 900 megawatts. (Courtesy Pacific Gas and Electric Company)

tapped by a well driven into the fissure or down to the porous layer, and the steam can be fed directly to a turbine.

The development of geothermal resources became an official national goal in the U.S. in December 1970 when Congress adopted the Geothermal Steam Act. The additional uses of geothermal fields (aside from the generation of electricity) have served as an impetus to this further exploration. The geothermal steam or hot water can be applied to desalting seawater and to heating houses and swimming pools. The hot water itself is a source of ex-

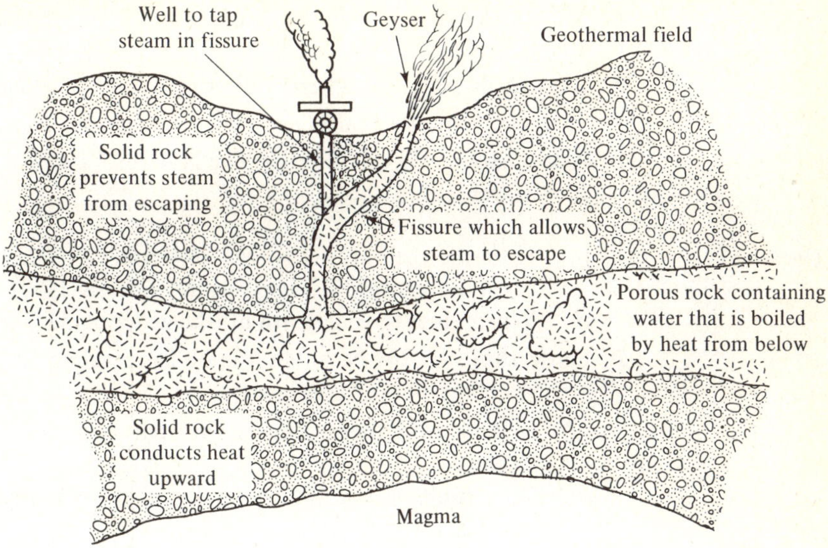

Figure 11-26 The source of geothermal energy is molten rock that has been pushed up into the earth's crust.

tractable minerals and can also serve to provide potable water. In the future we can look forward to the discovery of many more steam fields as better methods are developed to locate them.

Suggested reading

1 Holdren, John, and Philip Herrera, *Energy*. San Francisco: Sierra Club, 1971.

2 Schrodinger, Erwin, *What Is Life*? New York: Macmillan Co., 1946.

3 "Energy and Power," *Scientific American* (Sept. 1971).

4 "The Energy Crisis: Parts 1 and 2," *Bulletin of the Atomic Scientists*, (Sept. and Oct. 1971).

5 U.S.A.E.C. Understanding the Atom Series, Oak Ridge, Tennessee. There are many informative pamphlets in this series for the general public. The following are particularly relevant to this chapter: "Nuclear Power Plants," "Controlled Nuclear Fusion," and "Breeder Reactors."

6 "Report on Nuclear Energy," *Physics Today* (August 1973). The whole issue is pertinent to this chapter.

7 Culler, F., and W. Harms, "Energy from Breeder Reactors," *Physics Today* (May 1972).

8 MEINEL A., and M. MEINEL, "Physics Looks at Solar Energy," *Physics Today* (Feb. 1972).

9 BARNEA, JOSEPH, "Geothermal Power," *Scientific American* (Jan. 1972).

10 TAMPLIN, ARTHUR "Solar energy," *Environment*, **15**: 5 (June 1973).

11 ABRAHAMSON, DEAN, "Energy, The Nuclear Fast Breeder," *Environment*, **15**: (March 1973).

Questions

1 (*a*) Describe the various energy conversions necessary to convert the potential energy of raised water into electrical energy.
(*b*) Where does the energy to lift the water come from?

2 Show that a 3,200-pound automobile traveling at 60 miles/hr has a kinetic energy of 5.1×10^5 joules.

3 (*a*) Is there more thermal energy in a cup of hot coffee or in the Pacific Ocean?
(*b*) Why can't we use the thermal energy contained in the oceans to do useful work?

4 (*a*) What is the function of a heat engine?
(*b*) What is the formula for the maximum theoretical efficiency of a heat engine?
(*c*) Nuclear power plants operate at lower efficiency than fossil-fuel power plants because present nuclear reactors cannot be run as hot as boilers burning fossil fuels. The temperature of the steam produced by a boiling water reactor is around 350°C. Assume that T_1, the temperature of the cold reservoir, is 100°F. What is the maximum theoretical efficiency of a steam turbine whose input is from a nuclear reactor?
(*d*) How does this efficiency compare to that of a steam turbine which is fed steam from a fossil-fuel boiler?
(*e*) What else do we need to know in order to compute the overall efficiency of a nuclear power plant?

5 (*a*) How many joules of energy are used by a 100-watt light bulb which operates for 1 hour?
(*b*) If one-tenth of a kilowatt-hour of electric energy is used, how many joules of energy does this correspond to?
(*c*) What is the average power developed by a light bulb which converts 1,000 joules of electric energy into light and heat in 10 seconds?

6 (*a*) If the efficiency of a 1,000-megawatt power plant is 40%, what is the thermal power input to this plant?
(*b*) At what rate is waste heat produced?

7 (*a*) Why are fossil fuels considered to be a nonrenewable natural resource?

(*b*) Give at least two reasons for the varying estimates of the length of time a natural resource is expected to last.

8 (*a*) Describe the various energy conversions or transfers involved in obtaining electricity from fossil fuels.
(*b*) From U^{235}.
(*c*) From solar radiation. Use block diagrams.
(*d*) From raised water.
(*e*) Which of the above methods involves the least number of energy conversions or transfers?

9 (*a*) What is the mass of a U^{235} nucleus in kilograms?
(*b*) If, as a result of nuclear fission, 0.1% of the original mass of the uranium nucleus is converted into energy, how much energy results from each fission? In joules? In MeV?

10 (*a*) If 1 gram of U^{235} fissions, how much energy is released?
(*b*) How much U^{235} would be consumed (that is, undergoes fission) each day by a nuclear power plant producing 1,000 megawatts of power with an efficiency of 33%?

11 (*a*) How do breeder reactors differ from ordinary reactors and why are they being developed?
(*b*) What are some of the disadvantages of fast breeders as compared to ordinary reactors?

12 (*a*) Is it dangerous to hold a noncritical mass of U^{235} in your hand?
(*b*) How about Pu^{239}?

13 What are the functions of water in an ordinary nuclear reactor?

14 If you were dispensing research money, would you allot more for research on controlled fusion or solar energy? Why?

15 Make a rough estimate of the land area needed for a 1,000-megawatt solar power plant. Assume it is located where the incident solar radiation is normal to the earth's surface.

Chapter 12
Environmental and human health impacts of power production

Acquiring raw materials

Coal and uranium mining Extracting and transporting the fuels needed for power production has had a severe impact both on the environment and on the health of those individuals involved in these activities. Black lung disease and accidents have always plagued miners. Coal mining has been particularly damaging to the environment; acid drainage, soil erosion, and visual scars have resulted. Acid drainage results when compounds of sulfur (found near the coal) react with air and water to form sulfuric acid. It mostly occurs around coal mines and can be damaging to aquatic life when the acid runs off into streams. One of the areas in which a particularly damaging form of mining is taking place is the Four Corners region of the United States (where four states meet: Arizona, New Mexico, Utah, and Colorado). Figure 12-2 shows the aftermath of strip mining for coal. In this type of mining, the coal is near the surface and the overburden is stripped away with large earth-moving machines. The mining companies involved with this operation have undertaken to reclaim the land, but the results of their efforts will not be known for some years.

It has been found by personnel of the U.S. Bureau of Mines that about two million acres of land have subsided as a result of being undermined; that thousands of miles of streams and tens of thousands of acres of lakes have

Figure 12-1 Coal mining has always been a very dangerous occupation. As can be seen in this photograph of a miner applying rock dust to a working face area, it is often necessary to work in a stooped position with uncomfortable equipment such as a face mask and a helmet. (Courtesy U. S. Department of the Interior, Bureau of Mines)

Figure 12-2 About half of the coal mined in the U. S. is now strip-mined. Large shovels strip away overburden from shallow beds of coal. The 160-foot tall machine pictured here can lift 500,000 pounds 100 feet into the air. Most strip mining is being done in the western states such as Montana, Arizona, and New Mexico. Fewer people are put in physical danger in strip mining than in underground coal mining. (Courtesy U. S. Department of the Interior, Bureau of Mines)

been polluted as a direct result of mining and mineral processing. In addition, billions of tons of waste material from underground mining and processing are spread over millions of acres of land, and hundreds of uncontrolled fires are burning in mining waste piles, refuse banks, and abandoned mines. An underground coal mine fire near Laurel Run, Pennsylvania, has been burning for more than 50 years. Subsidence and poisonous gases have caused the abandonment of homes and businesses. Some of the toxic gases resulting from the fire are carbon monoxide, hydrogen sulfide, sulfur dioxide, and ammonia. There is no effective way of extinguishing all coal refuse bank fires.

The mining of uranium is a dangerous trade not only because of the hazards of mining in general, but also because of the presence of radioactive radon gas in the mines. Studies have shown a higher incidence of lung cancer among uranium miners than among other miners or the general population. Radon gas and its decay products can be inhaled and can reach the lungs. Since alpha particles (which have a short range) are emitted during the decay of radon to a stable isotope of lead, a large amount of energy can be absorbed in a small region of the lungs. The resulting ionization of lung tissue can produce lung cancer.

Another major problem arising from uranium mining has been the accumulation of large piles of tailings near the mines. Tailings are the sand-like remains of the uranium ore after it has been processed to remove the uranium. The tailings contain radium, which decays to radon gas. Radon gas can diffuse through the tailings, through soil, and even through concrete. During congressional hearings in 1971 the public learned that the town of Grand Junction, Colorado, was built on tailings from a nearby uranium mine. The controversy that ensued centered around these questions: What radiation level was high enough to warrant removal of the tailings from beneath structures? Who was financially responsible? For more details on Grand Junction see Chapter 6. In addition to the landfill problem, there is also the problem of these tailings blowing and washing into waters used for drinking or irrigation. If the radium concentration gets high enough, this could also constitute a public health hazard.

Oil and natural gas Obtaining oil and natural gas has become increasingly associated with offshore drilling of wells. Leakage and larger oil spills involved with drilling are a major problem because of the harmful effects of petroleum on marine organisms and birds, but the Department of the Interior claims that the risk of a large oil spill from drilling operations is very small because of recently strengthened control of the outer continental shelf program. They say that hurricanes can be a problem, but that there are safety features on all the wells to take this possibility into account.

Crude oil and petroleum products contain many substances* that are poisonous to marine life, and transportation of oil by ship has resulted in

*Toluene and benzene are two of the breakdown products of petroleum that are hazardous to marine life.

Figure 12-3 Offshore drilling rigs are becoming more numerous along the coast of California, Texas, and Louisiana. (Courtesy U.S. Department of the Interior, Bureau of Mines)

several disasters where tankers have collided or otherwise spilled their oil into lake, bay, or ocean water. The effects of these disasters, however, were not investigated immediately as were the effects of the spill in 1969 when an oil barge on the way to a power plant on the Cape Cod Canal ran ashore off West Falmouth, Massachusetts, in Buzzards Bay, and released between 650 and 700 tons of fuel oil into the coastal waters.

This small oil spill was studied immediately because it occurred near the Woods Hole Oceanographic Institute where both current information on local marine life and trained personnel were available for the investigation.

Massive immediate destruction of marine life occurred offshore during the first few days after the accident. A wide range of fish, crabs, lobsters, and other crustaceans and invertebrates were killed. Blue mussels that were juveniles in the polluted area at the time of the oil spill were generally sexually sterile the next season. It was found, contrary to previous expectations, that hydrocarbons taken up into the fat and flesh of fish and shellfish are not removed by natural flushing or by internal metabolic processes.

The hydrocarbons can then be passed from prey to predator where they may become a hazard to marine life and eventually to man.

In addition to its direct effects on marine life, it appears that oil may concentrate other fat soluble poisons, such as many insecticides and chemical intermediates. Dissolved in an oil film, these poisons may reach a concentration many times higher than that which occurs in the water column.

This oil spill was a small one compared to the Torrey Canyon tanker spill in 1967 which left 100,000 tons of crude oil in the ocean, and even though spectacular accidents, such as those at Santa Barbara, San Francisco and the Gulf of Mexico, made newspaper headlines, their contribution to the total oil pollution of the ocean is small. Far more oil reaches the oceans via routine discharges from tankers and other commercial vessels. As oil tankers increase in size, so does the potential damage they can cause in an accidental spill.

An alternative to the shipment of oil by tanker is to use an oil pipeline. There is a lot of opposition to this proposal in the case of Alaskan oil because of the possible effects of an oil leak on a fragile environment such as the tundra of Alaska. However, the United States government has approved construction of the Alaskan pipeline.

Energy conversion effects with fossil fuels

The effects of carbon dioxide and particulates on climates The combustion of fossil fuels for power production is responsible for about 25% of the particulates emission and a significant fraction of the carbon dioxide emission in the U.S. In Chapter 9 we discussed the ways in which the addition of carbon dioxide and particulates to the atmosphere can alter the climate. The atmospheric concentration of both of these substances has been increasing; this is true even in remote areas of the world. Carbon dioxide and particulates have opposing effects on the climate; CO_2 tends to warm the earth through the greenhouse effect, and particulates tend to cool the earth by absorbing and scattering incoming solar radiation. Volcano eruptions add particles to the atmosphere and also cause the earth to cool.

The average annual global temperature has been decreasing since 1940. Some scientists feel that natural causes such as volcanos have produced most of the temperature decrease, while others feel that the increasing concentration of particulates in the air is responsible. It is difficult to make quantitative predictions of the effects of adding CO_2 and particulates to the atmosphere. Particulate emissions from fossil-fuel power plants are being controlled, but industrial emissions are increasing. The only thing we can do about carbon dioxide emission is to burn less fossil fuels. The use of more nuclear power plants and other methods of power production (solar and fusion) will decrease the future output of particulates and CO_2. At this time, as a consequence of man's activities, we still do not know whether the earth will warm up or cool off.

Other combustion products Most of the other pollutants that arise from combustion of fossil fuels are of concern only within the region where they are emitted to the atmosphere. Some of the more important ones are sulfur oxides, carbon monoxide, nitrogen oxides, and hydrocarbons. The problems created by these pollutants (mostly in urban areas) include material damage to vegetation and structures, physical discomfort, loss of visibility, and public health hazards. (See Chapter 9.) Air pollution disasters have occurred in cities such as Donora, Pennsylvania, and London, England. (See Chapter 8.) In both episodes sulfur dioxide and particulates were implicated as the most harmful combination of pollutants present.

In the United States, power plants account for about 20% of the particulate emissions and about 70% of the sulfur oxides emissions. Natural gas is essentially sulfur free, but there is not nearly enough available to supply all the existing fossil-fuel power plants. Oil-fired power plants emit less sulfur oxides than coal-fired plants; however, there now appears to be a shortage of oil, low-sulfur oil in particular. As a result, power plants will begin to use more coal than they have in the past. Large deposits of low-sulfur coal in the western U.S. can and are being strip-mined. If there is enough low-sulfur coal so that only this type of coal is used in power plants, then it may be possible to meet the national air quality standards for sulfur oxides in the ambient air.

In principle, the amount of harmful by-products of combustion can be reduced by proper choice of fuel, effluent controls, and changes in the combustion process. However, the cost of all of these items is high, supplies of cleaner fuels are low, and some of the technology is not fully developed. For example, utility spokesmen claim that complete control of sulfur dioxide emission from coal- and oil-fired power plants has not yet been achieved.

The Environmental Protection Agency has established regulations to limit the emission of air pollutants from power plants. However, these new emission standards, which cover particulates, sulfur dioxide, and nitrogen oxides, are applicable to new power plants only. The NO_x and SO_2 standards vary, depending on whether the power plant is gas-, oil-, or coal-fired. These emission standards are applicable to an individual source and are distinct from the national air quality standards. It is possible for all of the power plants in some region to comply with the SO_2 emission standards, and still have the air quality standards exceeded because the concentration of power plants and other industry is too great.

Los Angeles County has taken a unique approach to the reduction of air pollution from power plants. New fossil-fuel power plants are not allowed in that air basin. Los Angeles now receives a significant amount of electricity from the Four Corners power plants, located in a sparsely populated region of the western U.S. The concentration of several coal-fired power plants has begun to pollute the air and reduce visibility over what used to be one of the clearest areas in the U.S.

The location of power plants far away from large urban areas and near the source of fuel is only a temporary solution since the air quality standards apply to rural areas as well as urban areas, thereby limiting the number of fossil-fuel power plants that can be concentrated in one area. In addition, the people living in these regions have become aware of the price they are paying (degradation of the environment through strip mining, adverse health effects from air pollution, reduced visibility, etc.) for their new-found economic well-being and the sacrifice they are making to spare Los Angeles further air pollution.

Energy conversion effects with hydroelectric power

The esthetic objection to hydroelectric projects is that they turn some of the most beautiful gorges and river valleys into lakes. When decisions concerning the location of a new power plant or the type of power plant to be built are being made, the loss of a rushing river will have to be weighed against the gain of more electricity and the addition of another body of still water. This delicate weighing of values is something that takes place in government agencies all the time. However, the public is not aware of the value judgments which go into everyday decision making.

Another effect of damming rivers is that sometimes the spawning grounds of migratory fish (such as salmon) of commercial and recreational importance are destroyed, or the fish are prevented from reaching them. Reservoirs lose water by evaporation, and the concentration of dissolved minerals in the water increases. Seepage from reservoirs may raise the water table and bring salts and minerals with it, thereby impairing the fertility of the soil.

Hydroelectric reservoirs eventually fill up with silt that is carried by rivers. Rivers which carry a large amount of silt, such as the Colorado, may fill up a large reservoir in one or two centuries. When that happens, the potential energy of the waterfall which occupies the dam site can still be converted to electric energy, but the available energy is then completely subject to the fluctuations in river flow. Therefore, hydroelectric power is not a source of energy of essentially limitless lifetime as is the case with solar energy.

When weighing the advantages and disadvantages of different methods of energy conversion, hydroelectric power compares very favorably. Hydroelectric projects usually have the added benefits of flood control and farm irrigation, and power generation is most efficient for hydropower.

Fission power, public health, and safety

Radioactive isotopes are the end product of nuclear fission. The fragments of fissioned atoms slowly accumulate in the reactor fuel. After many months to several years enough wastes accumulate to extinguish the chain reaction. This occurs when the fission products absorb too many of the fission neutrons

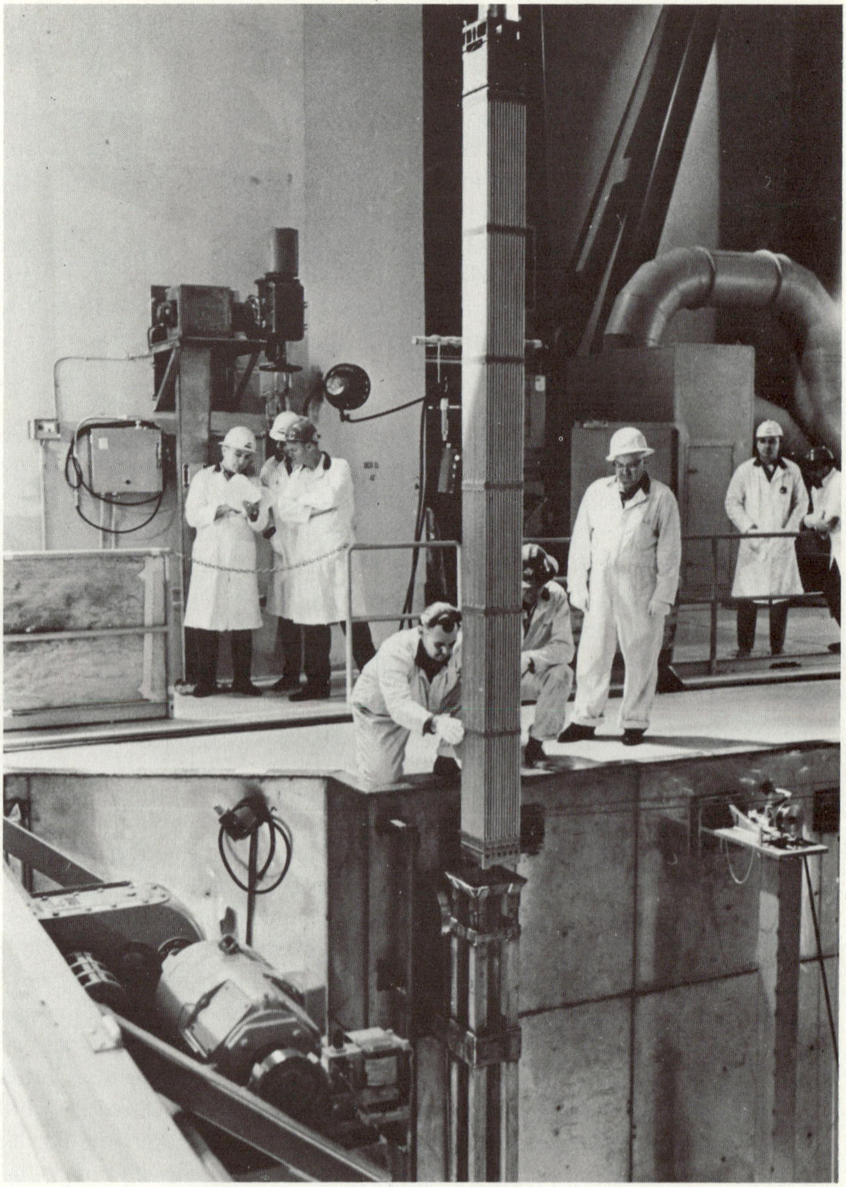

Figure 12-4 This is a fuel assembly, removed from a reactor vessel. This square assembly consists of a large number of metal tubes which are about 12 feet long and contain nuclear fuel. (Courtesy U.S. Atomic Energy Commission)

before they can collide with a U^{235} nucleus. Before the chain reaction is extinguished, the fuel (Fig. 12-4) is removed from the reactor and the radioactive fission fragments are separated from the fuel. The fission fragments range from a form of radioactive nitrogen, with a half-life of 7 seconds, to I^{129}, with a half-life of 17 million years.

We have already discussed the hazards of some of the more important fission products in Chapter 6. These are Sr^{90} which concentrates in bones, Cs^{137} which is found throughout the body, and I^{131} which concentrates in the thyroid gland. Two other fission products of importance are tritium and krypton 85 (Kr^{85}). Tritium, which is an isotope of hydrogen, replaces ordinary hydrogen in water and can, therefore, be found throughout the body. Krypton, which is a radioactive gas, is released into the air from reactors and from reprocessing plants. It does not remain in the body, but it can be an external hazard due to its emission of energetic gamma rays. If the emission of krypton from reprocessing plants is not reduced, it may become one of the more critical radioisotopes in the environment in the near future.

Fuel reprocessing and high-level radioactive wastes After fuel rods are removed from a nuclear power plant they are taken to a reprocessing plant where they are dumped into a nitric acid bath. The nitric acid dissolves the uranium and the fission products, but leaves the steel jacket of the rod intact. The acid solution containing the fuel and wastes then undergoes a complex process of chemical separation. Because of the extremely high levels of radioactivity, almost all procedures are carried out by remote control. Uranium and plutonium which are extracted and separated from each other, are reused for reactor fuel or fission bombs. The remaining liquid contains the fission wastes and is highly radioactive. It is stored in large underground tanks which must be cooled because of the heat given off in radioactive decay. There have already been leaks in some of these tanks, with large releases of radioactivity. These liquid wastes are now being converted to solid form. They must be disposed of in a place that insures their isolation from the environment for tens of thousands of years.

The Atomic Energy Commission has proposed putting these solid wastes in containers and then placing them in abandoned underground salt mines (Fig. 12-5). The Atomic Energy Commission's first choice, Lyons, Kansas, has been opposed by the Kansas Geological Survey. They pointed out that the heat given off by the radioactive isotopes could cause the salt to melt and recrystalyze, possibly creating fissures which would let in water. Salt was chosen as the dumping ground substance because of its high melting point and high heat conductivity, which would hopefully allow it to withstand the heat produced in radioactive decay. It has been assumed by the A.E.C. that since the salt mines still exist, water must not have been in them since the

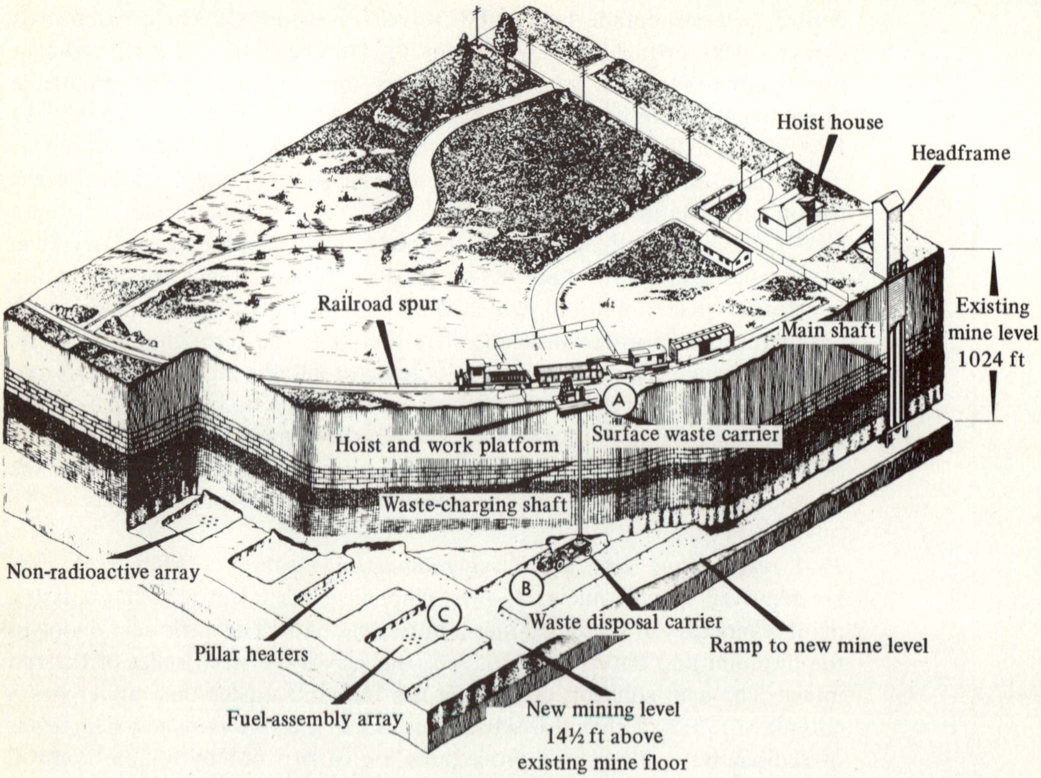

Figure 12-5 The Carey Salt Company mine at Lyons, KS, is the scene of a radioactive waste disposal experiment conducted by the Oak Ridge National Laboratory. (From "Radioactive Wastes," U.S. A.E.C.)

formation of the salt (not in large quantities, at least). However, the Geological Survey also pointed out that there were several old drilling sites nearby that could collect water, and that since the waste cannisters would last only a few years, there would be no way to retrieve the wastes if contamination of the water became imminent. At the present time, the salt mine proposal has an uncertain future.

Another A.E.C. proposal is to place the radioactive wastes in rockets and put the rockets in orbit around the sun. The dangers of both of these proposals are great, and as yet there is no solution that is acceptable to all parties involved. In the meantime we continue to accumulate more highly radioactive wastes which will be dangerous to all animals and plants in the biosphere for tens of thousands of years. Not only does the U.S. have to bury its own radioactive wastes, it is also responsible for the radioactive wastes of other nations to which it has sold nuclear reactors and reactor fuel.

Low-level effluents There are also low-level radioactive effluents which result from the routine operation of a nuclear power plant. These are in contrast to the high-level wastes just discussed, which accumulate in the reactor and are separated from the fuel at infrequent intervals. The low-level effluents are discharged to the atmosphere and to large nearby bodies of water. Some liquid wastes that have a low level of radioactivity consist of isotopes formed when impurities and corrosion products in the coolant water are bombarded with neutrons from the chain reaction (this is called neutron activation). Some examples of these radioactive isotopes are Co^{58}, Fe^{59}, and Mn^{51}. In addition, cracks in the fuel elements allow radioactive fission products to escape into the primary coolant. In current pressurized water reactors, portions of the primary coolant are periodically released to the environment. This results in the maintenance of a low level of tritium and other radioactive isotopes in the cooling water.

Tritium is produced in a nuclear reactor in many ways. Some of these are: fissioning of uranium, neutron capture by boron in control rods, and neutron activation of deuterium. Table 12-1 is a list of the concentrations in water of some of the more important radioactive isotopes found in the liquid effluent from a 1,000-megawatt pressurized water reactor. Also listed is the maximum permissable concentration (MPC) in water of the various isotopes and their half-lives. Tritium is the isotope which is nearest to its MPC, and its concentration is a factor of 1,000 less than its MPC. A proliferation of nuclear power plants could change this situation. In addition, concentration of the radioisotopes in food chains could result in a hazardous situation. This has occurred in the case of Zn^{65} released in H-bomb tests in the Pacific in 1956. Zn^{65} was concentrated in tuna fish to the point where 457 tons of tuna had to be destroyed because of high levels of radiation.

Gaseous fission products are also released to the environment. In a boiling-water reactor, when the steam is condensed for return to the reactor core, the gaseous fission products are separated and vented through a high stack. In general, gaseous wastes are held for an appropriate period of decay before being filtered and released through the stack. Some of the released radioactive gases are tritium and isotopes of krypton, xenon, and iodine.

Fuel reprocessing plants control the discharge of radioactive iodine by a combination of methods. First, the used fuel is stored from 90 to 120 days to allow natural decay of most of the iodine. Second, most remaining iodine is vaporized during the fuel reprocessing and is removed from other gases by activated charcoal or a chemical-reaction system that converts the iodine into a solid.

Table 12-1 LIQUID RELEASES FROM A 1,000-MW PWR

Isotope	*Half-life*	*Discharge concentration microcuries/cc*	*MPC microcuries/cc*
H^3	12.3 years	3.8×10^{-6}	5×10^{-3}
Mn^{54}	300 days	0.8×10^{-15}	8×10^{-4}
Co^{58}	71 days	2.4×10^{-14}	4×10^{-4}
Co^{60}	5.2 years	2.9×10^{-15}	1×10^{-4}
Sr^{89}	50.5 days	9.8×10^{-15}	7×10^{-5}
Sr^{90}	27.7 years	3.0×10^{-16}	4×10^{-7}
Y^{90}	64.8 hours	3.4×10^{-16}	3.0
Y^{91}	57.5 days	1.7×10^{-14}	0.2
Mo^{99}	67 hours	0.7×10^{-11}	8×10^{-4}
I^{151}	81 days	0.6×10^{-11}	2×10^{-4}
Cs^{134}	2.3 years	2.1×10^{-12}	9×10^{-6}
Te^{152}	78 hours	0.1×10^{-12}	5×10^{-4}
I^{133}	20.5 hours	0.7×10^{-11}	9×10^{-4}
Cs^{136}	13 days	0.9×10^{-13}	9×10^{-5}
Cs^{137}	27 years	3.4×10^{-12}	2×10^{-5}
Ba^{140}	12.8 days	2.3×10^{-15}	5×10^{-4}
La^{140}	40.5 hours	2.1×10^{-15}	2.0
Ce^{144}	290 days	0.8×10^{-14}	3×10^{-2}

(Reprinted from "Man's Impact on the Global Environment," by SCEP, by permission of The MIT Press, Cambridge, MA, Copyright 1970)

Low-level effluents and human health The Atomic Energy Commission feels that no public health hazard is produced by low-level effluents from nuclear power plants since the estimated dose per capita from nuclear power plants (0.01 mrem/yr) is only a very small fraction of the natural background dose (about 130 mrem/yr.). Many scientists, such as Drs. John Gofman and Arthur Tamplin of the Lawrence Radiation Laboratory in Berkeley, California, are concerned that the man-made radiation dose (excluding medical radiation) will be allowed to rise from its current value of 5 to 6 mrem/yr to the maximum allowable exposure of 170 mrem/yr. If this is allowed to happen, then (say Gofman and Tamplin) there may be as many as 24,000 additional cancer and leukemia cases each year in the U.S. after 30 years of exposure. (See Chapter 6.)

The A.E.C. feels that this result is incorrect and states that they have no intention of letting man-made radiation levels rise to anywhere near 170 mrem/yr. They have backed up their statements by reducing the standards for both gaseous and liquid radioactive emissions from new nuclear power plants by a factor of one hundred. These new standards do not apply to fuel reprocessing plants which are expected to increase in number along with new nuclear power plants. At this time the possibility of adverse health effects from low-level emissions from nuclear power plants is still a controversial subject which needs more clarification.

Reactor safety Another subject of great debate is that of reactor safety. A nuclear reactor, the heart of a nuclear electric power plant, is shown in

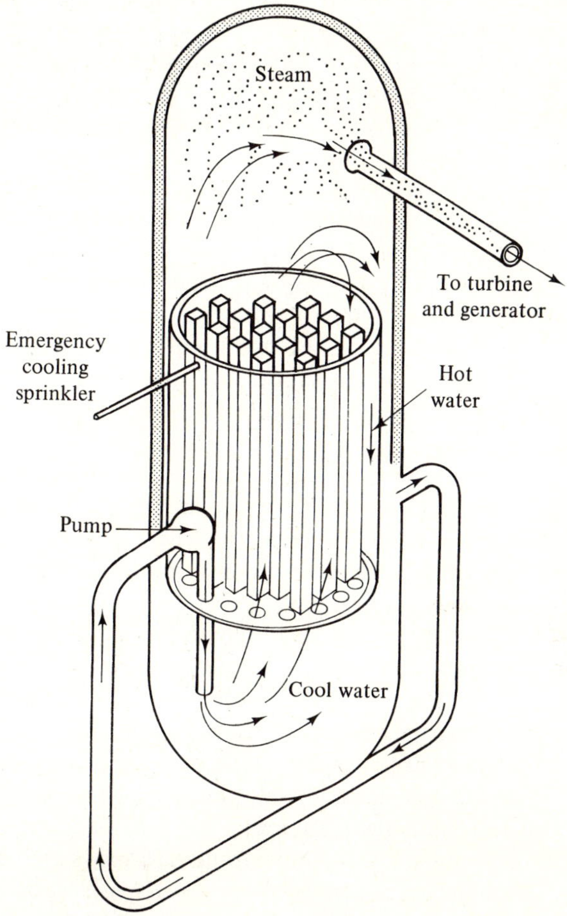

Figure 12-6 A nuclear reactor in simplified form. This illustration originally appeared in "Cooling Water," by Ian A. Forbes et al., in *Environment*, **14**: 1, 42. Copyright © Scientists' Institute for Public Information, 1972.

simplified form in Fig. 12-6. The reactor's fuel, or core, is represented by the long oblongs in the center of the drawing. The core is about 12 feet high and 15 feet across. Each oblong fuel assembly, roughly 6 inches across, is actually a bundle of 49 thin rods filled with uranium oxide pellets. In a large reactor there may be more than 700 such assemblies, containing 150 tons or more of uranium fuel. In the simplest design, the "boiling water" reactor in Fig. 12-6, cool water is pumped upward among the fuel assemblies which are heated by nuclear reactions. The water boils, and steam is drawn off to generate electricity while the remaining water is recirculated.

An ordinary nuclear reactor cannot explode like an atom bomb. The concentration of U^{235} in the fuel (about 3% of the total uranium in the fuel) is much less than in a bomb which is nearly 100% fissionable material. In addition, if an uncontrolled reaction does start, the control rods are automatically inserted into the reactor, stopping the chain reaction.

There is a possibility, however, that excess heating could melt the core and spill large quantities of radioactive material into the environment. If the flow of coolant to the reactor core was blocked for some reason, then the heat generated by the fission products could cause some or all of the fuel rods to melt (Fig. 12-6). An emergency cooling spray is therefore provided in case of emergency. If the many safety devices, which are built into all nuclear reactors, did not shut down the reactor quickly enough, then it is possible that the high temperature reached in the fuel might cause chemical reactions between reactor materials which would lead to a chemical explosion. Even if the series of events described above did occur, there would still be two physical barriers against the escape of radioactive substances into the environment. These are the steel reactor vessel itself (Fig. 12-7) and finally the spherical steel and concrete shell which is all that is seen from outside a nuclear ower plant.

Thus far there have been no radiation injuries or deaths resulting from the operation of licensed nuclear power plants in the United States. There have been, however, several accidents resulting in injuries and deaths to workers at experimental reactors. The A.E.C. feels that the probability of an accident releasing radioactivity at a commercial nuclear power plant approaches zero because of the many safeguards applied. Critics say that the probability of an accident is small but finite, and that if a large-scale release were to occur near a populated area, there would be a very significant loss of life, a large number of injuries, and induced cancers. Experience has shown that accidental releases from nuclear power plants are possible. A large amount of radioactive iodine was released from the stack of the Windscale power plant located in a rural part of England. Dairy cows in the area ate grass which had become contaminated with the radioactive iodine, and their milk had to be taken off the market for several days.

Figure 12-7 A steel reactor vessel being lowered into place. This 345-ton reactor is the heart of Unit 1 of Pacific Gas and Electric Company's Diablo Canyon nuclear project, 12 miles southwest of San Luis Obispo, CA. The 1,000-megawatt Unit 1 was scheduled for operation in 1975. (Courtesy Pacific Gas and Electric)

Designing additional safety features for nuclear reactors is a sensible procedure, but we should not allow this policy to give us a false sense of security. Although much time and effort have gone into the safety considerations for nuclear reactors, an accidental release of radiation is still possible. Since the consequences of such a release are grave, it would be best to locate all new reactors far from large population centers and far from earthquake faults.

Breeder reactors

If breeder reactors are used for commercial power generation, then the above mentioned problems will still be with us. Fuel reprocessing from breeder reactors will be an even more dangerous activity since much more plutonium will be handled than with ordinary reactors. Plutonium 239 (Pu^{239}) is an extremely toxic substance which has a physical half-life of 24,400 years. It decays by emission of alpha particles of 5.2 MeV. There are two hazards from this isotope of plutonium. If inhaled through the air it may become trapped in the lungs. As little as one millionth of a gram (10^{-6} g) of plutonium may be capable of producing fatal lung cancer. If plutonium is ingested, it will find its way to the bones where it concentrates with a biological half-life of 120 years. Therefore, it can be considered to remain in the bones of an individual for his lifetime at almost full concentration. There the danger is from bone cancer or leukemia.

The breeder program has had great difficulties since its inception around 1960. Because of the possibility of an atomic explosion, even greater safety precautions are necessary for breeder reactors than for ordinary nuclear reactors. Many nuclear scientists have stated that they would prefer to have breeder reactors located underground. There is a greater concentration of fissionable material in a breeder reactor and the moderator is removed because fast neutrons are more likely than slow neutrons to convert U^{238} into Np^{239}, which decays into Pu^{239}. If the temperature rises in the core of an ordinary reactor, the water loses some of its effectiveness as a moderator. The chain reaction slows down due to this effect of the moderator. This built-in automatic safety feature does not exist in a fast breeder reactor. The greater concentration of fuel in a breeder reactor produces higher core temperatures which require the use of liquid sodium as a coolant. Liquid sodium is highly reactive in the presence of water or air, so this adds another hazard to breeder reactor operation.

The Enrico Fermi Atomic Power Plant (Fig. 12-8) was the first breeder reactor to produce a significant amount of electric power (60 megawatts). In 1966, when it was run up to full power for the first time, it suffered a serious accident. Due to a blockage of the sodium coolant flow, there was a partial meltdown of its core. It is interesting that the cause of the accident was a piece of sheet metal that blocked one or two of the openings that admit coolant to the reactor core. About one year after the accident it was determined that the purpose of the sheet metal, which was not even included on engineering drawings, was to protect against one type of accident; unfortunately it was the cause of another kind of accident.

When the damaged fuel was removed from the core there was a small chance that it could be jarred into a critical mass too great to be controlled by the control rods, which were already in their maximum position. The result could have been an explosion violent enough to rupture the steel and concrete containment structure. Fortunately, this did not happen. After

Figure 12-8 The Enrico Fermi atomic power plant was the world's first large breeder nuclear power plant. It was located at Lagoona Beach on Lake Erie, about 35 miles southwest of Detroit. (Courtesy Atomic Energy Commission)

several more attempts to get the reactor functioning properly it was finally dismantled. There are no plans to reactivate it.

About breeder reactors, nuclear physicist Dr. Edward Teller (known as father of the H bomb) has written:*

> "In an accident involving a plutonium reactor, a couple of tons of plutonium can melt. I don't think anybody can foresee where 1 or 2 or 5% of this plutonium will find itself and how it will get mixed with some of the other material. A small fraction of the original charge can become a great hazard."

There is a feature of the nuclear power industry which is unique in the United States at least. The federal government has a legal obligation to provide 85% of the insurance coverage on nuclear power plants. Private

*Teller, Edward, "Fast Reactors: Maybe," *Nuclear News* (Aug. 1967), p. 21.

companies were not willing to undertake the whole premium. This obligation, which is assumed under the Price-Anderson Act, provides coverage of up to $560 million per accident. However, it has been estimated that a nuclear power plant accident could cause as much as $10 billion damage. In this event, those damaged by the accident would receive a return of only five cents on the dollar. The fact that private insurance companies are not willing to insure nuclear power plants must indicate that they feel the probability of an accident is an unknown, or that the damage would be too great in economic terms, or both.

Fusion power

The environmental and human health impact of fusion reactors will be less than for fission reactors. The main radioactive material which will be released from fusion reactors (routinely or accidentally) is tritium, which beta decays with a half-life of about 12 years. The low-energy electrons emitted by tritium can be easily shielded against; the only hazard would be the possible escape of tritium to the environment and its incorporation into living tissue. Fusion reactors will contain an intense flux of neutrons, so there is the additional hazard produced by neutron activation of structural materials. The great advantage of fusion reactors will be that no fission wastes will be produced.

The chance of an accident releasing radioactive material to the environment is much smaller in a fusion reactor than in a fission reactor. In a fission reactor great precautions are taken to keep the reactor from running too fast, whereas in a fusion reactor a great effort is made to get the reactor running at all. Any malfunction in a fusion reactor tends to slow the reaction very quickly by loss of confinement, a decrease in temperature, or both. The molten lithium in a fusion reactor is a chemical-explosion hazard as is the liquid sodium in a fission reactor. All factors considered, fusion reactors offer us a relatively safe, long-lived, and economical source of electricity, if the present technological problems can be overcome.

Solar power

The greatest environmental impact of solar energy conversion will be on land use. About 10,000 square miles of desert land would be needed for a continuous power production of 10^{12} watts, which is enough power to satisfy the present electric energy consumption in the U.S. The amount of land needed for solar power generation is greater than that needed for an equivalent amount of fossil-fuel or nuclear power. However, if the land used for obtaining the fossil or nuclear fuels were included, the disparity would not be so great. Land which is suitable for solar power plants is located in desert regions termed "ecologically fragile." An influx of population to the sources of electricity and employment could alter the deserts, and population restrictions would probably be necessary.

The climate may be altered in these desert regions by the absorption of more solar energy than would otherwise have been absorbed without the special absorbing surfaces of a "solar farm." Part of this excess energy will be shipped away in the form of electricity; however, this will account for only about 30%, at most, of the excess energy absorbed by a solar farm. Climate modification is a problem with any type of power production; the disadvantage with solar power is that all of the waste heat would be concentrated in several relatively small regions. Solar farms will have to be located in those areas where there is almost continuous sunshine. In the United States, this would probably be in southeastern California or western Arizona.

Waste heat: what can be done with it?

"Waste heat" results from the conversion of energy from one form to another. The efficiency of conversion of energy into useful work is never 100%; in fact, in many cases it is far less than 100%. For example, in the conversion of nuclear energy to electric energy, about 70% of the energy in the fuel appears as waste heat. It is released into a nearby body of water or into the surrounding air. If the waste heat is dumped into a body of water, it will eventually be transferred to the atmosphere by evaporation.

Many uses for waste heat have been proposed. In most cases they have been judged economically unfeasible. For example, the low-pressure steam discharged from a steam turbine could be used for space heating during the winter months. This is economically feasible when a power plant is located near a large city that requires heat such as New York City. Many power plants are not located near large population centers, and this will be more common in the future.

Another proposal is to use waste heat to desalinate salt water. If salt water is heated, the water can be boiled off and collected, leaving the salt behind. Desalination has been proposed for the Four Corners power plants, but the salt water would have to be transported to the plants and the salt disposed of afterwards. For power plants located in arid regions and near large bodies of salt water, desalination would be a good use for waste heat. Israel, which is located in an arid part of the world and near the Red and Mediterranean Seas, has put much effort into developing desalination plants in order to increase its supply of fresh water. The problem with desalination as an end use for waste heat is that most power plants are not located in arid regions where large quantities of fresh water are needed.

The production of fish or shellfish that thrive in warm water could also benefit from heated waste water. Long Island oyster farmers have found that their harvesting season is extended by the warming of the ocean by large power plants in the area. In California, Pacific Gas and Electric is succesfully using heated water from power plants to speed the growth of shellfish. Nationwide, however, there appears to be no beneficial use for all the waste heat from present power plants, let alone future ones. The question is: How can we limit the harm that waste heat can do the environment?

Example 12-1: A 1,000-megawatt power plant with an efficiency of 33% is located near an ocean. If all the waste heat is used to desalinate sea water, how much salt water can be converted to fresh water each day?

$$1{,}000 \text{ megawatts} = 10^9 \text{ watts} = 10^9 \text{ joules/sec}$$

If 1,000 megawatts is the electric power output, then we can calculate the thermal power input from:

$$\text{Efficiency} = \frac{\text{W}}{Q_2} = \frac{\text{Work Output}}{\text{Energy Input}}$$

$$0.33 = \frac{1{,}000 \text{ MW}}{Q_2} \qquad Q_2 = \frac{1{,}000 \text{ MW}}{0.33} = 3{,}000 \text{ MW}$$

From the conservation of energy law, $Q_2 = W + Q_1$ or the energy input, Q_2, equals the work done W plus the waste heat Q_1:

$$3{,}000 \text{ MW} = 1{,}000 \text{ MW} + Q_1$$

$$Q_1 = 2{,}000 \text{ MW}$$

So heat is rejected at the rate of 2×10^9 joules/sec.

Waste heat is generated at twice the rate at which electric energy is generated! We will suppose that we are in a desert region and the water temperature is 30°C. We need to add enough heat to some mass of water to raise its temperature to 100°C (the boiling point of water), and then enough energy to change it from a liquid to a gas (latent heat of vaporization).

We can use Eq. (4-3) from Chapter 4:

$$Q = mc\Delta T$$

If we add an amount of heat, Q, to a substance of mass, m, and specific heat, c, its temperature will rise by ΔT. If the substance changes state (for example, liquid to gas), then we must add this energy needed for it to change state to the right side of the above equation.

When water is already at its boiling point (100°C), it takes 540 calories to convert each gram to water vapor.

In order to desalinate sea water, we must increase its temperature to 100°C and then vaporize it. We collect the water later by allowing it to condense on a cool surface:

$$Q = mc\,\Delta T + 540 \text{ cal/g} \times m$$

where $Q = 2 \times 10^9$ joules/sec

$$c = 1 \text{ cal/g°C}$$

$$m = ?$$

$$\Delta T = 100°\text{C} - 30°\text{C} = 70°\text{C}$$

$$4.2 \text{ joules} = 1 \text{ calorie}$$

$$1 \text{ gallon} = 4 \text{ kg}$$

$$2 \times 10^9 \text{ joules/sec} = m(1)70 \text{ cal/g} + 540 \text{ m cal/g}$$

$$\frac{2 \times 10^9 \text{ joules/sec}}{4.2 \text{ joules/cal.}} = 610\, m \text{ cal/g}$$

$$m = \frac{2 \times 10^9}{4.2(610)} \text{ g/sec}$$

$$m = 0.78 \times 10^6 \text{ g/sec} = 780 \text{ kg/sec}$$

$$m = 780 \text{ kg/sec} = 195 \text{ gal/sec}$$

$$195 \text{ gal/sec} = 16.9 \text{ million gal/day}$$

This is enough fresh water to supply a city of about 100,000 people daily. We have assumed 100% efficiency in our conversion of salt water to fresh water which, of course, is never the case. So this calculation gives an absolute maximum for fresh water production.

Thermal pollution and its ecological effects

Large quantities of heated water are now dumped into rivers, lakes, and estuaries. The usual procedure is to pump water from a nearby source through a steam-condensing unit in order to condense the steam and recycle it through a boiler (Fig. 12-9). As the steam condenses to liquid water, heat is transferred to the cooling water. Depending on such factors as the flow rate, the temperature of the cooling water can be raised 10°F to 30°F over its initial temperature. A 1,000-megawatt power plant requires a water flow of about 600,000 gallons per minute when the water temperature rise is 15°F. This heated effluent is then discharged into the body of water from which it came.

Fish, as cold-blooded animals, are unable to regulate their body temperature and, therefore, are particularly sensitive to changes in their thermal environment. Most of the effects of temperature on the life processes of animals that live in the water stem from the metabolic changes which occur as a function of temperature. The speedup of metabolism that comes with increasing temperature, increases the animal's need for oxygen and, therefore, its respiration rate. At elevated temperatures, a fish's respiratory difficulties are compounded by the fact that the hemoglobin of the blood has a reduced

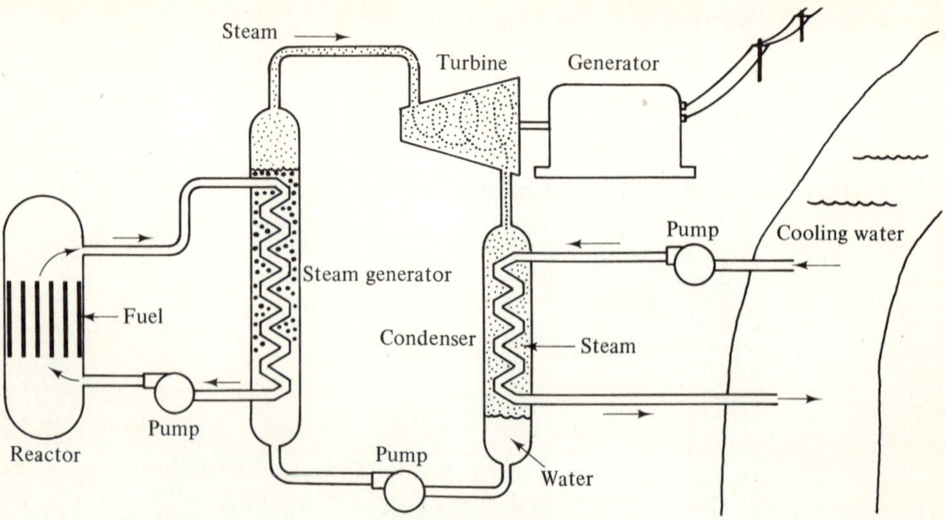

Figure 12-9 Cooling-water flow in a nuclear power plant. (From "Nuclear Power and the Environment," U.S. A.E.C. Understanding the Atom Series)

affinity for oxygen and thus becomes less efficient in delivering oxygen to the tissues.

This need for more oxygen at higher temperatures is compounded by the fact that the amount of oxygen dissolved in water depends on temperature (Fig. 12-10). The higher the temperature, the lower the oxygen content of the water. Adding waste heat to a body of water can lower the oxygen content below the level necessary to sustain many animals. Fish require at least 5 milligrams (mg) of dissolved oxygen per liter (about 1 quart) of water for well-being and 7 mg/liter in spawning areas; therefore, above 35°C the dissolved oxygen content may be dangerously low. At this time there are not many examples of adverse effects of thermal pollution on aquatic life, but there have been some suggestive studies.* As the number of power plants increases, a point will be reached when it is no longer possible to dump waste heat into rivers and lakes without obvious adverse effects. Most new power plants use cooling towers (they will be discussed shortly) to transfer the waste heat directly to the atmosphere rather than to a waterway.

Water temperature also affects the appetite, digestion, and growth of fish.

*Report of the Thermal Advisory Committee to the Maryland Department of Water Resources, Dec. 1969.

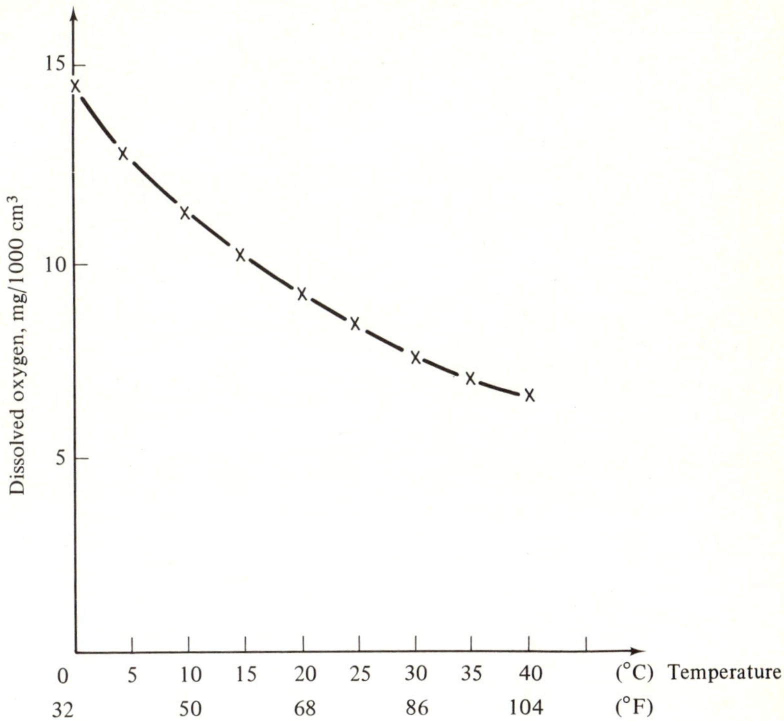

Figure 12-10 The concentration of dissolved oxygen as a function of temperature. The solubility in water of a gas decreases with increasing temperature. One liter is equal to 1,000 cm^3.

The effect of water temperature on the reproduction of aquatic animals may be most important. Temperature triggers the release of eggs of shellfish, such as oysters and clams, and exerts a precise control over the time it takes fish eggs to hatch. In general, increasing the temperature decreases the hatching time of fish eggs. However, excessive temperature can prevent normal development of eggs. The Oregon Fish Commission has declared that a rise of 5.4°F in the Columbia River could be disastrous for the eggs of the Chinook salmon. Standards for allowable temperature increases have been established to protect fish in their various environments. The optimal temperature for any water habitat depends not only on the preferences of individual species but also on the well-being of the system as a whole (Fig. 12-11).

Another effect of reducing the amount of dissolved oxygen in a body of water is that its capacity to assimilate organic wastes is lessened. Sewage

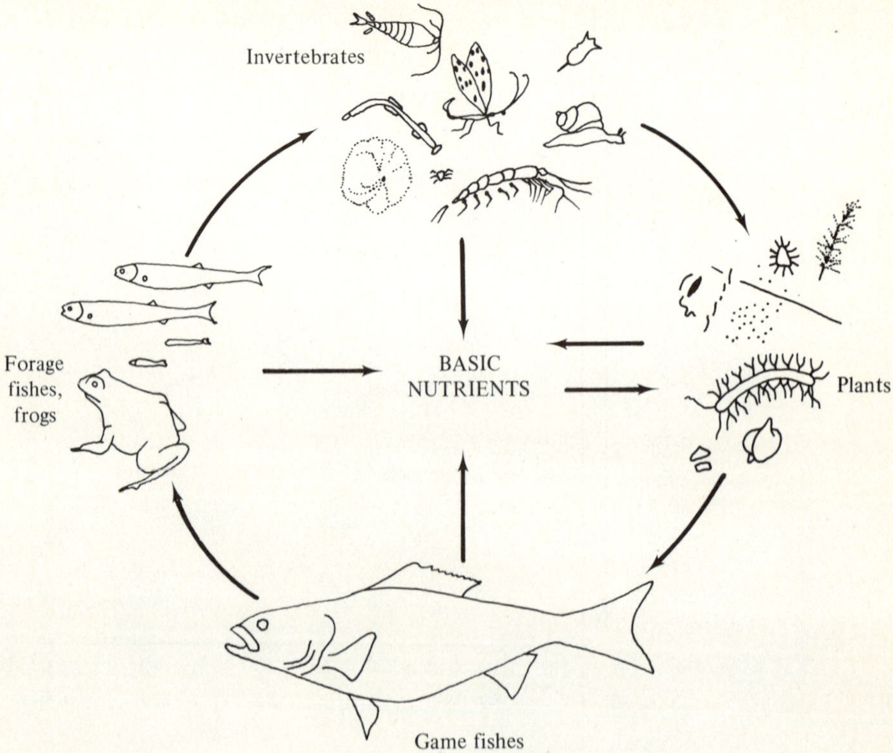

Figure 12-11 An aquatic ecosystem is more sensitive to changes in temperature than is an individual species of fish and plant. If any link in the food chain is affected by an environmental change (such as temperature), this will affect the proliferation of the other members of the food chain.

containing organic material is broken down by bacteria, using oxygen and releasing carbon dioxide in the process. Raw domestic sewage requires from 150 to 300 milligrams of oxygen per liter of water (mg/l) to be broken down. This amount of oxygen is called the biochemical oxygen demand (BOD) of the organic matter in the sewage. Sewage can be processed to the point where its BOD is only about 10 mg/l. If the temperature of a body of water rises to the point where the amount of dissolved oxygen is less than the BOD of the organic matter put into it, then some of the organic waste will be broken down by an anaerobic (without air) decomposition process with a different mixture of bacteria. Instead of releasing carbon dioxide in the decomposition process, anaerobic decomposition releases methane or hydrogen sulfide.

In these highly polluted situations the body of water turns dark and intense odors result (hydrogen sulfide has a very unpleasant odor). Recently, environmental groups persuaded a utilities company to build cooling towers for a new nuclear power plant (Davis-Besse) being built on Lake Michigan. The main consideration in their arguments was the avoidance of thermal pollution in the lake. Some bodies of water in the highly industrialized eastern states have reached the point where thermal pollution has probably been responsible for the absence of all aquatic life. Since toxic chemicals are also present in these waters, it is difficult to separate their effects from those of thermal pollution.

Another way of condensing the steam in a power plant is to use ocean water for cooling. The use of ocean water requires special pipes which can resist corrosion from salt water. This method is already being successfully used in nuclear power plants along the California coast. Ecologists are concerned about the possible adverse effects of a water temperature increase on marine life near the power plants. The higher water temperature may be more favorable to some species than others, thereby altering the distribution of species near the shore. Surprisingly, a large heat input into the ocean can increase the temperature near the shore since there is a lack of continuous mixing between water near the shore and the rest of the ocean. In order to alleviate this problem and for safety reasons, there have been proposals to place new nuclear power plants on platforms several miles out at sea. The use of cooling towers is a more ecologically sound method of dispersing waste heat.

Cooling towers

A 1,000-megawatt power plant may require about 600,000 gallons per minute for cooling purposes. Where water is scarce, cooling towers have been used to dissipate waste heat. For ecological reasons, cooling towers are now being considered even in localities where water is plentiful. They are enormous structures, typically about 400 feet high and as much as 400 feet wide at the base (Fig. 12-12).

There are two types of cooling towers, a wet type and a dry type. The wet type removes heat by evaporation. The hot water is broken into droplets in order to maximize the surface area and facilitate evaporation. The evaporated water molecules are removed by either a natural or forced-air draft through the tower [Fig. 12-13(a)]. Only a small percentage of the hot water is evaporated. The rest is collected in a basin at the bottom of the tower and is either recirculated to the condensers or pumped into a waterway. The towers for a 1,000-megawatt plant would eject from 20,000 to 25,000 gallons of evaporated water per minute. This is equivalent to a daily rainfall of an inch on an area of two square miles. On cold days such a discharge could condense into a thick fog, or even ice. The wet type of cooling tower is not very suitable for areas already subject to fog formation, such as the coastline of the western U.S. In England, the production of fog from wet cooling towers has already

Figure 12-12 The Trojan nuclear power plant of the Portland General Electric Company is shown under construction near Prescott, OR, 42 miles northeast of Portland on the Columbia River. The cylinder in the center will house a pressurized-water reactor. The large structure in the background is a wet cooling tower that will dissipate waste heat by evaporation and then recycle the cooling water. Note the size of the cooling tower as compared with the power plant and nearby trucks. (Courtesy U.S. Atomic Energy Commission. Photo by Jon Fardell for Portland General Electric Company)

become a serious problem. Wet cooling towers would also be ruled out where salt water is used as the coolant since the salt spray ejected from the cooling towers could destroy surrounding vegetation.

There is a variation of the cooling tower system that avoids these problems. In the dry tower method [Fig. 12-13(b)], the heated water travels

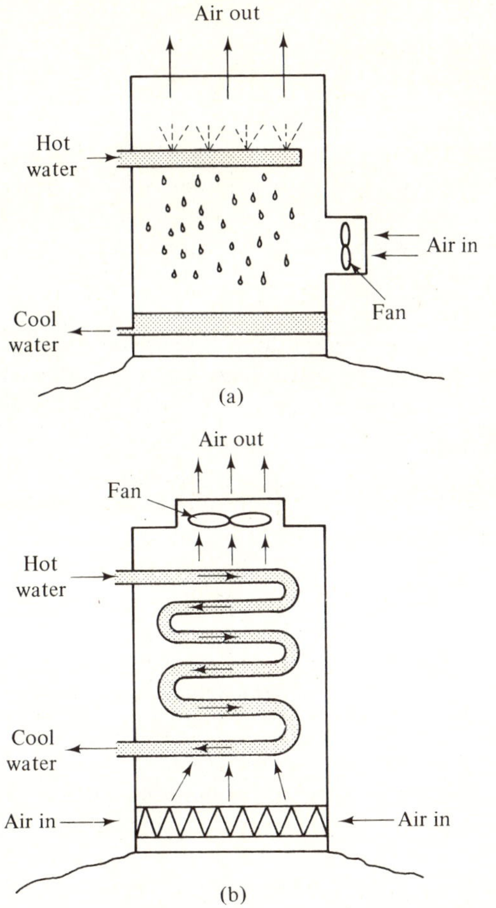

Figure 12-13 Cooling towers: (a) In a wet cooling tower, hot water is exposed to air moving up through the tower. Heat is removed by evaporation, and the cooled water is emptied into a waterway or recirculated through the plant. (b) In a dry cooling tower, hot water is channeled through tubing that is exposed to an airflow. The water gives up its heat to the cooler air without evaporating.

through closed tubing that is exposed to an air flow, and gives up its heat to the air without evaporation. Heat is transferred from the water pipes to the air by conduction and radiation, and is carried away by either forced or natural convection currents. Dry cooling towers are very expensive and have not yet been used in power plants in the United States. In a proposed nuclear power plant in Vernon, Vermont, it has been estimated that the cost of operation and amortization would be \$2.1 million per year for a dry system as compared to \$0.8 million per year for a wet system.

Climate alteration from waste heat

Another effect of the conversion of energy from one form to another is the addition of heat to the atmosphere. In all of the methods discussed for getting rid of waste heat from power plants the heat energy is eventually transferred to the atmosphere. When several power plants are located near each other, they may discharge enough waste heat into the atmosphere to affect local climate.

Cities on the average have a temperature 0.9°F to 1.4°F higher than the surrounding countryside. The effect is more marked in the winter when the minimum temperature may be 2°F to 3°F higher than that of the surrounding rural areas. Some of the factors tending to increase the mean temperature of cities are the direct energy inputs from power plants, air conditioners, and automobiles. The climate of cities is also affected by the alteration of wind speeds and the greater heat retention of buildings and pavements compared to trees and grass, so it is difficult to determine if man's heat input is significant in altering the climates of cities. Calculations of the thermal input from man's activities in the Los Angeles Basin have shown that this input is equal to about 5% of the solar energy reaching the surface. If energy consumption continues to increase in the U.S., we may reach the point where we have already modified the climate over the large continuous urban areas that are expected to exist by the year 2000.

We now make a rough calculation to determine the relationship between the thermal input from man's activities in the U.S. as a whole and the solar input. Averaged around the clock and over the total land area of the U.S., the solar input is about 18 watts per square foot. That is, 18 joules of energy (on the average) reaches each square foot of surface area each second.

The total U.S. energy consumption (1971) was about 7×10^{19} joules. All of this energy eventually ends up as heat.

The average power, P, used over the year is given by

$$1 \text{ year} = 3.15 \times 10^{7} \text{ sec} \qquad \bar{P} = \frac{E}{t} = \frac{7 \times 10^{19} \text{ joules}}{3.15 \times 10^{7} \text{ sec}} = 2.2 \times 10^{12} \text{ watts}$$

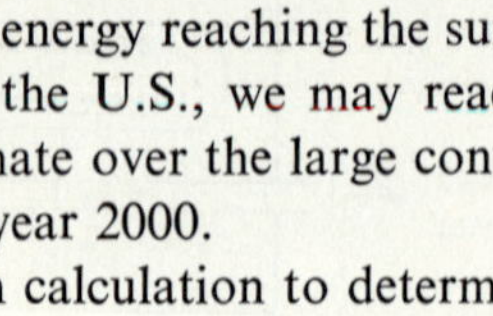

The land area of the U.S. can be estimated by considering the U.S. to be a rectangle 3,000 miles by 1,500 miles:

$$A = 3{,}000 \text{ miles} \times 1{,}500 \text{ miles} = 4.5 \times 10^6 \text{ miles}^2$$

We must now convert this area to square feet (ft^2):

$$A = 4.5 \times 10^6 \text{ miles}^2 = 4.5 \times 10^6 \text{ miles}^2 \times (5.28 \times 10^3 \text{ ft/miles})^2$$

$$A = 4.5 \times 27.8 \times 10^6 \times 10^6 \text{ ft}^2 = 125 \times 10^{12} \text{ ft}^2$$

$$\frac{\bar{P}}{A} = \frac{2.2 \times 10^{12} \text{ watts}}{125 \times 10^{12} \text{ ft}^2} = 0.018 \text{ watts/ft}^2$$

This means that in the U.S. we are now producing about 0.1% ($\frac{1}{10}$ of 1%) of the solar input. This is probably not enough power input to affect climate on a nationwide basis.

Suggested reading

1 Commoner, Barry, *Science and Survival.* New York: Viking Press, 1963.

2 "Environmental Quality," The Annual Report of the Council on Environmental Quality, available for each year beginning with 1970.

3 Wagner, Richard, *Environment and Man.* New York: W.W. Norton & Co. Inc., 1971.

4 Novick, Sheldon, *The Careless Atom.* Boston: Houghton Mifflin Co., 1968.

5 Lewis, Richard, "The Radioactive Salt Mine," *Bulletin of the Atomic Scientists,* **27**: 6 (June 1971).

6 Craig, Roy, "Cloud on the Desert," *Environment,* **13**: 6 (July/August 1971).

7 Brodine, V., "A Special Burden, *Environment,* **13**: 2 (March 1971), (discusses air Pollution).

8 Aaronson, Teri, "Problems Underfoot," *Environment,* **12**: 9 (Nov. 1970).

9 Blumer, M., H. Sanders, F. Grassle, and G. Hampson, "A Small Oil Spill," *Environment,* **13**: 2 (March 1971).

10 Clark, John, "Thermal Pollution and Aquatic Life," *Scientific American* (March 1969).

11 Novick, Sheldon, "Toward a Nuclear Power Precipice," *Environment,* **15**: 2 (March 1973).

12 Cicchetti, Charles, "The Wrong Route," *Environment,* **15**: 5 (June 1973). (Discusses the Alaskan Oil pipeline and Alternatives.)

13 "Oil and Gas Leases Sale," Impact Statement, Dept. of the Interior.

14 "Davis-Besse Nuclear Power Station," Impact Statement, Atomic Energy Commission.

Questions

1 (*a*) Why does the fuel have to be periodically removed from a nuclear reactor?

(*b*) Why is it so difficult to design containers to hold the high-level radioactive wastes?

2 Distinguish between the origins of low-level liquid and gaseous wastes?

3 Advertisements by public utilities often state that nuclear power plants are pollution free and, therefore, we should build more of them and less of the fossil-fuel power plants. Would you agree? Why?

4 If a government official stated that nuclear power plants are safe enough to be located near big cities, since they cannot explode like an atomic bomb, how would you react?

5 Why are breeder reactors considered to be more dangerous than ordinary reactors?

6 (*a*) If you were the president of a large insurance company would you offer insurance to a utility that owns a nuclear power plant?

(*b*) What about a utility that owns a fusion power plant?
How would you justify your decisions?

7 Do you feel that the advantages of solar power justify the use of such large amounts of land? Why?

8 Electricity enters your house or apartment from a central power plant. Explain how the incoming electric energy eventually gets converted to heat energy that does useful work along the way. Use as examples, several household items such as radios and power tools.

9 (*a*) Why is cooling water needed in all power plants?

(*b*) What is wrong with dumping the heated cooling water back into the body of water from which it came?

10 (*a*) What water temperature (in °C and °F) is necessary for the dissolved oxygen content in a body of water to be equal to 10 mg/liter? (See Fig. 12-10.)

(*b*) Is this temperature a likely one in the summer for the ocean? For rivers? For lakes? Choose bodies of water nearest your home.

(*c*) What may the effect be on these bodies of water if the oxygen concentration dropped below 10 mg/liter?

11 What is the main difference between the dry and wet cooling towers with respect to their methods of dissipating heat into the atmosphere?

12 Is climate alteration due to the release of heat energy into the atmosphere a problem at the present time?

13 If river water is pumped through a 1,000-megawatt power plant (efficiency of 33%) at the rate of 10,000 gallons per second, what will its temperature rise be?

Chapter 13

Energy consumption and the quality of life

Introduction

In this final chapter we discuss the relationship between energy consumption and the quality of life. In particular we look at the proposition that more energy consumption leads to a higher quality of life. While this thesis appeared to be true for a long period of time, its truth is not so apparent any longer. Machines can be used to reduce manual labor and to make living conditions more comfortable, but we need energy to run these machines.

Automobiles, dishwashers, power lawn mowers, etc., all require energy for their manufacture and operation. In order to maintain our comfort, we use large amounts of energy for heating and air conditioning. It is interesting to look at the household ownership of selected appliances or devices which many Americans now regard as necessities rather than luxuries. This information comes from the 1970 United States census (Table 13-1). We can also mention many more ways in which energy is needed to make our lives more convenient and enjoyable. In practically all cases the expenditure of energy is necessary, and the production (actually an energy *conversion* is what takes place) of this energy will usually have detrimental effects on the environment and man.

Table 13-1 ELECTRICAL HOUSEHOLD APPLIANCE OWNERSHIP IN U.S.

Appliance or device	*Family ownership*
Refrigerator	99.8%
At least one television	95.0%
Washing machine	91.9%
One automobile	80.0%
Clothes dryer	40.3%
Color television	37.8%
Air conditioner	36.7%
Two or more televisions	29.4%
Two or more automobiles	29.3%
Dishwasher	23.7%

Except for the energy cutback of the depression years, both population and energy consumption have been growing steadily in this century (Fig. 13-1). However, not all growth in energy consumption is due to the increase in population. While total energy consumption in the United States has increased by a factor of ten since 1900, the population has only tripled. Lately, energy consumption has been rising at a rate of almost 5% per year, about four times faster than the increase in population. On the average, each year Americans use more energy per capita than they did the year before. Because of this, even if the population leveled off, energy consumption would continue to rise. Obviously, this growth in energy consumption cannot continue indefinitely since our fuel supplies are not inexhaustable. In addition, it may not even be desirable to continue this rapid growth in energy consumption because of the adverse effects on our quality of life.

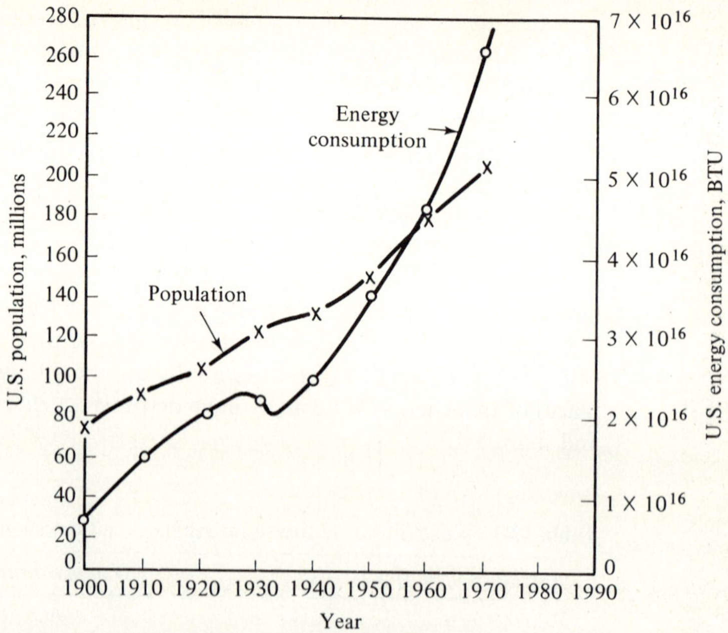

Figure 13-1 The U. S. energy consumption (o) has grown at a faster rate than the population (x), except during the depression.

Energy consumption and gross national product

An often-used indicator of the standard of living within a particular nation is its gross national product (GNP) per capita. We can obtain this figure by first adding the market value of the nation's production of goods and services for a particular year and then dividing that dollar value by the population of the nation during that year. For example, the United States with a population slightly greater than 200 million and a GNP which is above 1 trillion (10^{12}) has a GNP of about \$5,000 per capita.

Except for the depression years (1929–1933), there has been a steady increase in the GNP of the United States (Fig. 13-2). Since 1950 the annual growth rate has been around 6%. However, as we well know, the value of the dollar changes from year to year (generally downward, unfortunately), necessitating an adjustment in the growth rate. When inflation is taken into account (using 1958 as a base year), the real growth rate (since 1950) averages only around 3% a year (see Fig. 13-2, the dotted curve). When we plot the GNP in constant dollars (compared to 1958) we find that it has risen (from 1950–1970) by about a factor of two rather than the factor of $3\frac{1}{2}$ indicated by the solid curve.

In terms of 1958 dollars, the GNP went from \$355 billion in 1950 to \$720

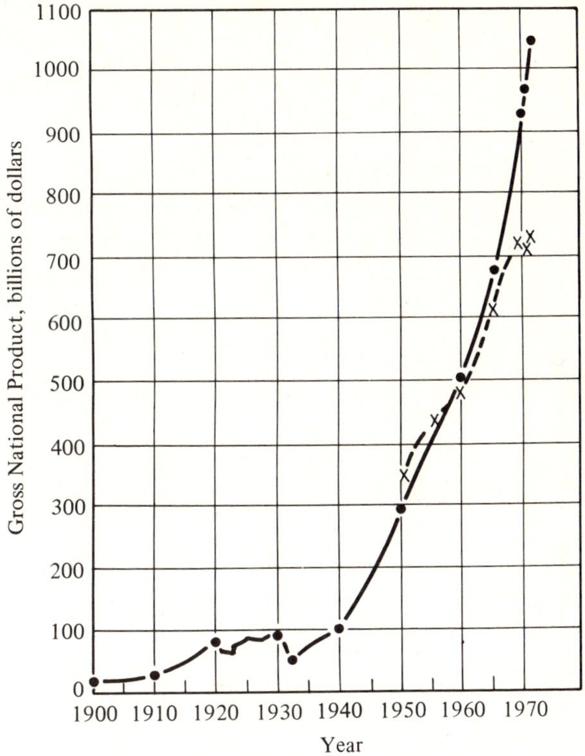

Figure 13-2 The gross national product (GNP) has risen in an exponential manner, except for two dips before 1940. These figures (solid curve) do not take inflation into account; that is, they are current dollars rather than constant dollars. When inflation is taken into account the dotted curve results. (The base year is 1958.)

billion in 1970. Since 720/355 is about 2, we have a doubling of the GNP in a period of 20 years. Looking at the dotted curve, which represents GNP in terms of the current value of the dollar (that is, the value it has during the year the GNP is computed), we find that the GNP rose from a value of $285 billion in 1950 to $974 billion in 1970. This represents a factor of 3.4 change in GNP over the same 20-year time period.

In the United States the real rate of growth of the GNP has fallen off rather sharply in recent years (Fig. 13-2, dotted curve), but the rate of growth of energy consumption (Fig. 13-1) has been holding steady. This has resulted in an increase in the amount of energy necessary to generate a fixed amount of GNP. Energy consumption actually generates GNP by providing the energy to produce the goods and services whose value is used in determining the GNP.

Figure 13-3 is a plot of energy consumption per dollar of GNP as a function of year. Gross national product in terms of constant 1958 dollars is used for the calculations.

The ratio of energy consumption to GNP has varied over the years. From 1920–1965 there was a rather steep decline (except for two short periods) in this ratio, indicating an increased efficiency in the use of energy. During this period the efficiency of electric power generation and transmission increased

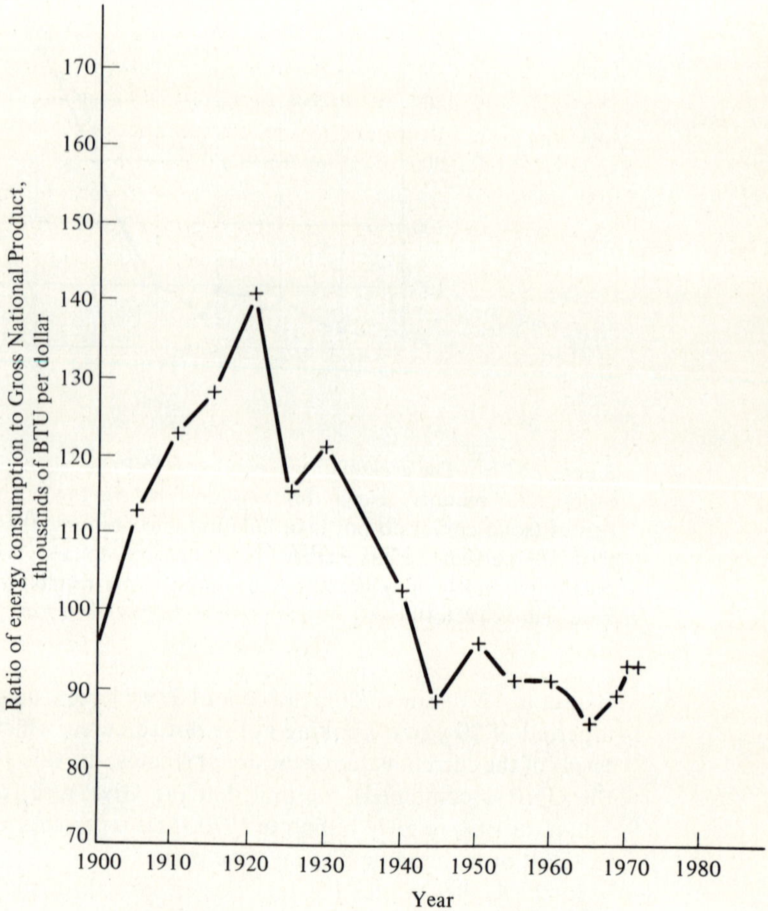

Figure 13-3 The ratio of energy consumption to gross national product as a function of time. GNP is in terms of 1958 dollars.

and the services sector of the economy, which does not use great quantities of energy, increased in importance. However, since 1965 the ratio of energy consumption to GNP has been rising from its low of 86,000 Btu per dollar. In 1970 and 1971 the United States consumed more energy (93,400 Btu per dollar) for each dollar of goods and services than at any time since 1951.

Much of this decrease in economic efficiency is due to the rapid growth of electricity consumption (a growth rate of about 9% per year). In many localities electricity is being subsituted for gas in home heating and cooking. These indirect uses of electricity are less efficient than the more conventional and direct gas heating or cooking. The fastest growing use of electricity has been air conditioning. The consumption of large amounts of electricity for air conditioning does not tend to increase the GNP as much as if the same amount of electricity were used to run machinery in a factory or office, for example. Therefore, the increasing electricity consumption for air conditioning raises the ratio of energy consumption to GNP.

There are certainly other possible explanations for the increase in the amount of energy necessary to produce a fixed amount of GNP. One of these is the decrease in the productivity of the American worker. Much has been written about assembly-line workers who are bored with their jobs and are less productive than they might otherwise be. The trend toward shorter working weeks may also contribute to lower productivity. This lower productivity of the average worker will lower the GNP and increase the ratio of energy consumption to GNP.

An assumption that is often made is that increasing energy consumption assures a nation of at least a proportionally increasing GNP. That is, if energy consumption increases by 10% in one year, then GNP should increase by at least 10% in that same year. Likewise, a decrease in energy consumption should result in at least a proportionate decrease in GNP. If the above statements were true, then the ratio of energy consumption to GNP, plotted in Fig. 13-3 as a function of year, would either stay constant or decrease with time. However, in the period from 1965–1970 this ratio was increasing, as it did during a few other time periods thus far in the 20th century (Fig. 13-3). Therefore, our original assumption is not a valid one. Not only will an increase in energy consumption not necessarily lead to a proportional increase in GNP, but it can even accompany a decrease or no change in GNP.

In Fig. 13-4, GNP per capita is plotted as a function of commercial energy consumption per capita for various nations. Each point represents the values of these two quantities for the nation indicated during one particular year. The data needed for this graph change each year and sometimes electricity consumption per capita is used instead of commercial energy consumption per capita. For these reasons, and also since we need to know only the general shape of the curve which fits the data, the axes are not labeled. Since it is possible to draw roughly a straight line through the points in Fig. 13-4, we

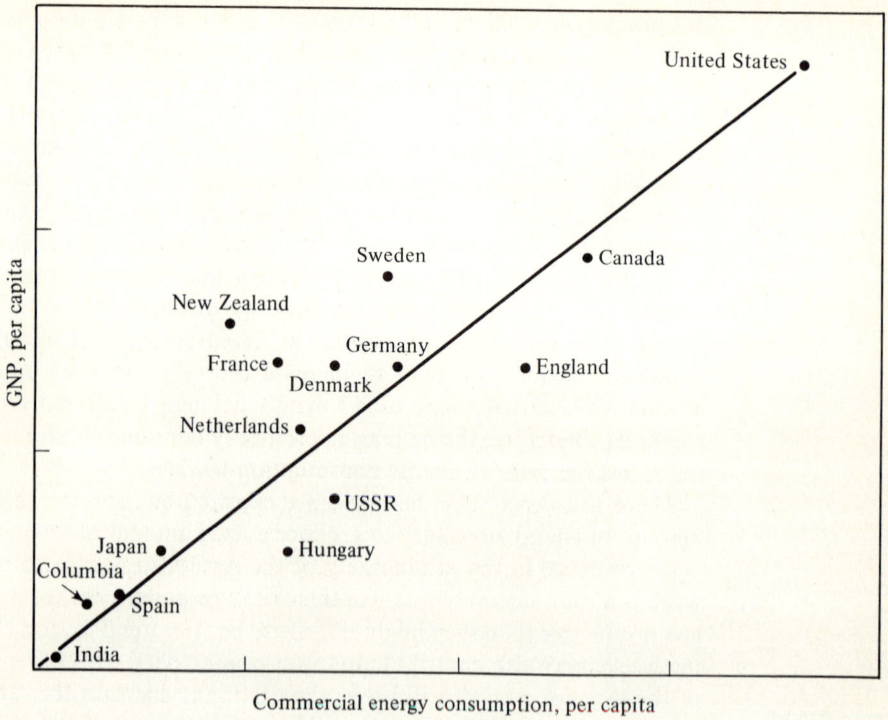

Figure 13-4 Gross national product (GNP) per capita as a function of commercial energy consumption per capita for several nations. If a straight line is drawn through the origin and the point representing the U. S., many of the other points fall close to this line.

can say that GNP is roughly proportional to energy consumption when each point on the graph represents a different country. This type of graph is probably the origin of the idea that GNP is proportional to energy consumption. However, the relationship we need to determine is between GNP and energy consumption for an individual country. Figure 13-4 cannot provide this information, but Fig. 13-3, which charts the ratio of energy consumption to GNP for one nation at a time (in this case, the U.S.), can and does illustrate the relationship between GNP and energy consumption.

By analyzing Fig. 13-4 we can see the inaccuracy of the assumption that GNP is proportional to energy consumption even when the data for different nations are plotted on the same graph. If we look at the points representing the USSR, Holland, and Denmark, we find that the energy consumption per capita is roughly the same for each of these nations. However, the GNP per capita is quite different for each of these countries. Denmark has a GNP per capita almost twice that of the USSR, and Holland a GNP per capita about $1\frac{1}{2}$ times that of the USSR. This is an example that shows how far wrong we might be if we assumed that energy consumption solely determines GNP.

The United States consumes almost 40% of the world's energy, but has less than 6% of the world's population. Looking at Fig. 13-4, we see that the United States has both the highest commercial energy consumption per capita and the highest GNP per capita. Many of the nations in the less industrialized parts of the world have a very low energy consumption and GNP per capita (see Fig. 13-4, Spain, India, and Colombia). Undoubtedly the governments of these nations will resist any effort, internal or external, to keep them from increasing their energy consumption per capita to the level of the more industrialized nations. The well-publicized affluence of the American consumer (see Table 13-1) is a goal desired by many people throughout less developed parts of the world. Therefore, even if the United States were to level off its energy consumption per capita, the world's consumption of energy (and, therefore, fuel) may grow at an even faster rate than at present. This should result in vigorous competition for the available fuel supplies, and eventually produce fuel shortages.

The greatest growth in energy consumption has been, and probably will continue to be, in electric energy. Those nations that can afford the expense will build many new power plants (mostly nuclear) during the remainder of this century. An important effect of increased energy (and fuel) consumption is the pollution which accompanies power production of all types. Pollutants such as carbon dioxide, dust particles, and waste heat can alter both local and global climates. If the rest of the world begins to approach the United States in energy consumption, modification of climate will be a very real possibility.

Unlimited growth of population or energy consumption cannot continue indefinitely. People need certain basic resources, such as air, food, and water, for their survival. Once these basic needs are satisfied, there will usually be a desire for consumer products such as clothes, refrigerators, automobiles, etc. In order to produce these goods the finite natural resources of the earth must be used. The world population is approaching 4 billion persons and is growing at about 2% per year. If growth continues at this rate, there will be about 6 or 7 billion people in the world by the year 2000. At some point it will no longer be possible to satisfy the basic needs of individuals, not to mention other less survival-oriented needs such as desire for quiet and privacy.

As already mentioned, the growth in energy consumption per capita also cannot continue indefinitely since there will eventually be a shortage of fuel needed for energy conversion and an unacceptable increase in pollution from the process of energy conversion. Hopefully, we may soon realize that unlimited growth is not only impossible, but it is also undesirable. Population control is not the whole answer to controlling growth; we must also limit our per capita consumption of energy and natural resources.

Gross national product and the quality of life

An assumption that is often made is that GNP is a valid measure of the standard of living or even the quality of life. It is easy to find fault with this assumption. For example, no one ever subtracts anything from the GNP because of detrimental effects of increases in goods and services. The loss of a scenic site to a power plant is not taken into account in the GNP, and neither is the crowding and noisiness of large cities. There are many more examples which show that the GNP is not measuring the quality of life. The main problem is that the indicators of the quality of life vary from person to person in the population. Therefore, an answer to the question, "Can we get along in an acceptable manner if we allow the energy consumption per capita to level off?" results in a weighing of values against each other. An instance of the type of question we will have to answer is: "Are we willing to give up the convenience of the automobile in order to reduce air pollution and noise in heavily populated areas?" The answer to this question depends on such factors as: How attached are we to the automobile? What type of mass transit is available? How convinced are we that air pollution is harmful to our health?

Many countries (Sweden, France, New Zealand) have a smaller ratio of commercial energy consumption to GNP (see Fig. 13-4). This may be due to more efficient methods of energy production, different degrees of emphasis on industrialization or services, and the country's efficiency in converting energy to work. Since GNP does not give a good indication of the quality of life, this fact may not have significance in any case. Japan, for example, also has a lower ratio of energy consumption to GNP than the United States, and the quality of life in cities such as Tokyo leaves much to be desired. Tokyo is crowded and noisy, and has severe air pollution problems. For a small fee, oxygen is dispensed on many of the city's busy streets. In addition, industrial pollution of waterways with toxic chemicals has not been well controlled. An infamous example of the disregard of the Japanese industrialists and government for the public welfare was the dumping of the toxic chemical mercury into Minamata Bay.

It doesn't appear that lack of fuels* (fossil or nuclear) will significantly limit energy consumption during the next hundred years, especially since before the end of that time period both fusion power and solar power should be developed. Therefore, if energy use is to be limited, it will have to be the

*An exception is gasoline. Some short-term shortages of gasoline have already occurred.

result of a conscious effort on the part of man. Since the conversion and use of energy are the causes of many environmental and public health problems, as already discussed, increasing energy consumption does not necessarily assure us of an improvement in the quality of our lives. This situation seems to be more true of the technologically advanced countries than for the less industrialized countries. Many people are becoming aware that there is more value in a fulfilling and harmonious life than in an ever-increasing supply of goods and services. It is felt by these people that it would be advantageous for the public health, for the state of the environment, and for the enhancement of the quality of our lives in general, for us to reduce our consumption of energy.

In order for the total energy consumption in the United States to begin leveling off, it is necessary that the population growth rate remain near zero. In the near future assuming that population growth in the rest of the world approaches that in the U.S., the distribution of population in the United States and in the world will be more important even than total population growth. A possible criterion for limiting further growth in metropolitan areas may be a limit on energy use per unit area. In the Los Angeles Basin (740 square miles) the thermal power from man's activities is already equal to about 5% of the thermal power reaching the surface from the sun. Projections using current growth rates show that by the year 2000, man's power input over 11,000 square miles of the Boston-Washington megalopolis will be equal to 50% of the solar power reaching the surface in winter and 15% in the summer. Our knowledge of the meteorological system is too meager to predict the results of such large energy additions, but significant changes are possible. In the long run, total energy consumed over the whole earth will be a significant factor in determining energy use worldwide. The effects of air pollutants and radiation on human health are the main considerations now and will remain so for some years. If and when these hazards are diminished, thermal effects on local and global climates will be the main considerations. Thermal pollution of waterways and climate alterations should be important factors in land use planning in the future.

Energy policy

Planning At present there is no central coordinating agency that decides how much power should be provided each year consistent with environmental and human health goals. The role of the utilities is to provide whatever amount of energy is demanded by society. The Federal Power Commission is charged with "assuring an abundant supply of electric energy throughout the U.S. with the greatest possible economy and with regard to the proper utilization and conservation of natural resources." I think that it seems clear that the F.P.C. is mostly interested in assuring an abundant supply of electric energy and avoiding brownouts or blackouts. Public health is hardly ever considered as is obvious in the building of the Four Corners power plants, probably the single largest source of air pollution in the world! The newly created Federal Energy Administration (1974) is mainly concerned with allocation of various fuels.

Many other federal agencies are involved in the production of power. The Department of the Interior approves leases to companies that seek oil and natural gas on federal lands (such as the offshore Louisiana lands). The department also approves constructions such as the Alaskan pipeline, roads through federal lands, and power transmission lines through federal lands. The Department of the Interior has several divisions: the Bureau of Land Management which can sell or lease federal land for mineral rights; the Bureau of Indian Affairs, which oversees and approves the leasing and sale of natural resources from Indian treaty lands as well as agreements that allow transmission lines to cross those lands; and the Bureau of Reclamation, which promotes the growth of regional economies by the optimum development of water and related land resources. The Bureau of Reclamation operates in three key policy areas: Its dams produce hydroelectric power, its water allocations control the location of regional power plants, and its water contracts stipulate the manner in which smokestack emissions will be controlled.

The A.E.C. licenses all nuclear power plants and must approve all construction plans. It regulates nuclear reactors, processing plants, and separation plants. The main problem is that the A.E.C. has been not only the regulator of all of the above, but also the promoter. They often tend not to consider alternatives in which they have no personal interest or with which they are unfamiliar. Even worse, they occasionally promote programs (like breeder reactors), which perpetuate their own existence, at the expense of less hazardous options. Some examples of options they do not seriously consider are the use of solar power and the reduction of electricity use.

There are many other agencies involved in the regulation of mining and power production; for example, state health departments, U.S.P.H.S., regional water quality districts, the Environmental Protection Agency, the Bureau of Mines, and the National Institute of Occupational Health and Safety. The main problems are that control and knowledge are too fragmented, and that there are too many agencies looking after their own narrow interests. What is needed is a central federal agency which uses an interdisciplinary approach to assess the need for more power, the methods of obtaining raw materials, land use, site location, method of power production, environmental and human health impacts, etc. Those people who already have expert knowledge in the above areas might be brought together into one group. The members of the group should include both government employees and members of interested citizen groups.

The U.S. Forest Service is a federal agency which uses an interdisciplinary approach in their planning. One of the factors they attempt to consider is the environmental impact of all the alternative methods of reaching a particular goal. The statement of a goal itself can offer opposing groups a chance to question the necessity of some very large undertakings. Public meetings are held before any decisions are made by the forest service and alternative proposals can be introduced at this time. A central energy agency might incorporate some of these ideas in its formulation.

There are many things which can be done immediately. For example, the growth of electricity use could be slowed down by elimination of waste by users, reduction of inefficient uses such as space heating by electricity, and by increasing the unit price of electricity as use increases. Since plenty of U^{235} is available if the more expensive U_3O_8 ore is used to obtain the fuel, the development of breeder reactors can proceed at a slower pace. They should remain experimental until a long safety record is built up, and they should be located far from large cities.

Federal money should be spent on research in pollution control for fossil-fuel plants and for development of solar power. A solar farm should be built even though the unit price of electricity from a solar farm may be several times higher than from a nuclear or fossil-fuel plant. The price of electricity should include such things as the degradation of the environment and human health. Both of these costs will be significantly lower for a solar farm than for fossil or nuclear power plants. An important question is: "Can we morally assign monetary values to human suffering and death?" If this is done by public agencies, then we should all be aware of the basis of the decision. When the A.E.C. states that the benefit of atomic energy is worth the risk, we should be informed how that decision was reached. The same is true of the government's decision to move rapidly toward development of fast breeder reactors. The public should be aware of the benefits and risks of breeder reactors. Only then can we, as an informed public, influence such decisions.

Impact statements The National Environmental Policy Act (NEPA) of 1969 has already had, and will probably continue to have, important effects on planning in the U.S. One of the provisions of the Act is that all agencies of the Federal government shall use an interdisciplinary approach in making decisions which may have an impact on our environment. A second provision is that such agencies include in every recommendation or report on proposals for legislation and other major Federal actions significantly affecting the quality of the human environment, a detailed statement by the responsible official on:

1 The environmental impact of the proposed action.

2 Any adverse environmental effects which cannot be avoided should the proposal be implemented.

3 Alternatives to the proposed action.

4 The relationship between local short-term uses of man's environment and the maintenance and enhancement of long-term productivity.

5 Any irreversible and irretrievable commitments of resources which would be involved in the proposed action should it be implemented.

The Act provides for the above impact statements to be made public. Therefore, interested citizen groups can learn of important proposals before they are implemented, and can then exert pressure on those parties carrying out projects they feel are damaging to the environment. The release of impact statements has led to the filing of many lawsuits against Federal agencies. The law suits are often concerned with improperly drawn impact statements and with possible environmental damage which may be caused by a proposed project. Essentially, all new nuclear power plant construction was temporarily stalled because the A.E.C. had not considered the effects of thermal pollution in their impact statements.

Projects have been stalled by the action of citizens in the past, before the NEPA. In 1968 a citizen-scientist coalition opposed a new nuclear power plant which was to be built on the shore of Lake Cayuga in New York State. Scientists from Cornell University provided the main argument against the power plant, which was that thermal pollution of the lake, whose water was to be used for cooling purposes, would result in eutrophication of the lake. That is, large algal blooms would occur, the fish population would change, and the lake would not appear as beautiful as it had earlier. In April 1969 the power company, New York State Electric and Gas, postponed its plans for the reactor pending "additional research on cooling systems for thermal discharge from the plant and for consideration of the economic effects of such systems." More recent nuclear power plant proposals include the use of cooling towers to cool heated water before its return to the body of water from which it came, or closed cycle cooling towers with no return of the cooling water.

Thus it appears that the NEPA coupled with high citizen interest have succeeded in convincing federal agencies to consider the important environmental impacts of their projects and to sometimes incorporate those features which will avoid harmful environmental impacts in their initial designs. This situation will tend to continue only as long as citizen interest remains high.

Suggested reading

1 COOK, EARL, "The Flow of Energy in an Industrial Society," *Scientific American* (September 1971).

2 EHRLICH, PAUL, *The Population Bomb*. New York: Ballantine Books, 1966.

3 LAPP, RALPH E., *The Logarithmic Century*. Englewood Cliffs, NJ: Prentice-Hall, Inc., 1973.

4 MEADOWS, DONELLA H., et al., *The Limits to Growth*. New York: Universe Books, 1972.

5 STARR, CHAUNCEY, "Social Benefits versus Technological Risks," *Science*, **165** (1969), 1232–1238.

6 BARKLEY, PAUL W., and DAVID SECKLER, *Economic Growth and Environmental Decay*. New York: Harcourt Brace Jovanovich, 1972.

7 BERG, CHARLES A., "Energy Conservation through Effective Utilization," *Science*, **191**: 40951 (July 13, 1973), 128.

8 MORRISON, MIKE, "The Realities and Unrealities of Energy Economics," *Sierra Club Bulletin* (May 1973).

9 HART, JOHN, "The Challenge of the Energy Crises, To Plan or To Panic," *Cry California*, **9**: 1 (Winter 73/74).

Questions

1 Make a list of the electric appliances or devices that you have in your house. Which of these do you consider to be essential?

2 If you start off with $1,000, and it grows at the rate of 10% per year (that is, your interest rate is 10%), after the end of two years you will have: (a) $1,200? (b) $1,020? (c) $1,210?

3 (*a*) By what percent did GNP (solid curve of Fig. 13-2) increase during the time period from 1940 to 1960?

(*b*) Can you estimate the average growth rate in GNP during this time? Explain how you arrived at your conclusion.

4 During the period from 1965–1970, the ratio of energy consumption to GNP rose by what percentage (see Fig. 13-3)?

5 A 20% increase in energy consumption for the U.S. in 1975 should result in how much of an increase in GNP? Explain.

6 Can you give any examples which illustrate that the GNP is not a very good indicator of your quality of life?

7 According to Fig. 13-4, would you say that the USSR or Sweden is more efficient in converting energy into work? Can you even make such a judgment from the data in Fig. 13-4? Explain.

8 Which criteria would you like to see a central energy agency use to make its decisions concerning energy?

9 Do you know of any instance in which the NEPA was used to halt a project which threatened to be damaging to the environment? If so describe the events which took place.

10 (*a*) Table 13-2 lists several home appliances, the rate at which they consume energy in watts, and the amount of time that they are used

Table 13-2

Appliance	*Average wattage*	*Amount of time operated*
Blender	386	40 hours
Coffee maker	894	100 hours
Dishwasher	1,201	300 hours
Oven, Microwave	1,500	180 hours
Toaster	1,146	33 hours
Clothes dryer	4,856	200 hours
Tooth brush	7	60 hours
Television (B&W)	237	1,500 hours
Television (color)	332	1,500 hours

each year on the average. Suppose that you use an electric toothbrush for about 60 hours each year and a toaster for about 33 hours each year. Which appliance would consume more energy during the year and how much energy would it consume?

(*b*) Which of all the appliances consumes the most energy during the year and how much energy does it consume?

(*c*) How much does it cost to run this appliance if the cost is 5 cents per kilowatt-hour?

Epilogue

It is difficult, if not impossible, to predict the duration and effects of the energy crisis, which began as this manuscript was being completed in the last months of 1973. There has been much debate concerning the probable cause of the energy crisis. Some people claim it is the fault of the oil companies for not keeping up with rising demand (purposely or because of lack of foresight), others say it is the fault of environmentalists for opposing construction of nuclear power plants and favoring the enforcement of strict air quality standards. In order to achieve these air quality standards, several actions were taken by the Environmental Protection Agency, for example, emission controls, which lowered gas mileage by as much as 10% on pre-1975 automobiles were required, and less polluting oil and natural gas were substituted for coal in some fossil-fuel power plants.

There is no doubt that making the environment more livable and healthy has cost us both money and energy. The cost in energy, however, is only a small fraction of the projected energy shortage for the next few years, and the benefits of cleaner air and more carefully designed nuclear power plants are great. The main cause of the energy shortage is the rapid increase in per capita energy consumption. In the United States, per capita energy consumption has grown from about 28,000 kilowatt-hours in 1900 to its present value of around 100,000 kilowatt-hours, a factor of 3.5 increase! Therefore, on the average each person in the U.S. uses 3.5 times more energy than an individual used in 1900 and, of course, there are many more of us now—over 200 million compared to 80 million in 1900.

One result of this energy crisis seems to be apparent. Our way of life in the U.S. will be greatly affected, although not necessarily for the worse. No longer will electricity and other forms of energy be treated as inexhaustible, almost free commodities by the American consumer. We will have to give up such luxuries as large gas-guzzling automobiles and unlimited use of air conditioners and space heaters.

Another result of the energy crisis is that pressure is being applied by various groups and individuals to suspend certain environmental protection laws. The years ahead will provide a significant test for the environmental movement and also a significant opportunity. The need for energy conservation can be used as an opportunity to slow the growth of per capita energy consumption, recycle valuable resources, reduce air pollution from automobiles by building more small, energy-efficient models, and build more energy-efficient mass transit systems instead of highways.

Appendix A
Mathematical review

Mathematical review

Very small and very large numbers are used in this book. For example, the diameter of a hydrogen atom is about 0.0000000001 meter and the speed of light in a vacuum is 300,000,000 m/sec. It is very convenient therefore, to use scientific notation for very large or very small quantities. That is, it is much neater to write 0.0000000001 m as 10^{-10} m and 300,000,000 m/sec as 3×10^8 m/sec. We have expressed these two unwieldy numbers in a simpler form as powers of ten., The following table shows how we express numbers as powers of ten:

$10^0 = 1$	$10^{-1} = 0.1$
$10^1 = 10$	$10^{-2} = 0.01$
$10^2 = 100$	$10^{-3} = 0.001$
$10^3 = 1{,}000$	$10^{-4} = 0.0001$
etc.	etc.

In general, 10^x is equal to 1 followed by x zeros, and 10^{-x} is equal to 1 preceded by x decimal places. The number 31,000 can be expressed as a power of ten as follows: $31{,}000 = 3.1 \times 10^4$:

$$3.1 \times 10^4 = 3.1 \times 10{,}000 = 31{,}000$$

The number 0.0002 can be expressed as 2×10^{-4} since 2×10^{-4} equals $2 \times 0.0001 = 0.0002$. The same procedure can be used to express any small or large number as a power of ten.

Examples: Express the following in scientific notation:

1 300,000,000	*4* 0.000036
2 0.0000000001	*5* 0.001
3 93,000,000	*6* 5,400

Express the following in regular notation:

7 3×10^{10}	*9* 6×10^{-6}
8 10^{-8}	*10* 4.1×10^5

The answers are at the end of this appendix.

Mutiplication and division

Sometimes we need to multiply or divide numbers expressed as powers of ten. In general:

$$(10^a) \times (10^b) = 10^{a+b}$$

The symbols a and b are called exponents, and when powers of ten (or powers of any number) are multiplied, we add the exponents together. Therefore, $10^3 \times 10^2 = 10^{3+2} = 10^5$.

When we divide, the exponents are subtracted from each other:

$$\frac{10^a}{10^b} = 10^{a-b}$$

So $10^2/10^4 = 10^{2-4} = 10^{-2}$

By using these two rules we can evaluate complicated expressions such as:

$$\frac{(6.6 \times 10^{-34})(3 \times 10^{10})}{(1.5 \times 10^{-8})}$$

The numbers in front of the powers of ten are treated separately as follows:

$$\frac{6.6 \times 3}{1.5} = 6.6 \times 2 = 13.2$$

The powers of ten are handled in this manner:

$$\frac{(10^{-34})(10^{10})}{10^{-8}} = \frac{10^{-34+10}}{10^{-8}} = \frac{10^{-24}}{10^{-8}} = 10^{-24+8} = 10^{-16}$$

The final answer is the product of the separate answers: 13.2×10^{-16}.

Examples: Evaluate the following:

11 $10^7/10^2$

12 $10^3/10^{-7}$

13 $10^8 \times 10^{-3}$

14 $\dfrac{10^3 \times 10^{-8}}{10^2}$

15 $\dfrac{10^{-2} \times 10^{-4}}{10^{-3}}$

16 $\dfrac{(6 \times 10^{-3})(4 \times 10^{-5})}{10^2}$

Illustrative example: Example 1 in Chapter 2 could have been done by using scientific notation. The time, t, that it takes light to travel from the sun to the earth is $t = d/v$, where d is the distance between the sun and the earth, and v is the speed of light in a vacuum:

$$t = \frac{93{,}000{,}000 \text{ miles}}{186{,}000 \text{ miles/sec}} = \frac{9.3 \times 10^7 \text{ miles}}{18.6 \times 10^4 \text{ miles/sec}}$$

$$t = 0.5 \times 10^3 \text{ sec} = 500 \text{ sec}$$

Logarithms

Taking the logarithm of a number is the opposite of raising it to some power. For example,

$$10^2 = 100$$

and

$$\log_{10}(100) = 2$$

When we compute the logarithm of 100 to the base 10($\log_{10} 100$) we are asking: To what power must 10 be raised to get 100, or what value of x satisfies $10^x = 100$? The answer is 2.

Logarithms of numbers such as 10, 100, 1,000, etc., are easily obtained. Here are some examples which you should do if you are not already familiar with logarithms:

17 $\log_{10} 1{,}000 =$

18 $\log_{10} 10 =$

19 $\log_{10} 10^7 =$

20 $\log_{10} 10^3 =$

If we are trying to find the logarithm of a number such as 32, we are asking: What value of x satisfies this equation: $10^x = 32$?

The answer to this problem is not obvious, but we can use Table A-1 which lists logarithms that have already been calculated:

Table A-1

N	*log N*	*N*	*log N*	*N*	*log N*	*N*	*log N*	*N*	*log N*
0	$-\infty$								
1	0	11	1.0414	21	1.3222	31	1.4914	41	1.6128
2	.3010	12	1.0792	22	1.3424	32	1.5051	42	1.6232
3	.4771	13	1.1139	23	1.3617	33	1.5185	43	1.6335
4	.6021	14	1.1461	24	1.3802	34	1.5315	44	1.6435
5	.6990	15	1.1761	25	1.3979	35	1.5441	45	1.6532
6	.7782	16	1.2041	26	1.4150	36	1.5563	46	1.6628
7	.8451	17	1.2304	27	1.4314	37	1.5682	47	1.6721
8	.9031	18	1.2553	28	1.4472	38	1.5798	48	1.6812
9	.9542	19	1.2788	29	1.4624	39	1.5911	49	1.6902
10	1.000	20	1.3010	30	1.4771	40	1.6021	50	1.6990

From the table, $\log 32 = 1.5051$.

Examples:

21 $\log_{10} 2 =$

22 $\log_{10} 20 =$

23 $\log_{10} 200 =$

Exponential functions

Many natural processes can be described mathematically by using exponential functions. For example, radioactive decay (Chapter 3) is an exponential process. If we start off with N_0 radioactive atoms at time $t = 0$, then the number, N, of radioactive atoms remaining as a function of time, t, is given by

$$N = N_0 e^{-\lambda t}$$

where λ is a constant which depends on the particular isotope that is decaying and e is equal to 2.718. In this case, we are working with powers of e rather than powers of ten. The only difference in working with powers of e is that it is more difficult to do calculations, and we must resort to tables that list the values of e raised to various powers. For example, if $\lambda t = 1$, then $N = N_0 e^{-1}$. In Table A-2 we can find the value of e^{-1}, which is 0.3679. Therefore, after $1/\lambda$ seconds there are $0.3679 N_0$ radioactive atoms left.

Table A-2

x	e^x	e^{-x}	x	e^x	e^{-x}
.00	1.0000	1.0000	2.0	7.3891	.1353
.10	1.1052	.9048	2.5	12.182	.0821
.20	1.2214	.8187	2.0	20.086	.0498
.30	1.3499	.7408	3.5	33.115	.0302
.40	1.4918	.6703	4.0	54.598	.0183
.50	1.6487	.6065	4.5	90.017	.0111
.60	1.8221	.5488	5.0	148.41	.0067
.70	2.0138	.4966	6.0	403.43	.0025
.80	2.2255	.4493	7.0	1096.6	.0009
.90	2.4956	.4066	8.0	2981.0	.0003
1.00	2.7183	.3679	9.0	8103.1	.0001
1.50	4.4817	.2231	10.0	22026.	.0000

Answers to questions in Appendix A

1 3×10^8

2 1×10^{-10}

3 9.3×10^7

4 3.6×10^{-5}

5 1×10^{-3}

6 5.4×10^3

7 30,000,000,000

8 0.00000001

9 0.000006

10 410,000

11 10^5

12 10^{10}

13 10^5

14 10^{-7}

15 10^{-3}

16 2.4×10^{-11}

17 3.0

18 1.0

19 7.0

20 3.0

21 0.3010

22 1.3010

23 2.3010

Appendix B

Basic electricity

Basic electricity

Electrical phenomena are considered by most people to be mysterious in nature. We plug a lamp into an electrical outlet and it "magically" lights up. With the presentation of a few basic electrical principles it should be possible to gain a rough understanding of what is going on when we use various appliances and devices around the house. A good place to start is with a discussion of the nature of conductors and insulators.

Metals, as most of us know, are classified as good conductors of electricity, whereas glass, wood, and rubber are classified as poor conductors, or insulators, of electricity. In general, we call any material through which charge flows easily, when it is subject to an electrical force, a *conductor*. A material that offers appreciable resistance to the flow of charge is called an *insulator*. A fraction of an inch of rubber insulation placed between two current-carrying wires can prevent the flow of charge between them.

Whether a substance is an insulator, conductor, or semiconductor* depends on how tightly the atoms of that substance hold their electrons. Metals, unlike other solids, contain many free electrons; that is, electrons which are loosely bound to their atoms and are free to move about if an electrical force is applied. On the other hand, the electrons in insulators are tightly bound to their atoms and are not free to move about.

We said earlier that charge can flow in a conductor if an electric force is impressed upon it, but we did not mention the source of the electric force (usually called an electromotive force). In most cases the source of electromotive force (emf) is a battery or electric generator. When this source is hooked up to a light bulb, an electric field is set up in the hookup wire and electrons can flow through the circuit, thus causing the light bulb to give off light. Electromotive force is measured in volts; a standard flashlight battery has an emf of 1.5 volts (Fig. B-1).

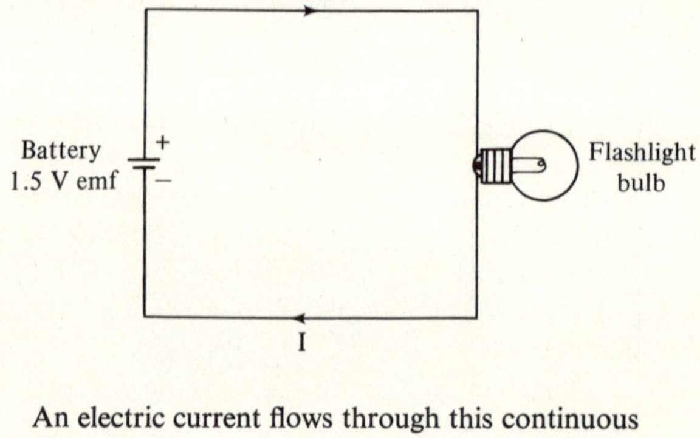

Figure B-1 An electric current flows through this continuous circuit composed of a battery, a light bulb, and some connecting wire.

*A semiconductor is a material that is normally insulating, but which becomes conducting when certain impurities are added.

The flow of charge is called an electrical *current*. The rate of electrical flow is measured in amperes, where 1 ampere of current is equal to the flow of 1 coulomb* of charge per second past any point in an electrical circuit. The greater the emf (or voltage), the greater the current in a circuit. We can think of voltage in the same way that we think of pressure in a water pipe (Fig. B-2): the greater the water pressure, the greater the flow of water, and similarly, the greater the voltage, the greater the electrical current.

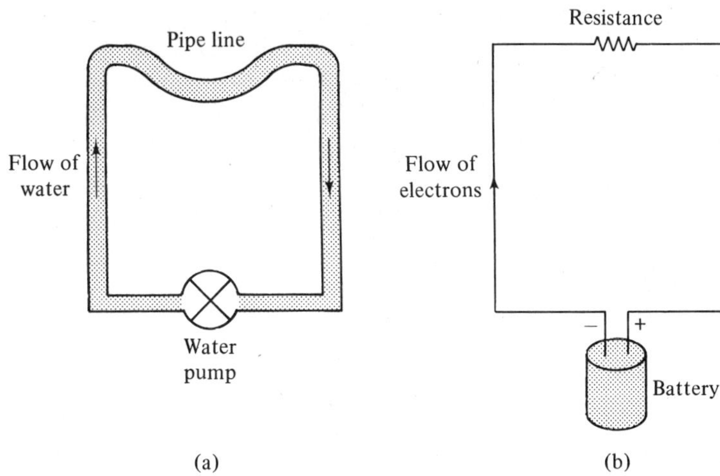

Figure B-2 An analogy can be made between water flowing through pipes and electrons flowing through wires.

There is another factor that affects the rate at which charge flows in an electrical circuit. It is called *resistance*. A wide, short pipe offers less resistance to the flow of water than a long, narrow one. The same is true of a metal wire in terms of the flow of electrons; a long, thin wire offers more resistance to the flow of electrons than a short, fat one. Other factors affecting resistance are the nature of the conducting material and its temperature. On a microscopic level, the source of resistance is the collisions that take place between the electrons and the atoms making up the wire. For most conductors, a decrease in temperature causes a decrease† in resistance since the atoms move about less at low temperatures. Some metals and their compounds show an abrupt decrease of resistivity at very low temperatures (like several degrees

*One coulomb of charge is equivalent to 6×10^{18} electrons.
†For semiconductors the opposite is true.

above absolute zero). A current in a loop of such material (called superconducting), can persist for hours or days after the source of emf is removed. There is a relationship (called Ohm's law*) between the current, I, voltage, V, and resistance, R, in an electrical circuit:

$$\text{Current} = \text{Voltage/Resistance} \qquad I = V/R$$

The magnitude of the current in a circuit is proportional to the applied emf or voltage, and inversely proportional to the resistance. If emf is expressed in volts and current in amperes, then resistance will be in ohms.

A typical 100-watt light bulb has a resistance of about 100 ohms. From Ohm's law we find that if 100 volts is applied to this light bulb, a current of 1 ampere is produced. Ohm's law is strictly true only for metallic resistors through which a steady current is passing.

Power and heat

Electricity is normally used to run electric motors, produce light, or produce heat. Most of us have probably noticed that a light bulb emits both heat and light. In fact, about 90% of the electric energy supplied to a light bulb is converted to heat energy and only 10% to light. A small electric motor (such as in an electric drill) converts about 60% of its input energy into mechanical energy and the rest appears as heat energy. This tendency of electrical devices and appliances to warm up when a current is passed through them is even more obvious in irons, toasters, and space heaters, which convert almost all of the electrical input energy into heat.

James Joule, an English physicist of the 19th century, found that the rate of production of heat energy in any electrical device is proportional to the square of the current passing through it. The proportionality factor is the resistance of the device. Therefore,

$$\text{Power} = (\text{Current})^2 \cdot (\text{Resistance}) = I^2R$$

When current is expressed in amperes (A or amps) and resistance in ohms, then the rate at which energy is dissipated (the power) is in joules per second or watts; 1 joule per second equals 1 watt.

An electric toaster typically has a resistance of about 15 ohms. When it is connected to a 120-volt electrical outlet it draws a current of about 8 amps:

$$I = V/R = 120 \text{ volts}/15 \text{ ohms} = 8 \text{ amps}$$

*After George Ohm, a German physicist of the 19th century.

Therefore, the rate at which energy is dissipated in the form of heat is:

$$\text{Power} = (8 \text{ amps})^2 \cdot (15 \text{ ohms}) = 64 \cdot (15) = 960 \text{ watts}$$

If we were to run such a toaster for 1 hour, then the amount of energy dissipated would be equal to the power multiplied by the length of time the device was run:

$$\text{Energy} = \text{Power} \times \text{time} \qquad E = P \times t$$

$$E = 960 \text{ watts} \times 1 \text{ hour}$$

$$960 \text{ watts} \cong 1 \text{ kilowatt, so } E \cong 1 \text{ kilowatt} \times 1 \text{ hour} = 1 \text{ kilowatt-hour.}$$

A kilowatt-hour represents the amount of electrical energy converted to heat by the toaster in 1 hour. The cost of electric energy usually runs somewhere between a few cents and 10 cents per kilowatt-hour. If it were 10 cents per kilowatt-hour, then it would cost 10 cents to run a toaster for 1 hour. It would also cost 10 cents to run a 100-watt light bulb for 10 hours. We can check to see that the units are really energy units in the above equation:

$$1 \text{ kilowatt-hour} = 1{,}000 \text{ joules/sec} \times 3{,}600 \text{ sec} = 3.6 \times 10^6 \text{ joules}$$

The equation for power can be written in a different way if Ohm's law is substituted into the power equation:

$$P = I^2R \qquad \text{Ohm's law} \quad I = V/R$$

Rewriting,

$$P = I(IR) = IV$$

Often electrical circuits contain elements other than resistors, such as capacitors and inductors, which make them more complex to analyze. We do not discuss such circuits in this book.

Transmission of electricity

Electricity is generated at a central power plant and then transmitted on overhead power lines to various cities. Additional power lines (overhead or underground) carry the electricity to our homes. This electrical energy is transmitted in the form of an *alternating current* which means that the direction in which the electrons move inside the wire reverses itself periodically. In a 60-cycle per second alternating current the electrons move back and forth about their equilibrium position 60 times each second. When a battery is used as a source of emf the electric current in a circuit is a *direct current*; electrons flow in only one direction.

The electricity which enters our homes is mostly in the form of an alternating current since this type of current can be easily transformed* from one voltage to another. Power may be transmitted from a power plant to a city at about 100,000 volts stepped down to about 2,000 volts in the city, and finally stepped down to 120 volts for safe home use. In Chapter 11 the process of power generation is discussed.

For those readers who are interested in learning more about electrical and magnetic phenomena, all of the introductory physics texts mentioned at the end of Chapter 1 contain several chapters on these subjects.

*In a device called a transformer input voltage can be increased (stepped up) or decreased (stepped down) with a corresponding decrease or increase in current output. Thus the power in is equal to the power out except for some small power losses in the core of the transformer.

Physical constants and conversion factors

Important constants

speed of light	c	3.00×10^{8} meters/sec
charge of electron	e	1.60×10^{-19} coulomb
electron rest mass	m_e	9.11×10^{-31} kilogram
Planck's constant	h	6.63×10^{-34} joule·sec
proton rest mass	m_p	1.67×10^{-27} kilogram
Stefan-Boltzmann constant	σ	5.67×10^{-8} watt/m^2·°K^4
gravitational constant	G	6.67×10^{-11} newton·m^2/kilograms^2

Conversion factors

Length

1 Angstrom (Å) = 10^{-10} meter = 10^{-8} centimeter
1 centimeter = 0.39 inch
1 meter = 3.28 feet
1 kilometer = 0.62 mile
1 mile = 5,280 feet = 1.61 kilometers
1 inch = 2.54 centimeters

Mass

1 kilogram = 2.21 pounds

Volume

1 meter^3 = 1,000 liters = 10^6 centimeters^3 = 35.31 feet^3

Velocity

60 miles/hour = 88 feet/sec = 26.82 meters/sec
1 mile/hour = 1.61 kilometers/hour

Force

1 newton = 10^5 dynes = 0.23 pound

Pressure

1 atmosphere = 760 mm of mercury = 14.7 pounds/inch^2

Work and energy

1 joule = 10^7 ergs = 0.24 calorie
1 calorie = 4.18 joules
1 Btu = 252 calories = 1,054 joules
1 kilowatt-hour = 3.60×10^6 joules

Power

1 watt = 1 joule/sec = 0.0013 horsepower = 0.057 Btu/minute

Answers to even numbered quantitative questions

Chapter 1

6 (*a*) 11.6×10^{-3} gram = 0.0116 gram
(*b*) 0.24×10^{-4} pounds = 0.000024 pound

Chapter 2

6 28.8°F

10 12% N_2, 86% O (atomic oxygen), 2% He

Chapter 3

4 2.0×10^{-15} joules

14 50%

18 (*a*) 410 inches or about 34 feet
(*b*) 30.6 feet

Chapter 4

4 (*a*) about 2.5°C
(*b*) about 29 minutes

12 (*a*) 580 kilocalories per day
(*b*) Yes, on the order of 1/4 to 1/3 of the total metabolic rate for most adults.

20 A temperature decrease of about .50°C from 37°C to 36.50°C doubles the heat production rate.

Chapter 5

14 ${}_{54}Xe^{131}$ atomic number is 54 and mass number is 131.

Chapter 6

16 20%

Chapter 7

2 If we assume that the height of each apartment is 3 meters, then it takes only about 0.001 second (one-thousandth of a second) for the sound to travel to the manager's apartment.

6 95% of the population

8 The fundamental natural frequency would be 14,176 cycles/sec. Therefore, if the ear canal were filled with water, we would be able to hear higher frequency sounds than usual.

Chapter 8

4 (*a*) 20 pounds of sulfur
(*b*) about 40 pounds of SO_2

8 540 μg = 0.00054 g

10 speed equals 64 ft/sec $\cong$ 43.6 miles/hr

12 60%

Chapter 9

2 80%

Chapter 10

14 Energy used in bicycling is 63 kcal, in walking 150 kcal, and in driving 5,460 kcal. Therefore, in terms of energy usage, bicycling is the most efficient method of covering short distances of a few miles.

16 2 minutes if passengers enter and exit train at same part of the 1-minute waiting period at station.

18 70 trains

Chapter 11

4 (*b*) Maximum theoretical efficiency of a heat engine $1 - T_1/T_2$ where T_1 is the exhaust temperature and T_2 is the input temperature in degrees Kelvin.

(*c*) The efficiency equals 50% for a turbine in a nuclear power plant.

(*d*) From a calculation in this chapter, the maximum theoretical efficiency for a turbine in a fossil-fuel power plant was 62%.

(*e*) We also need to know the conversion efficiency of nuclear energy in the fuel into heat before we can calculate the overall efficiency of a nuclear power plant.

6 (*a*) 2,500 megawatts power input

(*b*) 1,500 megawatts of waste heat

10 (*a*) 9×10^{10} joules

(*b*) 2.9 kilograms of U^{235} undergo fission each day in a 1,000-megawatt nuclear power plant assuming continuous operation.

Chapter 12

10 (*a*) 15°C or 59°F

Chapter 13

2 (*c*)

4 5.7%

10 (*a*) Toaster consumes 37.8 kWh of energy.

(*b*) Dryer consumes about 970 kWh of energy.

(*c*) $48.50

Index

A

B

C

X